The History of Mathematical Tables

FROM SUMER TO SPREADSHEETS

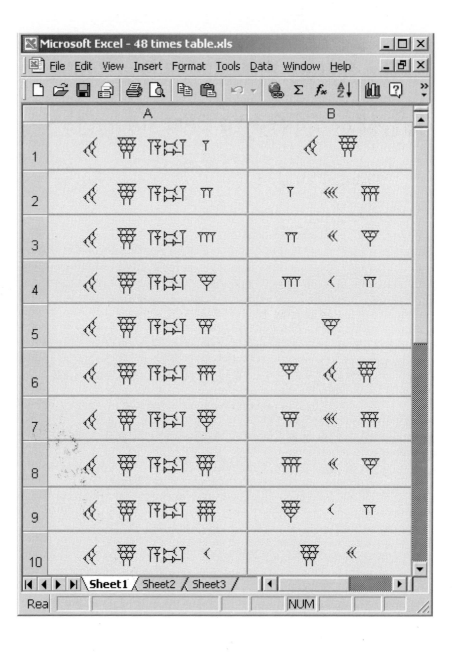

The History of Mathematical Tables

FROM SUMER TO SPREADSHEETS

Edited by
M. Campbell-Kelly
M. Croarken
R. Flood
and
E. Robson

OXFORD
UNIVERSITY PRESS

OXFORD
UNIVERSITY PRESS

Great Clarendon Street, Oxford OX2 6DP

Oxford University Press is a department of the University of Oxford.
It furthers the University's objective of excellence in research, scholarship,
and education by publishing worldwide in

Oxford New York

Auckland Bangkok Buenos Aires Cape Town Chennai
Dar es Salaam Delhi Hong Kong Istanbul Karachi Kolkata
Kuala Lumpur Madrid Melbourne Mexico City Mumbai Nairobi
São Paulo Shanghai Taipei Tokyo Toronto

Oxford is a registered trade mark of Oxford University Press
in the UK and in certain other countries

Published in the United States
by Oxford University Press Inc., New York

© Oxford University Press

The moral rights of the author have been asserted
Database right Oxford University Press (maker)

First published 2003

All rights reserved. No part of this publication may be reproduced,
stored in a retrieval system, or transmitted, in any form or by any means,
without the prior permission in writing of Oxford University Press,
or as expressly permitted by law, or under terms agreed with the appropriate
reprographics rights organization. Enquiries concerning reproduction
outside the scope of the above should be sent to the Rights Department,
Oxford University Press, at the address above

You must not circulate this book in any other binding or cover
and you must impose this same condition on any acquirer

A catalogue record for this title is available from British Library

Library of Congress Cataloging in Publication Data
(Data available)
ISBN 0 19 850841 7

10 9 8 7 6 5 4 3 2 1

Typeset by Newgen Imaging Systems (P) Ltd., Chennai, India
Printed in Great Britain
on acid-free paper by
Biddles Ltd., www.biddles.co.uk

In memory of
JOHN FAUVEL
1947–2001

Contents

Introduction	1
1. Tables and tabular formatting in Sumer, Babylonia, and Assyria, 2500 BCE–50 CE ELEANOR ROBSON	19
2. The making of logarithm tables GRAHAM JAGGER	49
3. History of actuarial tables CHRISTOPHER LEWIN AND MARGARET DE VALOIS	79
4. The computation factory: de Prony's project for making tables in the 1790s IVOR GRATTAN-GUINNESS	105
5. Difference engines: from Müller to Comrie MICHAEL R. WILLIAMS	123
6. The 'unerring certainty of mechanical agency': machines and table making in the nineteenth century DORON SWADE	145
7. Table making in astronomy ARTHUR L. NORBERG	177
8. The General Register Office and the tabulation of data, 1837–1939 EDWARD HIGGS	209
9. Table making by committee: British table makers 1871–1965 MARY CROARKEN	235

10.	Table making for the relief of labour DAVID ALAN GRIER	265
11.	The making of astronomical tables in HM Nautical Almanac Office GEORGE A. WILKINS	295
12.	The rise and rise of the spreadsheet MARTIN CAMPBELL-KELLY	323

Biographical notes   349
Index   353

Metric System. (See Tables, pp. 218-19)

The Metre is 1/10-millionth of the Equator to Pole surface-distance; the **gram**, a cube centimetre (millilitre); the **litre**, a cube decimetre, or 1000 cube cm., or 1 kilogram, of water at 39° F. The **are** is a sq. decametre; the **stere**, a cube metre, or kilolitre, or 1 million grams (tonne) of water.

Latin prefixes, deci-, centi-, milli-, denote sub-divisions.
Greek prefixes, deka-, hecto-, kilo-, myria-, denote multiples.
*, †, ‡, §, show square and cube of length-measure with same sign.

Lineal:—	*metres*	*Brit. equivt.*
1 millimetre	$\frac{1}{1000}$	0·039 ins.
1 centimetre ‡	$\frac{1}{100}$	0·394 ,,
1 decimetre †	$\frac{1}{10}$	3·937 ,,
Metre* ..	**1**	**39·370 ins.**
1 decametre §	10	32·808 feet
1 hectometre	100	328·084 ,,
1 kilometre	1000	1093·61 yds.
1 do. (abt. $\frac{5}{8}$m.)		**0·621 mile**

Square:—	*ares*	*Brit. equiv't.*
1 milliare	$\frac{1}{1000}$	1·076 sq. ft.
1 centiare*	$\frac{1}{100}$	10·764 ,,
1 deciare	$\frac{1}{10}$	107·639 ,,
Are § ..	**1**	**119·599 sq. yds.**
1 decare ..	10	0·247 acre
1 hectare	100	2·4711 acres

Cube:—1 **Stere*** (cub. metre) = 35·31 cub. feet (metric ton water); millistere † (·001 stere) = 61·02 c. in.

Troy:—1 gram = ·03215 Troy oz.

Weight:	*grams*	*Brit. Equiv't*
1 milligram	$\frac{1}{1000}$	0·015 grains
1 centigram	$\frac{1}{100}$	0·154 ,,
1 decigram	$\frac{1}{10}$	1·543 ,,
Gram ‡ ..	**1**	**15·432 grains**
1 decagram	10	5 drms. 18 grs
1 hectogram	100	3½ oz. 12 grs
1 kilogram †	1000	2 lb 3¼ oz 10 grs.
1 myriagram	10,000	22 lb 0¾ oz.
1 Quintal	100,000	220·46 lbs.
1 **Tonne** or millier		**0·984 ton**, or 2204·622 lb. or 1·102 short tons.

Capacity:	*litres*	*Brit. equiv't.*
1 millilitre ‡	$\frac{1}{1000}$	16·9 minims
1 centilitre	$\frac{1}{100}$	0·35 fluid oz
1 decilitre	$\frac{1}{10}$	3·52 ,,
Litre † (1000 cc)	**1**	**1·7598** pints
1 decalitre	10	2 gall. 1½ pts.
1 hectolitre	100	2¾ bush. (22 gls)
1 kilolitre (cb. metre)		220·00 galls.

Weights of Materials. D, Density or Specific Gravity, compared with water taken as 1. **Weight** in lbs. (gases, ozs.):—of a cube inch, **In.**; of a cube foot, **Ft.** Different samples vary greatly, especially wood

Material	D.	In.	Material	D.	In.	Material	D.	Ft.	In.
Alumin'm	2·7	·10	Lead ..	11·4	·41	Ash, Oak, Teak	·84	45-58	·028
Brass ..	8·5	·31	Mercury	13·6	·49	Beech, Birch	·72	43-58	·027
Brick, abt	1·8	·07	Nickel ..	8·7	·31	Cork ..	·24	15	·009
Celluloid	1·4	·05	Platinum	21·5	·78	Elm, Mahog'y	·64	35-45	·023
Coal (*solid*)	1·3	·05	Sand ..	1·6	·06	Pine, Larch	·56	31-35	·020
Copper ..	8·9	·32	Silver ..	10·6	·38	,, Yellow	·50	28-33	·018
Ebonite	1·2	·04	Slate ..	2·9	·10	Walnut ..	·67	30-50	·024
Glass ..	2·7	·10	Steel ..	7·9	·28	Benzol ..	·80	50	·029
Gold ..	19·3	·69	Stone, abt	2·5	·09	Meth. Spirit	·82	51	·029
Ice	0·93	·03	Tin ..	7·4	·27	Paraffin oil	·80	50	·029
India r'br.	0·9	·03	Water (*pure*)	1·0	·036	Petrol (Avn ·68)	·74	46	·027
Iron (*cast*)	7·2	·26	,, (*sea*)	1·03	·037				
,, (*wrt*)	7·7	·28	Zinc ..	7·2	·26				

Gases: Ft. Acetylene, 1·10 oz; Air, 1·22; Carb. Acid, 1·86; Coal Gas, 0·61; Hyd'gn, 0·08; Oxygen, 1·35.

Imperial Coinage. Diameter in ins.; weights in grains, &c.

Gold:—	Ins.	Grs.	**Silver** *cnt'd*	Ins.	Grs.	**Copper:**	Ins.	Oz.
Sovereign	0·86	123·2	Florin	1·12	174·5	Penny	1·21	$\frac{1}{3}$ oz
Half-Sov.	0·75	61·6	Shilling	0·92	87·2	Halfp'ny	1·00	$\frac{1}{5}$ oz
Silver:—			Sixpence	0·76	43·6	Farthing	0·80	$\frac{1}{10}$
Half-cr'n	1·27	218·1	Threep'ny	0·64	21·8			

Money lying at undrawn Interest. When it doubles itself:—

Interest	2½%	3%	3½%	4%	4½%	5%	5½%	6%
Simple	40 yrs.	33⅓ yrs.	28·57 y.	25 yrs.	22·22 y.	20 yrs.	18·18 y.	16⅔ yrs.
Comp'd	28·07 y.	23·45 y.	20·15 y.	17·67 y.	15·75 y.	14·21 y.	12·95 y.	11·89 y.

Introduction

Tables have been with us for some 4500 years. For at least the last two millennia they have been the main calculation aid, and in dynamic form remain important today. Their importance as a central component and generator of scientific advance over that period can be underestimated by sheer familiarity. Like other apparently simple technological or conceptual advances (such as writing, numerals, or money) their influence on history is very deep. The history of tables now deserves, and is ready, to be brought forward from the narrow floodlights of particular special studies into the open sunlight.

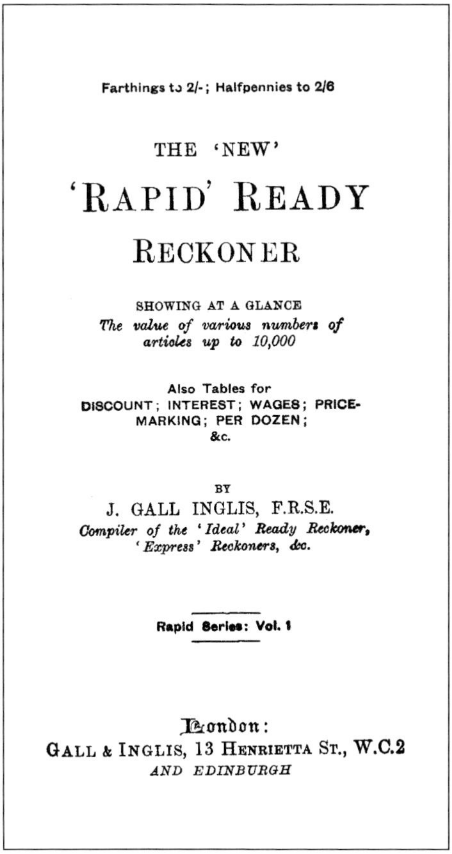

Farthings to 2/-; Halfpennies to 2/6

THE 'NEW'

'RAPID' READY

RECKONER

SHOWING AT A GLANCE
The value of various numbers of articles up to 10,000

Also Tables for
DISCOUNT; INTEREST; WAGES; PRICE-MARKING; PER DOZEN; &c.

BY

J. GALL INGLIS, F.R.S.E.
Compiler of the 'Ideal' Ready Reckoner, 'Express' Reckoners, &c.

Rapid Series: Vol. 1

London:
GALL & INGLIS, 13 HENRIETTA ST., W.C.2
AND EDINBURGH

Fig. 0.1 A page from J. Gall Inglis's *The rapid ready reckoner*, published in London in the early twentieth century. Ready reckoners were pocket- or handbag-sized books of tables, designed to aid everyday calculations, especially in multi-base Imperial systems of weights, measures, and coinage, in the days before electronic computing devices. This page shows four different kinds of table: (i) The Metric System (empirical data for use as a calculation aid); (ii) Weights of Materials (presentation of theoretical data); (iii) Imperial Coinage (presentation of empirical data); (iv) Money Lying at Undrawn Interest (theoretical data for use as a calculation aid).

The issues turn out to be very interesting. From the earliest times there has been a range of different kinds of table, from the representation of mathematical functions to documents summarizing empirical values. What they have in common is an expression of complex information in a two-dimensional form. From the start, issues of design and legibility jostle with issues of abstract information processing. The structure of tables, the transition from one-dimensional to two-dimensional layout in the location of information has a far greater significance than might naively be expected. A number of diverse roles are involved in the panorama of activities associated with tables, among them theoretician, constructor, scribe, printer, and consumer.

The insights brought to bear by recent historiography enable us to position the history of tables as a fascinating confluence of different aspects of human endeavour. In the history of information processing, for example, we can see table making in the modern era as lying at the historical junction between the factory-based production of physical goods and the office-based processing of information. The wide range of consumers of tables, whether scientific users such as astronomers, mathematicians, physicists, and medical statisticians, or professional and trade users such as engineers, actuaries, or navigators, indicates something of the challenge to historians to make sense of an extremely rich field of understanding.

The history of this extraordinary human invention falls into four periods. From around 2500 BC to 150 AD, the story is one of the invention of the table as a concept and its realization in a number of forms for different purposes. Over the next millennium and a half the second period saw some of the great achievements of the human mind, in the astronomical and trigonometric tables which lay at the heart of progress in the hard sciences leading up to the scientific revolution. The third period, from the early seventeenth to the mid-nineteenth centuries, was the heyday of work on logarithm tables which formed the basis of calculation needs for the industrial revolution. The fourth period, from the mid-nineteenth century up to the present, has seen a number of developments in the production of a range of ever more sophisticated tables for physical, mathematical, industrial, and economic purposes, as well as the development of technology to help in their calculation. The story is by no means over, as the development of spreadsheets, dynamic tables in computers, shows that there is still a lot of life in the deep idea of presenting information on a two-dimensional tabular screen.

In recent decades there has been unprecedented historical activity on each of these four periods, work of high quality which has appeared in specialist journals in different areas (such as economic history, Oriental studies, and the histories of technology, astronomy, computing, information science, mathematics, and science). It is now timely to seek to synthesize the results of these studies for a broader audience and present for the first time a historic sweep of a fascinating and unexpectedly important human invention.

INTRODUCTION

These paragraphs were written by John Fauvel at the start of our planning meetings for 'Sumer to Spreadsheets: The History of Mathematical Table Making', the summer meeting of the British Society for the History of Mathematics held at Kellogg College, Oxford in September 2001, of which this book is the final outcome. John died unexpectedly in May 2001. Not wanting to dilute his voice amongst our own, we have chosen to let him have the first word, and simply to expand on three of the themes he mentioned: the different types of table; the communities of table makers and consumers; and the production of tables.

Types of tables

Tables facilitate the selection, categorization, calculation, checking, and extraction of data—but there are as many different applications of mathematical tables as societies which used them. The twelve chapters of this book cover a bewildering variety: from Sumerian tables of squares (Chapter 1) to late twentieth-century spreadsheets (Chapter 12), from natural logarithms and other purely mathematical functions (Chapters 2, 4, 9, and 10) to astronomical ephemerides (Chapters 1, 7, and 11) and statistical data (Chapters 3 and 8). How can we begin to make sense of this enormous range?

Take, for example, the ready reckoner. These little books, often published with titles such as *The pocket ready reckoner*, were small, inexpensive sets of commercial tables. They were a common calculating aid in the Western world, but particularly in Britain because of the difficulty of computing sterling amounts and quantities in imperial weights and measures (both of which were non-decimal). They were particularly common—almost essential—in retail emporia that unit-priced goods. Imagine walking into a draper's at the turn of the twentieth century and buying $13\frac{1}{2}$ yards of curtain material at 5s 9d a yard. This was not an easy calculation, even in an age when schools taught commercial arithmetic. However, the appropriate table gives the almost instant answer of £3 17s $7\frac{1}{2}$d. The ready reckoner became a commodity item during the late Victorian period, when low-cost publishing made them very affordable. They continued to be used up to about 1970, when currency decimalization (in Britain) and inexpensive electronic cash registers and calculators made them obsolete.

The ready reckoner was not just for calculating purchases in retail establishments, though that was certainly its dominant use. It was a compendium of tables that any citizen—from cleric to jobbing builder—might find useful to have about their person. Here are examples of some of the typical tables they contained: Perpetual calendar, Table to calculate Easter day, Metric equivalents of imperial weights and measures, Tables for commissions and discounts, Interest tables, Stamp duty, Postal rates, Wage tables, Currency conversion tables, Areal, liquid, and dry measures.

Classification and taxonomy offers one way of making sense of the universe of tables. First, we can think about what goes into tables: are they computed from empirical or theoretical data? Or to put it another way, to what extent are the contents of the table determined by social convention and/or objective mathematical formulae? Second, we can consider what tables are used for: are they aids to further calculation (such as the dreaded four-figure tables of school mathematics before the pocket calculator made them obsolete) or do they present self-contained information in a final form (for instance a railway timetable)? In short, it should be possible to make a rough and ready 'table of tables', categorizing them as to whether they are computed from empirical or theoretical data, and whether they are primarily used to calculate other things or are an end in themselves. For example, the tables from *The rapid ready reckoner* shown in the frontispiece to this Introduction might be classified like this:

	Computed from empirical data	*Computed from theoretical data*
Used as calculating aid	The Metric System	Money Lying at Undrawn Interest
Used to present data	Imperial Coinage	Weights of Materials

But categorization is not straightforward. At one end of the empirical–theoretical spectrum there is census data, as described by Edward Higgs in Chapter 8. Not only do census designers have to choose the types of information they wish to elicit from the population, but they have to then categorize the relatively free-form responses they receive into discrete categories for tabulation. However carefully designed the process is, censuses do not produce objective results but reflect the ideals and prejudices of the commissioning society, as embodied in the minds and work of the census officials. At the opposite extreme there are tables of mathematical functions, such as those manufactured by de Prony and his team in late eighteenth-century Paris (see Ivor Grattan-Guiness's description in Chapter 4). Sines, logarithms, or number theoretical series—provided they are calculated accurately—are (and have been) reproducible across time and space regardless of the language, number base, or personal belief system of the tabulator. Along the spectrum in between are myriad tables which are constructed from a mixture of empirical and theoretical data, the socially constructed and the value-free. How, for instance, should we consider astronomical ephemerides, which tabulate the predicted movement of the major heavenly bodies over the coming year(s)? They depend on a celestial coordinate system and on a mathematical model describing lunar, solar, and planetary motion, both of which are socially defined and open to constant revision and refinement, as Arthur Norberg explains in Chapter 7. But from another viewpoint an ephemeris is no more than the tabulation of a mathematical function, which makes it in some sense more objective than a census table.

On the calculation-aid/data-presentation spectrum we see a similar range of possibilities. Logarithms, of course, are the archetypical calculating aid, as Graham Jagger shows in Chapter 2. Ephemerides were calculating aids too, enabling navigators to locate their position on the Earth's surface by the night sky. On the other hand astronomers could also use them like a railway timetable, looking up the times and locations of celestial events that they wished to observe. Census tables and tables of vital statistics were, on the whole, meant to be read as final reports rather than as raw material for further calculations—although this was necessary too for instance to enable civil servants to plan a community's needs. In Chapter 3 Chris Lewin and Margaret de Valois explain how actuaries used tabulated mortality data to create and develop the concept of life assurance.

When we move on to look at dynamic tables in the form of spreadsheets (Chapter 12), we now see a form of table that can both show empirical or derived data and then manipulate that data and display the results all in one and the same 'table'. Here we see an interesting paradox: on the one hand computers have been the death of the printed table-as-calculating-aid but conversely computerized spreadsheets have given new and vigorous life to the still ubiquitous table-as-data-presentation format.

Communities of table makers and users

Table making is a community-based activity. Communities of table makers and table users are as old and diverse as tables themselves. However, three particular groups feature across many of the chapters of this book: the astronomical, mathematical, and actuarial communities.

Astronomers have proved to be the most enduring community of mathematical table makers and table users as generation after generation has built upon the work, and tables, of its predecessors. Claudius Ptolemy (c.85–165 CE), working in the famous library at Alexandria in Egypt, was author of the 13-volume astronomical work *Syntaxis* or *Almagest* which revolutionized astronomical thought.[1] He selected the most useful of the astronomical tables from the *Almagest*, edited and republished them with instructions on their use as the *Handy tables*. They were translated from Greek into Latin, Arabic, Persian, and Sanskrit and had a circulation larger even than the *Almagest*. The *Handy tables* were originally copied by hand in manuscript form but editions in print form continued to be published in Western Europe for centuries. Their longevity, wide distribution and influence among astronomers worldwide mean that Ptolemy's *Handy tables* can justifiably claim to be the first mass produced mathematical table.

Another famous table maker who was the product of a scholarly institution was al-Khwarizmi (c.780–850), of the royal House of Wisdom in Baghdad. While he is best known as the father of modern algebra and algorithms, al-Khwarizmi's collection of astronomical tables and accompanying instructions, or *zīj*, is the earliest surviving astronomical treatise in Arabic. It had a long life: Adelard of Bath, for instance, translated it into Latin in the early twelfth century—and it was still used in the Jewish Geniza in Cairo in the nineteenth century![2]

In the second half of the eleventh century, an important group of astronomers gathered in Toledo in Muslim Andalusia (southern Spain) to produce the *Toledan tables*—some parts of which derived from the work of Arab astronomers such as al-Battani and al-Khwarizmi and other parts from Ptolemy. The *Toledan tables* became highly popular throughout Europe until the fourteenth century.[3] Another collaborative medieval table-making effort was the *Alfonsine tables*, assembled between 1263 and 1272 by a group of 15 astronomers at the instigation of Alfonso X of Leon and Castile. The *Alfonsine tables* gained popularity in Paris in the 1320s, were further developed in England, and had a major influence on European astronomy in the following centuries.

The Renaissance saw huge advances in both theoretical and observational astronomy in which the invention of the telescope and the gradual adoption of the Copernican view of the solar system played a fundamental part. The international nature of the astronomical community allowed for astronomers across Europe to learn from one another and to build upon each others' work. The work of Brahe, Kepler, Newton, Mayer, and others led to increasingly accurate tables for predicting the positions of the Sun, Moon, planets, and stars. The interplay between changing astronomical theory and the generation of new astronomical tables is well described by Norberg in Chapter 7.

Ephemerides, or tables of predicted positions, were highly collaborative works. They were constructed from the best available astronomical tables, often by a team of computers who were sometimes part of the astronomical research community and sometimes simply hired hands. It is here that we begin to see the division between those who actually computed the tables and the editors in whose name the tables were published. International collaboration in the construction of national ephemerides was common. For example Nevil Maskelyne, editor of the *Nautical almanac* 1767–1811, not only checked the *Nautical almanac* tables against those of the French *Connaisssance des temps* but also kept up a vigorous astronomical correspondence with his French counterpart Jerome Lalande throughout the intermittent Anglo-French wars of the period.[4] By the 1930s national ephemeris publishers throughout Europe had divided up the necessary table-making work between the various international offices in order to reduce the workload and avoid duplication of effort. Even today international cooperation is common and, as George Wilkins explains in Chapter 11, the British and US almanacs are now jointly computed and published.

While astronomers have been the predominant makers and users of mathematical tables over the last four millennia, another important, and often linked, group is the mathematical community. In Chapter 2 Jagger describes Napier's invention of logarithms and the subsequent publication of logarithm tables by Briggs, Gunter, and others. Jagger's narrative demonstrates how, following Napier's initial work, the mathematical community worked to develop the use of logarithms as a calculating aid and to prepare the necessary tables.

Mathematical tables of the higher functions began to appear at the beginning of the eighteenth century and the increasing importance of higher mathematical functions to mathematical physics throughout the nineteenth century led to a rapid increase in the number of tables published. Mathematicians began to be concerned by the lack of coordination and unwitting duplication of table-making effort within the mathematical and mathematical physicist communities. In response the British Association for the Advancement of Science set up a Mathematical Tables Committee in 1871 to produce a survey of existing mathematical tables and to coordinate the creation of new ones. As described by Mary Croarken in Chapter 9, over the next ninety years the British Association Mathematical Tables Committee became a central focus for mathematical table making in Britain. At its inception the Mathematical Tables Committee was highly representative of the communities it served; later on in its history we see a growing gulf between the mathematical table maker and the user communities. In several instances Mathematical Committee members were making tables primarily because they enjoyed the process of making tables and not wholly for the benefit of the target users.

A similar separation can be found in the Works Project Administration Mathematical Table Project created in 1930s New York and described by David Grier in Chapter 10. Here the table makers were not willing volunteers but unemployed workers desperately in need of work. Arnold Lowen, the project's leader, sought to gain approbation for the project within the American mathematical community, albeit with limited success. Here the table makers were intellectually, culturally, and socially separate from the community of users they sought to serve.

Another table-making community, the actuaries, is discussed by Lewin and de Valois in Chapter 3, and touched on by Higgs in Chapter 8. It is often assumed that actuarial table makers had little in common with either the mathematical or astronomical communities, their products being used in different spheres of human endeavour. However, this assumption proves to be much too simplistic. The skills for producing actuarial tables were equally applicable to making astronomical or mathematical tables and *vice versa*. For example Edmund Halley, Astronomer Royal 1720–1742, was author of a highly influential life table published in 1693.[5]

Perhaps the most significant contact between the actuarial, mathematical, and astronomical communities occurred in early nineteenth-century Britain. In 1820 the Astronomical Society of London was founded with the aim of standardizing astronomical calculation and collecting and distributing data. Its founding members included Francis Baily and his brother Arthur, successful London stock brokers; their colleague Benjamin Gompertz; currency exchange expert Patrick Kelly; the financier Henry Thomas Colebrooke; West India merchant Stephen Groombridge; the mathematician and astronomer John Herschel; and Charles Babbage, the mathematician son of a London banker and briefly actuary of the Protector Assurance Society. The predominance of businessmen, most of whom had practical experience of actuarial table making as well as an interest in astronomy and mathematics, is remarkable.

Adversity too has shaped mathematical table-making communities. During both World Wars mathematicians and astronomers came together and applied their table-making skills to the preparation of ballistics and other tables required by the armed services. During the First World War it was primarily mathematicians such as Karl Pearson in Britain and Oswald Veblen in the US, but in the Second World War the net was stretched wider. In Britain the staff of the Nautical Almanac Office were joined by graduate mathematicians to create the Admiralty Computing Service which produced almost one hundred tables of direct relevance to the war effort.

Table making as a communal enterprise has thus been part of the history of mathematical tables from ancient times. Today so many businesses, universities, and individuals run Microsoft Excel spreadsheet software that the community of table users with the ability to share spreadsheet data on a world wide basis is truly worldwide.

Making tables

A theme that surfaces in many of the chapters of this book concerns the way in which tables were actually calculated. One can discern five distinct styles of table making: the solitary table maker; communal computing; computing bureaus; mechanized computing; and, finally, computerization. Although this is the order in which these five different styles evolved, solitary and communal, manual and mechanized computing styles have co-existed throughout the history of table making.

The working methods of the early table makers are shrouded in mystery. We do not know the extent to which tables were solitary or communal endeavours, but both modes clearly existed. The solitary table maker is a heroic figure, though the tradition is perhaps exaggerated. For example, according to D. E. Smith, it was said that when Napier and Briggs first met—the latter having completed the tables the former had begun—'almost one quarter of an hour was spent, each beholding the other with admiration, before one word was spoken.'[6] However, as Jagger writes in Chapter 2, Napier at least had a little help from his friends. But the lone table maker continued to work long into the twentieth century. For example, Emma Gifford, 'an extremely cultured lady but by no means a mathematician', was occupied for several years in the solitary task of compiling and publishing her 500 page volume of *Natural sines*.[7]

However, from the earliest times big table-making projects had to employ several human computers. One of the best known examples of such a group endeavour is the *Nautical almanac*, published continuously from 1767. As described by Wilkins in Chapter 11, the Astronomer Royal Nevil Maskelyne employed a network of computers, many of whom served for decades. The computers were all freelance, located all over England, and their work was entirely coordinated though the postal system.[8] Such clear evidence of the way computing is organized is rare; much more usually the historian is left with just a tantalizing hint. For example, as explained in Chapter 1 by Eleanor Robson, table making in Mesopotamia was very much a collective activity, but the surviving artefacts are necessarily silent about how the work was actually organized. The *Opus Palatinum* of Rheticus published in 1596 (and the source book for Gifford's *Natural sines*) was said to be the work of eight computers. Leibniz hints at the existence of major table-making projects, noting in 1745 that his newly invented calculating machine 'has not been made for those who sell vegetables or little fish, but for observatories or halls of computers.'[9] In the twentieth century, as Croarken explains in Chapter 9, the British Association Mathematical Tables Committee made use of small calculating teams as well as solitary calculation.

In 1790, Gaspard Riche de Prony established the Bureau du Cadastre to produce a new set of tables for the French ordnance survey. The Bureau marked a transition from small, informal computing teams to a large-scale, highly organized bureaucracy. De Prony's innovation was path-breaking for two reasons. First, the sheer scale of the activity: the new French tables of logarithms and trigonometrical functions were of an unprecedented precision and were computed *de novo* (unlike earlier tables which had generally been compiled from previously existing tables). Second, de Prony consciously adopted an industrial metaphor. He had been influenced by Adam Smith's classic *Wealth of nations* (1759), in which he gave his famous description of a pin-making factory where, by division of labour, a community of specialized workers was vastly more productive, per person, than one solitary worker who undertook all the processes of pin manufacture. In an evocative passage, quoted by Grattan-Guinness in Chapter 4, de Prony decided 'to manufacture logarithms as one manufactures pins.' The Bureau de Cadastre was in every sense a computing 'factory'. The Bureau employed three classes of labour: a small advisory group of eminent analysts; an executive group of between six and eight professional mathematicians; and a large number of between 60 and 80 relatively unskilled computers. By employing the method of differences, which required only the operations of addition and subtraction, it was possible to use the lowest grade, and therefore the most economical, computing labour. Organized computing bureaus became the norm for large-scale table-making projects. In the 1830s, for example, the Nautical Almanac Office's network of freelance computers was replaced by a permanent office of about a dozen computing staff. The processing of census data was another example of the large-scale employment of low grade staff for table production. In the case of the British Census, as explained by Higgs in Chapter 8, Census tabulation used the 'ticking method' which was extremely tedious but required only the lowest form of clerical life.

The computing bureau style of operation remained popular for major table-making projects until after the Second World War. The best documented was the WPA Mathematical Tables Project established in New York in the 1930s depression, described by Grier in Chapter 10. The WPA project was established partly as a make-work project for the unemployed, but also as a genuinely needed computing service. At its peak the project employed 200 computers, and it came into its own with the computation of LORAN tables in the early years of the war. The WPA project was by no means the only such computing organization in wartime America. For example, the Army's Ballistics Research Laboratory and Moore School of Electrical Engineering at the University of Pennsylvania employed a team of 100–200 female computers (each equipped with a desk calculating machine) to make ballistics tables. This tide of table making, which threatened to overwhelm existing methods, led the Moore School to become the birthplace of the modern computer.[10]

The difference engine was the most radical departure from manual methods of table making. Although table makers sometimes used calculating aids (such as logarithm tables or desk calculating machines), the difference engine was intended to entirely remove the human element, and once set up an engine would churn out results indefinitely. As Mike Williams explains in Chapter 5, Charles Babbage was the most important figure in the history of the difference engine. He was one of the foremost economists of his day, and thoroughly conversant with the principle of the division of labour.[11] However, in the 1820s human labour was starting to be challenged by machinery and the pin-making factory was giving way to the pin-making machine. For reasons discussed by

Williams, and Doron Swade in Chapter 6, Babbage failed to complete a full-scale machine—being diverted by the analytical engine. However, he inspired several imitators, notably Georg and Edvard Scheutz in Sweden. A copy of their machine was purchased by the General Register Office to make a new life table, although the machine required much coaxing, and did only a fraction of the job. Other difference engines followed; by Martin Wiberg in Sweden, George Grant in the United States, and Christel Hamann in Germany. The difference engine concept was never really commercially viable, however, and special-purpose machines had died out by the First World War. By contrast, after the 1890 US Census, Herman Hollerith recognized that census data processing was not a big enough market for his punched card machines, and so he developed equipment for ordinary businesses. Thus, by the time the British census came to use Hollerith machines in 1911, census tabulation was only one of many applications for punched-card machines. It seems that table making was too specialized and too narrow an activity to support a dedicated technology. This was clearly understood in the late 1920s by L. J. Comrie who, as described by Williams in Chapter 5, adapted commercial calculating machinery for the *Nautical almanac*, ushering in an era of mechanized table making at last.

The emergence of the digital computer first transformed the way tables were made, and then undermined their very existence. Computer power enabled an explosion in the production of tables—not all of them of obvious utility. For example the Harvard Mark I computer produced some fifty volumes of Bessel function tables, many of which were barely used, and earned the computer the nickname 'Bessie'.[12] At first it was thought that computers, by giving more individuals and institutions the capability to do computing, would encourage the production of many new tables. However, when it became clear that computers could calculate function values on-the-fly, the need for tables—whether on punched cards or in printed form—evaporated. In Chapter 9 Croarken describes rather poignantly how one institution, the British Association Mathematical Tables Committee, came to terms with the fact that the life's work of its members was no longer needed.

From Sumer to spreadsheets

The history of tables is not just an arcane corner of the history of science (mathematics, computing, or astronomy). We have already touched upon some aspects of economic history in the way that labour and machinery were organized to construct tables. We have seen that tables have a social history too, or at least their producers and users do. The communities they worked in, and the support those communities gave or denied them, played a major part in how tables were developed, published, and circulated. But we can also examine the internal structure of tables, as greatly overlooked evidence for cognitive history. They tell us much about how people have selected, classified, and manipulated quantitative data at different times and places. Tables also speak to us about the history of literacy and numeracy. The material culture of tables—from clay tablets, papyri, and manuscripts to printed books and shrink-wrapped software packaging—sheds new light on the history of the book and the transmission of knowledge.

While the *list* has been hailed as a major breakthrough in cognitive history, most famously by the anthropologist of literacy Jack Goody,[13] the table as a pre-modern phenomenon of structured thought has been completely neglected. Indeed, for Goody (non-numerical) tables are no more than a means for modern scholarship to present 'the communicative acts of other cultures, non-literate and literate... a way of organising knowledge about classificatory schemes, symbolic systems, human thought...' whose 'fixed two-dimensional character of may well simplify the reality of oral communication beyond reasonable recognition, and hence decrease rather than increase understanding'.[14] For Goody, then, tables are a means by which Western academia imposes inappropriate, simplifying order on the complexities of other cultures. Higgs's chapter on the General Register Office's tabulation of data in nineteenth-century Britain does indeed show that 'the construction of tables involves decisions about what is important and what is not, and what should be collected and presented, and what can be ignored' but that 'there is nothing necessarily sinister about such processes of truncation'.

Robson's chapter on the uses of numerical tables in many aspects of literate life in ancient Iraq (Chapter 1) shows that tables are by no means an invention of modernity but have been in lively, inventive, and constructive use for millennia. It is not difficult to assemble examples from other ancient cultures. Over a hundred arithmetical tables are known from the world of Classical Antiquity, written in Demotic Egyptian, Coptic, and Greek,[15] and a similar number of astronomical tables have been recovered from the first five centuries CE, 'with links not only to the Greek theoretical astronomy of Hipparchus and Ptolemy but also backward to the Babylonians and forward to the astronomers of Byzantium, Islam, and the Latin West.'[16] This list could perhaps be extended indefinitely.

The two-dimensional structure of the table was elaborated remarkably quickly, in the blink of an eye in historical terms, but it took millennia for the amazing power of the deceptively simple table to reach its zenith. In Chapter 1 Robson traces the functional development of administrative tables in early Mesopotamia over a period of just 50 or 60 years in the nineteenth and eighteenth centuries BCE. Categorization and data selection come first: these first tables were no more than multi-column lists of different named types of empirical data. Column headings and row labels, were an integral part of tables right from their inception, even though it was to be millennia before titles, headings, and subheadings were applied to works of connected prose, whether literary epics or royal decrees. Thus quantitative and qualitative information could be separated, producing a great efficiency both in recording and retrieving structured information. The scribe no longer had to waste time and effort in describing each item accounted for, but could simply tabulate it in the correct row and column, while the supervisor could take in at a glance the finished document and the pattern of information it contained.

Within very short order—a decade or so—the computational power of tables was discovered. First, one could use the very structure of tables as an aid to calculation. Columns could be totalled; products could be found across rows; cross-checking for errors became easier. Second, tables could be used to store basic functions used for more complex calculations, from integers and their squares or cubes to reciprocals.

When we substitute, say, paper for clay tablet, Bessel functions for squares of the integers, or EU farming subsidies for tables of goats and sheep, one can see that the similarities between tables ancient and modern are much greater than their differences. Right to the present day, the table can be seen as an elaboration of the basic theme of homogeneous values arranged in rows and columns. This remains true even in the age of the spreadsheet. User surveys tell us that spreadsheets are most commonly used to write lists and to keep simple accounts: two basic human needs, unchanged through the millennia.

Tables are quintessential cultural objects of the civilizations that created them, improved and perfected by each succeeding generation. For example, in logarithmic tables, residual errors were gradually eradicated, values truncated or extended, and intervals reconsidered according to the dictates of the day. It is likely that, at least in theory, an ancestral line could be drawn from the logarithms of Briggs to the four-figure tables used in schools in the early twentieth century. Tables have followed the fashions of the day with regard to typeface, the use of rules and 'white space', and even the colour of paper. Indeed, as Swade describes in Chapter 6, such attention to typographical niceties made the printing part of Babbage's difference engine almost as complex as the calculating part. This trend continues even into the modern spreadsheet—where competing brands are distinguished not by their function, but by their user interface, and each 'upgrade' combines the best of the old leavened with a little of the new.

As Martin Campbell-Kelly has noted in the closing chapter, the two-dimensional table is almost an historical necessity—suggested by the two dimensional writing surface common to all civilizations. We know that different civilizations independently invented (or perhaps, like the integers, discovered) the table. Would the civilizations of another planet use tables? We like to think so.

Notes

1. G. J. Toomer, 'Ptolemy', in *The Dictionary of Scientific Biography*, New York, 1970, pp. 186–206 and *idem, Ptolemy's Almagest*, Princeton University Press, Princeton, 1998.
2. J. D. North, *The Fontana history of astronomy and cosmology*, Fontana, London, 1994, pp. 184–5.
3. J. D. North, *The Fontana history of astronomy and cosmology*, Fontana, London, 1994, pp. 207–13.
4. S. L. Chapin, 'Lalande and the longitude: a little known London voyage of 1763', *Notes and Records of the Royal Society* 32 (1978), 165–80 and D. Howse, *Nevil Maskelyne: the seaman's astronomer*, Cambridge University Press, Cambridge, 1989, p. 47.
5. E. Halley, 'An estimate of the degrees of mortality of mankind...', *Philosophical Transactions of the Royal Society* 17 (1693), 565–610 and 654–6.
6. D. E. Smith, *A source book in mathematics*, vol. 1, Dover Publications, New York, 1959.
7. E. Gifford, *Natural sines*, vol. 2, Haywood, Manchester, 1914. Quote taken from A. Fletcher *et al.*, *An index of mathematical tables*, 2nd edn, Blackwell, Oxford, 1962, p. 818.
8. M. Croarken, 'Tabulating the heavens: computing the Nautical Almanac in eighteenth century Britain', *IEEE Annals of the History of Computing*, forthcoming.

9. Translation from *The works of Charles Babbage,* vol. 2 (ed. M. Campbell–Kelly), Pickering, London, 1989, p. 181.
10. H. H. Goldstine, *The computer: from Pascal to von Neumann,* Princeton University Press, Princeton, 1972.
11. Indeed, he refined the concept into a more nuanced variation known as the Babbage Principle, which advocated that manufacturers should purchase no more than the lowest quality labour that could perform any given task. M. Berg, *The machinery question and the making of political economy, 1815–1848,* Cambridge University Press, Cambridge, 1980.
12. I. B. Cohen, *Howard Aiken: portrait of a computer pioneer,* MIT Press, Cambridge, Mass., 1999.
13. J. Goody, *The domestication of the savage mind,* Cambridge University Press, Cambridge, 1977.
14. J. Goody, *The domestication of the savage mind,* Cambridge University Press, Cambridge, 1977, pp. 53–4.
15. See the catalogues by D. H. Fowler in *The mathematics of Plato's Academy: a new reconstruction,* 2nd edn, Oxford University Press, Oxford, 1999, pp. 268–76 and 'Further arithmetical tables', *Zeitschrift für Papyrologie und Epigraphik* 105 (1995), 225–8.
16. A. Jones, 'A classification of astronomical tables on papyrus', in *Ancient astronomy and celestial divination* (ed. N. M. Swerdlow), MIT Press, Cambridge, Mass., 1999, pp. 229–340, quote from p. 335.

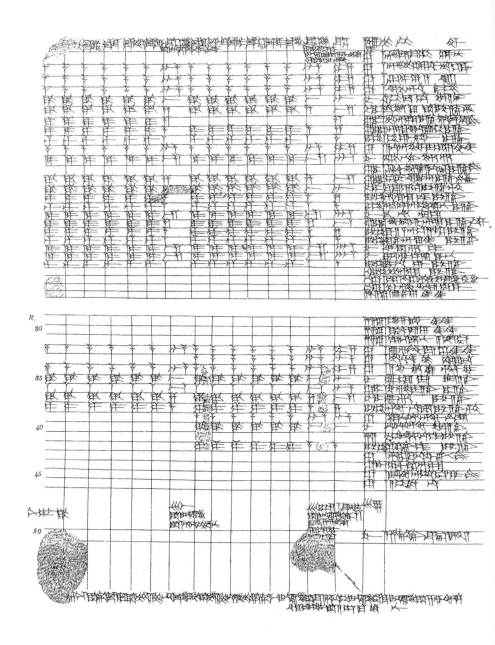

1

Tables and tabular formatting in Sumer, Babylonia, and Assyria, 2500 BCE–50 CE

ELEANOR ROBSON

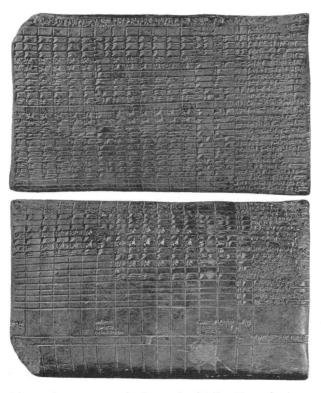

Fig. 1.1 Month-by-month wage account for the temple of Enlil at Nippur, for the year 1295 BCE. This tablet, recording the monthly salaries of forty-five temple personnel, exhibits most of the classic features of Mesopotamian tables. Column headings at the top of the table give the month names. There are also subtotals for each person every six months and a yearly total. Names and professions are given in the final column. Eighteen of the individuals listed get no payment for all or half the year; most of these are classified (in the penultimate column) as dead or absconded. There are explanatory interpolations under two headings and under their totals, and a summary at the end. (CBS 3323, from Nippur; now in the University Museum, University of Pennsylvania. A. T. Clay, *Documents from the temple archives of Nippur dated in the reigns of the Cassite rulers*, 3 vols., The University Museum, Philadelphia, 1906, pls. 25–6.)

While most of this book is concerned with the development and production of standardized mathematical tables for use as reference tools, this chapter will focus more on the invention and evolution of the numerical table as an information storage device. This was not a one-off event, with clearly traceable consequences across an ever-widening arena of functions, contexts, and cultures. Rather, even within the single cultural milieu of ancient Mesopotamia we shall see a fitful pattern of invention, partial adoption, disappearance, and re-invention time after time over the course of some two and a half millennia. Broadly speaking, documents with tabular formatting could be found in three distinct Mesopotamian locales: in the large institutional administrative archives of Sumer and Babylonia; amongst the detritus of scribal schooling, especially in mathematics and metrology; and, later, in the scholarly libraries attached to the great temples of Assyria and Babylonia. We shall examine all three in turn.

But first, if we are to talk about tables clearly and effectively we need a consistent terminology to describe them. I shall define a *formal* table as having both vertical and horizontal rulings to separate categories of information; *informal* tables, on the other hand, separate quantitative and qualitative data by spatial arrangement only, without explicit delimiters. Documents with no tabular formatting at all we might call *prose-like* or *prosaic*. *Headed* tables have columnar headings, while *unheaded* tables do not. Some tables are preceded by *titles* or introductory *preambles*; others are followed by *summaries* and/or *colophons*. In Mesopotamian tables, any qualitative or descriptive information is almost invariably contained in the final right-hand column (*row labels*), or interrupts the table as an *explanatory interpolation*. Mesopotamian tables have at most two *axes of organization*: the horizontal axis, along which different types of numerical information are categorized, and the vertical axis, down which the data is attributed to different individuals or areas. While some tables have two *axes of calculation* others exhibit just one, usually vertical, axis of calculation, or none at all. Both calculation and organization run, as a rule, left to right and top to bottom, following the direction of cuneiform script.

Tabular formatting in administrative records

There are no tables to be found in the earliest written records of Sumer. The 5600 documents from late fourth-millennium Uruk and its neighbours 'serv[ed] the accounting needs of a complex administration including offices of the fisheries, of herded animals and animal products, of field management, grain production and processing, and of labor'.[1] They were formatted not into lines of ordered text but into boxes or 'cases', each of which represented one sense unit or accounting unit. The spatial organization of the cases on the surface of the tablet could be quite complex,[2] reflecting often sophisticated accounting procedures,[3] but nowhere do we see the separation of quantitative and qualitative data into separate cases. This earliest writing comprised just number signs and ideograms, representing whole words or ideas, which could be displayed in any order within each case. But once writing evolved the capacity to represent syllables and thereby approximate the sounds of speech, it was no longer determined solely by the needs of accounting. As literacy spread into other social domains—for recording legal decisions, royal deeds, and even poetry—it became increasingly imperative to order the written symbols to follow the structure of Sumerian as a spoken language. The visual logic of the earliest accounts was thus irrevocably lost in the shift to linear organization of writing during the first half of the third millennium BCE.

Nevertheless, accounting continued to dominate the written record: a recent estimate has put the total administrative output at some 97 per cent of all surviving tablets from the third millennium.[4] The bureaucracy of the twenty-first century BCE kingdom of Ur was particularly prolific: some 45 500 administrative records have so far been published from its archives in the ancient cities of southern Iraq.[5] The accounts are almost exclusively prosaic, in both senses of the word: they record the day to day transfers of goods, livestock, and personnel from the responsibility of one official to another, and those records are made linearly across the surface of the tablet. This, for instance, is an extract from a monthly summary of four different types of ovids under the responsibility of the central livestock depot at Puzrish-Dagan:

7 nanny goats 3 billy goats: day 10
5 sheep, 5 kids, 5 billy goats
7 nanny goats: day 13
4 kids, 1 nanny goat 1 billy goat: day 14
2 sheep, 3 kids 2 nanny goats
4 billy goats: day 15
3 sheep, 15 kids 3 billy goats, 9 nanny goats: day 16
3 sheep, 5 kids, 1 billy goat
6 nanny goats: day 18
12 sheep, 11 billy goats, 7 nanny goats:
day 20[6]

Early Mesopotamia: Sumer and Babylonia

Mesopotamia is the classical Greek name for the Land Between Two Rivers, the Tigris and Euphrates, corresponding more or less to the modern state of Iraq. Ancient national, regional, and cultural boundaries were very fluid, however, and at times Mesopotamia incorporated neighbouring lands which now belong to modern Turkey, Syria, and Iran.

Mesopotamia was first settled some eight thousand years ago, as the Gulf receded and people moved south into the gradually more habitable marshlands it left behind. The shift towards urban living was a long and complex process, but by the middle of the fourth millennium BCE the cities of Mesopotamia were the largest, wealthiest, and most sophisticated in the world. More or less politically autonomous, and each centred around the temple of an anthropomorphic deity, the cities drew their strength from local religious power and from long-distance trade of agricultural and animal products in exchange for metals and luxury goods.

Driven by the need to manage complex economies and large populations of people and livestock, the temple administrators of the city of Uruk developed an accounting system which, by about 3300 BCE, had matured into the world's first writing. It encoded the Sumerian language—related to no other known language, living or dead—with marks in the surface of hand-held clay tablets. Originally impressed number signs and incised pictograms or more abstract symbols, representing whole words, were arranged in boxes, or *cases*. During the Early Dynastic period (*c.* 3000–2400 BCE) the incised script developed into wedge-shaped or *cuneiform* characters, while retaining the ancient impressed notation for numerals. At the same time, cases evolved into ordered lines of characters, which evolved the capacity to represent syllables and thereby approximate the sounds of speech. This enabled the scribes of Sumer to record a wide variety of documents and texts: they were no longer restricted to keeping accounts, although this was still a core activity. A second language also came into widespread use in the course of the third millennium. Akkadian is a Semitic language, indirectly related to Hebrew and Arabic, and bears no resemblance at all to Sumerian.

The last third of the third millennium BCE saw some major shifts in Mesopotamian socio-political structure. For the first time, large territorial states came into being, centred on one city but aiming to unify the whole land. The kingdoms of Akkad (*c.* 2350–2250 BCE) and Ur (*c.* 2100–2000 BCE) were each created by the ambitions and abilities of individual, long-lived and charismatic kings, Sargon and Shulgi, who seem to have constructed their empires through sheer force of personality. Inherently unstable, these kingdoms both have histories of rapid unification, centralization, and standardization followed by unsteady but inexorable decline over the next two or three generations.

> **Early Mesopotamia: Sumer and Babylonia** *cont.*
>
> The fall of the kingdom of Ur at the end of the third millennium saw the political landscape of Mesopotamia revert to its earlier configuration of small city states vying for land, water rights, and trade routes. The city of Nippur, religious centre of Enlil, the 'father of the gods', changed hands between the southern kingdoms of Isin and Larsa at least half a dozen times in two centuries. In the early eighteenth century Hammurabi, king of the hitherto unimportant city of Babylon, managed to outwit and outmanoeuvre his rival dynasts in this real-life game of Monopoly, often playing one ruler off against another. By the end of his forty-year reign (*c.* 1792–1750 BCE) vast swathes of southern Mesopotamia had come under Babylonian influence or control. Hammurabi also made sweeping religious changes, culminating in the 'retirement' of Enlil in favour of his personal deity Marduk as head of the pantheon. However, none of Hammurabi's successors appears to have had his longevity, intelligence, or stamina, so that as before Mesopotamia was united only for a matter of decades before a series of contractions and retrenchments. By the end of the sixteenth century Babylon was little more than a city state again.
>
> For the city of Nippur under Hammurabi's successor Samsu-iluna we can closely identify the curriculum used in three or four different scribal schools, and the variations between them. Scribal schools were for the most part very small indeed, housed in the home of the teacher, whose primary profession was often that of a temple administrator or priest. They catered for up to half a dozen boys at any one time, each at different stages of their education. Much of the learning was through repeated copying and rote memorization.

It is not an easy task to total each category of animal, or to see at a glance which days there are no entries for: given that this document would have been compiled from at least thirty daily records and might itself have been one of up to five hundred tablets from which an annual account had to be compiled, this non-tabular format was gravely inefficient.

A tiny number of tables are known from the bureaucratic archives of Ur, one of which is also from Puzrish-Dagan.[7] It too is an account of sheep and goats:

3	3	3	2	1	lambs
93	93	93	6[2]	31	first-rate sheep
6	6	6	4	2	billy goats
102	102	102	68	[3]4	

Fig. 1.2 Compare these two accounts of sheep and goats. On the left the non-tabular account, c. 2050–2025 BCE, makes no visual separation between numerical and descriptive data, nor between different categories of data. On the right is the earliest known tabular account, c. 2028 BCE. Row labels giving the type of animal are on the far right; columnar totals are in line 4. Line 5 appears to contain column labels in the form of abbreviated personal names. Summary information at the bottom of the tablet is largely missing. The date is on the top of the reverse. (AUAM 73.0639 and AUAM 73.0400, from Puzrish-Dagan; now in the Horn Archaeological Museum, Berrien Springs, Michigan. M. Sigrist, *Neo-Sumerian account texts in the Horn Archaeological Museum*, vol. 1, Andrews University Press, Berrien Springs, Michigan, 1984, no. 63, no. 56.)

In the fifth row the columns are labelled with the abbreviated names of the officials responsible; at the bottom is the date on which it was drawn up. It has most of the features of later administrative tables: two axes of organization, one, vertical axis of calculation, row labels to the right of the table—but no headings, so that we cannot tell why the three varieties of livestock were further categorized.

This appears to be a reasonably successful attempt at table making—its structure is clear, the additions are correct—so why didn't the idea catch on? We can put forward a tentative hypothesis, based on its probable archival context. Puzrish-Dagan, a purpose-built depot near Nippur,[8] functioned as the central livestock management centre for the kingdom of Ur for over thirty years (c. 2056–2020 BCE). Almost 12 000 documents from its archives have been published, all of them excavated illicitly and therefore without context. Hundreds of tablets from each year of its operation have survived. Our table comes from the penultimate active year of Puzrish-Dagan, after which date the number of surviving records declines dramatically, from an annual average of 220 in the preceding decade to single figures. Maybe the scribes were simply under too much pressure preventing institutional collapse—part of the general 'relatively sudden retreat of the central administration'[9] at that time—to experiment further with new data storage formats.

It was not until the mid-nineteenth century BCE that tables began to be used with any consistency or continuity.[10] The first phase in their development is attested by a group of around two hundred tabular lists from the temple of the god Ninurta in Nippur. Over four hundred tablets record regular food offerings (bread, flour, and beer) made to Ninurta and other deities and divine objects which were then redistributed to temple personnel.[11] The ancient dates on the documents themselves span some eighty years and six changes of political allegiance between the kingdoms of Isin and Larsa between 1871 and 1795 BCE. At least three different methods were used to record the offerings and redistributions, the most successful and enduring of which spanned at least twenty years (*c.* 1855–1836) and one change of government. The archive's scribe ruled the obverse of the tablet vertically into six columns, heading them 'bread', 'shortbread', 'fine flour', '*utu*-flour', 'beer', but leaving the final column blank. In this final column he listed the names of the divine recipients and in the others the quantities of goods they received. He made no calculations, horizontal or vertical, and on the reverse he never tabulated the redistributions but simply listed them prosaically, aligned with the names of the human beneficiaries.

Outside the Nippur offerings archive we can detect three phases in the development of administrative tables,[12] the first of which was a period of experiment and evolution (*c.* 1850–1795). The tables from the immediately following decade (the mid-1790s to 80s) exhibit by contrast a relatively rigorous standardization. Almost all of them are in landscape format, and all are headed uniformly, with the final heading (over the row labels) reading MU.BI.IM 'its name'. None has an introductory title or preamble but almost all have two axes of calculation and interlinear explanatory interpolations. Several groups of related tables can be detected, in which the scribes sometimes experiment with the format and layout of the documents. There is then a twenty-five year gap in the documentary record, until after the conquest of Larsa by Babylon. At this point (*c.* 1758 BCE) the tables begin to be preceded by general introductory matter, and the tablets are predominantly in portrait format.

Very few tables are attested after the mid-1720s for over 70 years until the last half of the seventeenth century BCE. All are much simpler organizationally than their precursors, with at most one axis of calculation, usually simple columnar additions. Almost all are in portrait orientation, reflecting the fact that they have no more than three data columns—whereas the earlier administrative tables regularly had eight or nine. These late tables are almost all from Sippar and the region around Babylon, whereas their predecessors are mostly from Larsa and the south.

After the mid-seventeenth century BCE the Mesopotamian historical record falls silent for about three hundred years. Our next evidence for tabular documentation comes once again from Nippur, and shows a remarkable similarity to what had gone before.[13] An administrative archive of some 600 tablets, spanning the years *c.* 1360–1225, comprises 'records of the receipts of taxes or rents from outlying districts about Nippur; of commercial transactions conducted with this property; and the payment of the salaries of the storehouse officials as well as of the priests, and others in the temple service. In other words they refer to the handling and disposition of the taxes after they had been collected.'[14] Some 33 percent of the records in that archive are tabular. They exhibit all the featural complexity of the eighteenth-century tables—headings, totals, subtotals, two axes of calculation, titles and preambles, explanatory interpolations, even the final column heading MU.BI.IM 'its name'—some four hundred years after the last datable table from eighteenth-century Nippur. We can only speculate how this extraordinary stability was maintained, as although drawings of all the documents in the archive were published nearly a century ago, no analytical work has yet been done on them. Nevertheless, we can get a sense of the strength of the written tradition, and a hint of the huge gaps in all aspects of our knowledge of Mesopotamian history.

So far we have had nothing to say about northern Iraq: that is simply because there is no evidence for administrative tables there before the first millennium BCE. Even in the period of Assyrian world domination, and in the face of massive documentation, the evidence for tables is very slim. Administrators in the successive capital cities of Ashur, Kalhu, and Nineveh used tables simply as columnar lists, in order to separate numerical data of different types: wine from beer, present from absent.[15] Most were headed, with columnar totals at the end of the table. No surviving exemplars exhibit horizontal calculations or any operations more complex than simple addition. Just eighteen of the 450 published administrative records from Nineveh are tables, some 4 per cent of the total. The evidence for first-millennium Babylonia is even scarcer.

Tempting though it is to attribute the lack of tables in the first millennium BCE to some ill-defined decline in bureaucratic ability, a more satisfying explanation is at hand. From the late second millennium onwards, Aramaic became increasingly common as a written language. For whereas the Assyrian and Babylonian dialects of the Akkadian language were still recorded with the increasingly cumbersome and recondite cuneiform script on clay tablets, Aramaic used an alphabet of just 22 letters, written freehand with ink on parchment or papyrus—neither of which organic media survives at all in the archaeological conditions of Iraq. All three languages, however, were also written on waxed wooden or ivory writing boards, several specimens of which have survived,[16] and for which there is ample documentation in the cuneiform record. The durability, tradition and tamper-proof qualities of tablets ensured their continuing use for legal documents and religious literature, whereas writing boards and ink-based media were particularly suited to keeping accounts. Cuneiform tablets had to be inscribed while the clay was fresh; the new media, on the other hand, could be drawn up, corrected and added to over time, like modern-day ledgers. It is highly likely, then, that the vast majority of tabular accounts from the first millennium BCE perished long ago with the boards on which they were written.

Formal and informal tables in school mathematics

The first securely datable mathematical table in world history comes from the Sumerian city of Shuruppag, *c.* 2600 BCE. The tablet is ruled into three columns on each side with ten rows on the front or obverse side. The first two columns of the obverse list length measures from *c.* 3.6 km to 360 m in descending units of 360 m, followed by the Sumerian word sá ('equal' and/or 'opposite'), while the final column gives their products in area measure. Only six rows are extant or partially preserved on the reverse. They continue the table in smaller units, from 300 to 60 m in 60 m steps, and then perhaps (in the damaged and missing lower half) from 56 to 6 m in 6 m steps. While the table is organized along two axes, there is just one axis of calculation, namely, the horizontal multiplications. Around a thousand tablets were excavated from Shuruppag, almost all of them from houses and buildings which burned down in a city-wide fire in about 2600 BCE,[17] but sadly we have no detailed context for this tablet because its excavation number was lost or never recorded.[18]

While at first sight this table appears simply to list the areas of square fields, in fact it is akin to much more abstract multiplication tables. Prior to the invention of the sexagesimal place value system in the twenty-first century BCE there was no concept of abstract numeration: numbers were not thought of as independent entities but as attributes of concrete objects—the length of a line, for

Later Mesopotamia: Babylonia and Assyria

The disintegration of Babylonian kingdom in the late seventeenth century BCE marked the end of the old order. Major new communication, transport, and military technologies were developing that would henceforth enable huge empires to grow and flourish over centuries. Horses enabled people and ideas to travel faster than ever before, while glass, and later iron, became luxury must-have commodities that stimulated long-distance trade and contact between elites all over the Middle East from Turkey to Iran to Egypt.

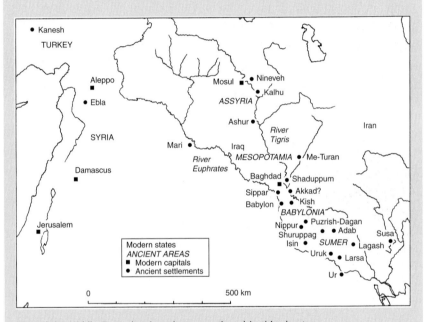

Map of the Middle East, showing places mentioned in this chapter

In Babylon the Kassite dynasty (c. 1550–1150) had taken power. Although the ruling family had its origins somewhere outside Mesopotamia they quickly adopted the languages, religion, and customs of their native predecessors. The old city temples retained their traditional wealth and status too, remaining significant employers, producers, and consumers in the Babylonian economy. It was a time of literary and scholarly innovation, in which traditional texts were revised, expanded, and interpreted anew. While the Kassites ruled over southern Iraq, the north of the land belonged to the Syrian empire of Mitanni. From the mid-fourteenth century onwards, though, the city of Ashur on the Tigris began to assert its independence, and gradually built up

Later Mesopotamia: Babylonia and Assyria *cont.*

a kingdom which by *c.* 1200 BCE was large enough to attempt a take-over of Kassite Babylonia. By the early first millennium Assyria held vast tracts of the Middle East in its power, extracting heavy tribute from those lands that were not yet under its direct control. In the heyday of the Assyrian empire (*c.* 900– 650 BCE) its capital moved up the Tigris, first to Kalhu and then to Nineveh, where Sargon and his successors Sennacherib, Esarhaddon, and Ashurbanipal ruled the Middle East from enormous palaces built at the expense of their subject peoples, decorated with detailed depictions of their conquests and filled with booty from their annual campaigns. Cuneiform survived the coming of alphabetic Aramaic in the early first millennium, in large part because of its continued high status. Assyrian attempts to usurp Babylon as the cultural centre of Mesopotamia were never entirely successful, but Ashurbanipal managed to create a definitive library of traditional Mesopotamian works in his palace at Nineveh, in part through the theft of tablets from Babylonia and the employment of captive Babylonian scribes to write for him.

Nineveh fell to the Babylonians and Medes in 612 BCE. Babylon enjoyed a brief period of independence and prosperity under Nebuchadnezzar and his successors, but in 539 BCE it became subsumed into the unprecedentedly enormous Persian empire. Although this event marked the definitive loss of Mesopotamian political autonomy it was by no means the end of its cultural history: the last datable cuneiform tablet is an astronomical record from Babylon written in 75 CE, some 600 years later. It had survived political and cultural conquest by Persians, Greek-speaking Macedonians, and the later Parthians. By the end, though, the temple of Marduk in Babylon was the last refuge of a small handful of traditionalist priests and scholars, the guardians of a late flowering of the cuneiform culture that had blossomed for over three thousand years.

---

instance, or the quantity of sheep in a flock. Therefore in the third millennium BCE to square a number meant, at some level, to construct a square area from two equal lengths. The table was not the only means available for expressing numerical relationships, though: a linear version has survived from the city of Adab.[19] It too is in three columns, but these function more like newspaper columns than the columns of a table. Odd lines of the text give length measurements from 1 to 12 cubits (*c.* 0.5 m–6 m), followed, as expected, by the word sá. The even lines give the areas that result from squaring the lengths. Once again, we have no exact information about its original context, but the style of its handwriting suggests that it is perhaps a few generations younger

Fig. 1.3 The world's oldest datable mathematical table, from Shuruppag, c. 2600 BCE. The first two columns contain identical lengths in descending order from 600 to 60 rods (c. 3600–360 m) and the final column contains the square area of their product. The sequence continues on the reverse, and probably finished at 1 rod (6 m). (VAT 12593, from Shuruppag, now in the Vorderasiatisches Museum, Berlin. H. Nissen, P. Damerow, and R. K. Englund, *Archaic bookkeeping*, University of Chicago Press, London and Chigago, 1993, fig. 119.)

than the tablet from Shuruppag (c. 2550 BCE). It is signed by someone called Nammah—about whom, frustratingly, we know nothing else at all.[20]

Oddly enough, we find a strong disinclination towards the truly tabular format in later school arithmetic, where one might expect to find tables galore. While there are many thousands of published and unpublished documents from the early second millennium BCE that are normally characterized as mathematical tables, on closer inspection most of them turn out to be prosaic lists of equivalences rather than formally laid out tables.

As usual, the evidence is patchy and difficult to date accurately. The earliest exemplars of the second-millennium arithmetical tradition are actually from the tail end of the third millennium BCE, at the time of the first sexagesimal numeration. We have just one or two administrative tablets with the sexagesimal workings still showing,[21] and a handful of reciprocal tables, or lists of inverses, which are highly likely to have originated in the administrative archives of the twenty-first century kingdom of Ur.[22] They list sexagesimally regular numbers up to 60, or sometimes beyond, with their inverses. Their informally tabular appearance derives from the fact that each line is left–right justified, as are the lines on all well written cuneiform tablets. Little attempt was made to align the numbers themselves.

The same type of reciprocal list is very well attested in later centuries, when it belonged at the head of the standard series of multiplications taught in scribal schools. The students were required to memorize it, like all other elementary school subjects, through repeated copying and revision, first as individual sections and then as long extracts from the whole list. In general, more verbose forms of the list were copied on first exposure to a new section, while a terser—and often much more untidy—version, consisting of the numbers only, was repeated for revision. The whole sequence could comprise up to forty lists, each of up to 25 lines long—some 1000 lines in total. Different multiplications were omitted in different schools, it appears, but the sequence was universally the same, with head numbers from 50 down to 1;15. The multiplicands were always 1–20, 30, 40, and 50, sometimes followed by the square and square root of the head number. At around the same point in the curriculum, students copied out a similar sequence of metrological units, from the capacity, weight, area, and length systems respectively. The series ran from tiny units to enormous ones, and at its maximum extent had about 600 entries.[23]

Less frequently extra-curricular arithmetical lists were copied, such as reciprocal pairs halved and doubled from an initial pair 2 05 ~ 28 48;[24] squares and inverse squares of integers and half-integers to 60;[25] and inverse cubes from 1 to 30.[26] The lists are usually informally tabular, like the multiplication lists, although occasionally formal columnar layouts are adopted, but never with column headings. The so-called coefficient lists, or lists of geometrical, metrological, and practical calculation constants, are similarly

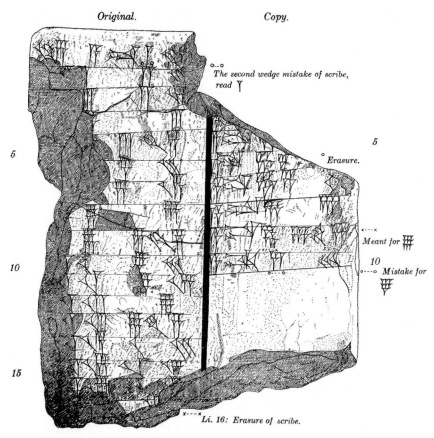

Fig. 1.4 A fifteen times multiplication 'table' for multiplicands 1–20, 30, 40, and 50, from early second-millennium Nippur. The teacher's copy is in the left hand column, the more poorly executed student's copy in the right. The student's copy is unfinished. The tablet broke in antiquity, and parts of it are now missing. (CBS 2142, from Nippur, now in the Collection of the Babylonian Section, University Museum, Philadelphia. H. Hilprecht, *Mathematical, metrological and chronological tablets from the Temple library of Nippur*, Philadelphia: Department of Archaeology, University of Pennsylvania, 1906, no. 94.)

informal, making a visual separation between the values and the names of the constants, but breaking columnar boundaries whenever necessitated by long names or explanations.[27] Only one is ruled into more than two columns, and only one other is headed.

Scribal students were expected not only to memorize arithmetical facts but also to put them into practice. Around a hundred published or partially published rough calculations and diagrams, mostly on small round or square tablets from Ur and Nippur, show the students' numerical solutions to mathematical problems.[28] As one might expect, certain standard layouts were used for particular problem types and students were encouraged to use columnar rulings to group data types.

Relatively little contextual detail is known about the metrological and arithmetical lists and tables, despite their superabundance in archaeological museums. Because there is a general (mis-)perception that they are all identical they are heavily under-represented in publications, so that it is very difficult to quantify the corpus, or give a good account of its chronological and geographical range. Nevertheless a few general statements can be made: around 430 exemplars have been published (*c.* 300 standard arithmetical lists, *c.* 50 non-standard ones, *c.* 50 metrological lists, *c.* 20 tabular calculations, *c.* 10 coefficient lists), representing perhaps 5 or 10 per cent of such material in accessible museums. They are known from almost all early second millennium urban centres in and around southern Iraq. We can estimate the date of several groups of these tablets, either based on direct archaeological information, or from clues gleaned from other tablets in the same museum accession lots. The earliest datable tablets are probably from the 1790s–80s (Larsa, Ur, and Uruk) but there are also groups from the 1750s–40s (Isin, Larsa, Mari, Nippur, Shaduppum, Ur) and the 1640s (Sippar, Susa).

Extraordinarily, the sophisticated bureaucratic ability to display and manipulate numerical data was almost never exploited in the elementary training of future scribes. The arithmetical and metrological lists could easily have been organized into multi-columnar headed tables, with two axes of organization and calculation, given the current clerical facility with tabular layouts. Yet there is only one known mathematical cuneiform tablet which is conspicuously indebted to administrative tabular practice. Plimpton 322 has achieved such iconic status as the Mesopotamian mathematical tablet *par excellence* that it comes as quite a shock to realise how odd it is.[29] Its fame derives from its mathematical content: fifteen rows of four extant columns containing sophisticated data relating to Pythagoras' theorem. The fact that this data is laid out in a landscape-oriented headed table, with a final heading MU.BI.IM ('its name') for the non-numerical data, has gone completely unremarked. These, of course, are formal features of administrative tables from Larsa during the period of rigorous standardization in the 1790s–80s BCE. Clearly, then, there was potential for innovations in data-management to cross over from administration into mathematics; it is puzzling that it happened so rarely.

At the moment we cannot even begin to date Babylonian tablets written in the later second millennium BCE, unless the scribes dated them for us, so badly do we understand the changes in writing conventions at that time.

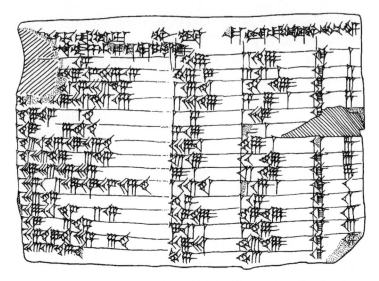

Fig. 1.5 Plimpton 322, the only true table in Old Babylonian mathematics, from Larsa c. 1800 BCE. The second and third columns of the table list the diagonals and short sides of 15 right triangles, while the final column is simply a line count. The heading on the first surviving column, which reads, 'The holding-square on the diagonal from which 1 is torn out, so that the short side comes up', gives important information about how the table was computed. There is still much speculation about the what the missing column(s) on the left of the tablet might have contained. (Plimpton 322, from Larsa, now in the University of Columbia Rare Book and Manuscript Library, New York. O. Neugebauer and A. Sachs, *Mathematical cuneiform texts*, American Oriental Society, New Haven, 1945, pp. 38–41. Drawing by E. Robson.)

Thus, in the absence of secure archaeological evidence, we cannot say with any confidence at all whether any mathematical tables have survived from Kassite Babylonia. Early publications often ascribe a Kassite date to multiplication lists that we would now confidently date to the eighteenth century,[30] while one of the coefficients lists may be from any time between fourteenth and the sixth century BCE.[31]

Assyrian metrological and arithmetical lists and tables are equally thin on the ground. A large metrological list has been discovered at Kanesh, an Assyrian trading post in eastern Turkey, c. 1850 BCE. It is now rather damaged but originally ran from 1 shekel (c. 8 g) to 100 talents (c. 3 metric tonnes).[32] The earliest tablet from Ashur itself, dating to c. 1200 BCE, gives the standard reciprocals, the complete set of multiplications from 50 to 1;15, and squares of integers to 30, in the classic early second-millennium format.[33] There is only one list of Assyrian capacity and weight measures, the extant part of which runs from 1 ban (c. 10 litres) to 70 000 'ass-loads' (c. 7 million litres) and 2 shekels to just over

1 mina (c. 16–500 g).³⁴ A small table gives the ratios between length units from 6 grains = 1 finger (c. 1.7 cm) to 30 USH = 1 double-mile (c. 10.8 km).³⁵ There is even less from Nineveh, the later Assyrian capital: one enigmatic table of 3-place reciprocals in base 70,³⁶ and a few lexical tables giving the names of unit fractions in Sumerian and Akkadian, all from Ashurbanipal's palace library.³⁷

Several thousand school tablets of the first millennium BCE are also known from the Babylonian cities of Kish, Nippur, Sippar, Ur, Uruk, and Babylon itself, of which a tiny number are mathematical or metrological.³⁸ The earliest are from a group of fifteen school tablets found in the archive of the city governor of Nippur, c. 755–730 BCE. Both list capacity measures from 1 ban to 1 gur (c. 10–300 litres); the first is on the back of a letter, while the second has been disguised as a ration list by attaching a fictitious name to each entry and giving an (erroneous) total.³⁹

About a century later are the two hundred or so tablets from a badly preserved building in Kish, where 'almost every room contained tablets which had been stored in large jars, arranged round the room according to contents, primarily syllabaries and religious texts.'⁴⁰ Sadly the exact disposition of the tablets within the jars was never recorded. Thirteen of those tablets contain extracts from the standard list of weights and measures, none of which is truly tabular, inscribed on the same tablets as snippets from elementary syllabaries and word lists.⁴¹ At the other end of the scholastic spectrum is a beautifully inscribed mathematical table. Whereas most mathematical lists and tables are known to us from several dozens, hundreds, or even thousands of exemplars, this document is unique. It is the only identified table that combines both squares and square roots in a single entry. Each line reads 'n times n is n^2, whose square root is n', where $n = 1, 1\frac{1}{2}, 2, 2\frac{1}{2}, 3, \ldots, 59, 59\frac{1}{2}$. The vertical lines of the tabular formatting appear to have served more as aids to producing a beautifully written document than as conceptual separators of different classes of data. This suggests it was written by a professional scribe and designed as a reference document for a library. The tablet is signed at the bottom by one Bêl-bani-apli ('the lord [god Marduk] is creator of an heir').⁴²

Tables for scholarship and astronomy

While little use was made of tabular formatting for mathematical training, from the early second millennium BCE onwards it became ubiquitous in other aspects of schooling and scholarship. The table's horizontal axis of organization enabled inter-columnar relationships to be expressed between words, phrases, or longer texts. One of its first uses was in the school sign-lists now called Proto-Ea and Proto-Diri, attested from at least the eighteenth century BCE, which enumerate the different Sumerian readings each cuneiform sign or sign-combination can take. Each has a simple bi-columnar structure, with no headings. The first five lines of Proto-Ea go as follows:

a2-a	A	'The sign A can be read as "aya"
ia	A	The sign A can be read as "iya"
du-ru	A	The sign A can be read "duru"
e	A	The sign A can be read "e"
a	A	The sign A can be read "a" '.[43]

An optional further column to the right gave one or more Akkadian translations for each Sumerian sign combination. In the course of time the standard lists of Sumerian nouns also acquired columns of Akkadian translations; then and now they are known by their first line, $UR_5.RA$ = *hubullu* (a type of loan). Outside Mesopotamia proper, such lists also accrued Egyptian, Hittite, Hurrian, or Ugaritic translations and could comprise up to five columns of text.[44] Headings were never used, however. It is a real irony—and testament to our collective blindness to document formatting—that these intrinsically tabular documents are nowadays called 'lexical *lists*' while their list-like arithmetical counterparts are always known as 'multiplication *tables*'.

A tiny percentage of urban Mesopotamians trained as scribes, and a tiny percentage of scribes became neither bureaucrats nor amanuenses but scholarly experts, Akkadian *ummânu*. From the early second millennium BCE onwards, Mesopotamian scholars carried one or more of the following titles:

Tupsharru   'scribe/celestial diviner'. Experts in interpreting celestial (and other) portents.

Barû   'haruspex/extispicer/diviner'. Experts in extispicy [divination by the entrails of sacrificed goats] and lecanomancy [divination by the patterns of oil on water].

Ashipu   'exorcist/healer-seer'. Experts in magical manipulation of the supernatural.

Asû   'physician'. Experts in curing diseases by drugs and physical remedies.

Kalû   'lamentation chanter'. Experts in soothing angered gods.[45]

Each scholarly discipline had its own corpus of specialist literature, running to a hundred or more standard tablets of omens, prescriptions, or laments. Indeed, the scholarly corpus is so vast, complex, and esoteric that only a tiny proportion of it has been published and studied thoroughly. Tabular documents were used sporadically throughout the corpus, whether, for instance, for adding commentary to core works or for drawing up tables of propitious and unpropitious days of the month. We, however, will focus on the works of the celestial diviners—more commonly if anachronistically referred to in the modern literature as astronomers or astrologers.

Most of our contextualized knowledge of the men themselves comes from three separate locales. At the Assyrian court at Nineveh, *c*. 750–612 BCE, a select inner circle of scholars wrote regularly to the king about what the future portended for matters of state and the well-being of the royal family. A great many of their letters and reports survive, which directly cite the scholarly reference works, thereby enabling us to see how they worked in practise.[46] We also have tablets by and about the celestial diviners attached to two later Babylonian temples: the Resh of Anu in Uruk (*c*. 400–100 BCE and perhaps later) and the Esangila of Marduk in Babylon (*c*. 200 BCE–50 CE and probably earlier).[47] By this late stage in Mesopotamian history, Babylon was ruled by a succession of outsiders (Persians, Greeks, Parthians) with their own belief systems; the professional locus of the scholars, who could no longer rely on state patronage, had thus become the temples.

Until at least the second century BCE the full Akkadian title of the celestial diviner was *tupshar Enuma Anu Ellil*, 'scribe of "When [the gods] Anu [and] Ellil"', after the first line of the huge compendium of celestial omens that constituted their core reference work:

When Anu, Ellil, and Ea, the great gods, in their sure counsel had fixed the designs of heaven and earth, they assigned to the hands of the great gods the duty to form the day well and to renew the month of mankind to behold.[48]

The series runs to 68 or 70 tablets (it exists in different versions) which were divided into four sections: lunar omens (tablets 1–22), solar omens (23–29), weather omens (30–49), and omens from stars and planets (50–70). There is strong evidence to suggest that parts of it at least go back to the early second millennium, but the earliest surviving sources come from the Assyrian city of Kalḫu, *c*. 720 BCE—including a 'deluxe' edition written on 16 ivory writing boards for Sargon, king of Assyria, which was found down a palace well with only a few fragments of its wax writing surviving.[49] Fortunately the title and destination of the work had been inscribed in cuneiform on the front cover! Other copies found in Assyria had been looted from Babylonian libraries in 647 BCE.

Only one of *Enuma Anu Ellil's* 70-odd tablets is tabular. Tablet 14 in fact comprises four tables, all related to lunar motion:

A. Duration of lunar visibility in the equinoctial month, [according to] the tradition of Nippur
B. Duration of lunar visibility in the equinoctial month, [according to] the tradition of Babylon
C. Seasonal variation in length of day and night
D. Monthly variation in lunar visibility and new moon; and in lunar visibility at full moon.[50]

The first two tables have 30 entries each, the last two 24. They each end with one-line summary 'headings' but are otherwise very informal, with verbose, list-like entries that are broken into sense units on the surface of the tablet. Here, for instance, is the third entry from each of the four tables:

(A)	On day 3	the moon	is present for 15 (USH).
(B)	On day 3:	1/2 mina	6 shekels.
(C)	In month Ayyar, day 15:		the day watch is $2\frac{2}{3}$ mina, the night watch is $2\frac{1}{3}$ mina.
(D)	In month Ayyar, day 1:		the moon's period of visibility is 10 USH.[51]

All four model the moon's visibility using what are now called 'linear zigzag functions' whereby the variable increases or decreases by a constant amount over fixed intervals, and attains fixed maxima and minima. Plotted on a graph over time such functions trace regular zigzags, hence their modern name. Although these are the earliest known *tabular* representations of linear zigzag functions, the schemes on which they are based can be traced right back to much shorter descriptions in mathematical and astronomical texts of the early second millennium BCE, about a thousand years earlier.[52]

At around the same time that *Enuma Anu Ellil* was being copied for the king of Assyria, scholars began to make systematic records of celestial phenomena without always taking omens from them. These observations were collected into daily, monthly, and annual Diaries, as they are now called, from at least 652 BCE. They formed a continuous tradition, much of which is now lost, that would last for around seven hundred years and eventually (indirectly, and in translation) comprised much of the observational basis for Ptolemy's *Almagest* of the second century CE. The monthly Diaries typically contain the following data:

A statement of the length of the preceding month; the time interval between sunset and moonset on the first day of the month; time intervals between sun/moonrise/set in the middle of the month; the time interval between moonrise and sunrise on the morning of the moon's last visibility; the dates on which the moon approached the various Normal Stars and the watch of the night in which this occurred; and the date and description of lunar and solar eclipses. For the planets they record dates and position among the stars of first and last visibility, direct and retrograde motion and stationary points, and conjunctions with Normal Stars. [...] The dates of the solstices and equinoxes and of the heliacal rising of Sirius are recorded. [...] Weather conditions are regularly reported, and occasional astronomical phenomena such as meteors and comets (including Halley's Comet in 164 and 87 BCE).[53]

These diaries were key to the development of mathematical astronomy, some time around 500 BCE. They enabled scholars 'to devise consistent and powerful sets of rules (embodied in "Procedure Texts") that allowed them to generate tables (called by modern scholars "Ephemerides") permitting one to foretell the longitudes and times of the occurrences of lunar and planetary phenomena and the longitudes of these bodies in the intervals between these occurrences.'[54] Alongside those Procedure Texts and Ephemerides increasingly accurate and sophisticated mathematical tables were developed, presumably to help scholars with their calculations. Not surprisingly, many of them come from Uruk and Babylon, where the celestial scholars were most active.

Some collections of lunar and planetary data are also set out in a ruled tabular format. Each cell of the table contains a description (often on more than one line) of a particular observed (or sometimes predicted) planetary or lunar event. Successive events are given in the cells going down a column, and going along each row we get events separated by the characteristic planetary or lunar periods (e.g. 8 years for Venus, 18 years for the Moon). The similarities between events separated by these periods can then be easily viewed. There are never any headings to these tables. Frequently, columns continue from one side of the tablet to the other. Sometimes, however, the tablets turn sideways so that the rows continue from one side to the other.[55]

In many ways the dozen or so extant metrological lists and tables of the first millennium BCE are the inverse of their second-millennium counterparts: they now followed the order of capacity, weight, area, and length but more often reversed that order. Tables tended to put the sexagesimal (or more

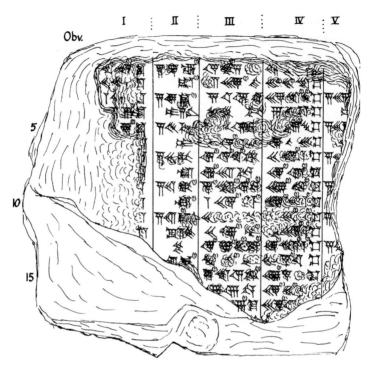

Fig. 1.6 The latest datable Ephemeris, from Babylon, recording lunar and solar eclipse possibilities for the years 12 BCE–43 CE. It is a fragment of a larger tablet in which columns are grouped into threes: date (here cols. II, V), longitude of Moon (col. III), and magnitude of eclipse (cols. I, IV). (BM 34083 = Sp. 181, from Babylon; now housed in the British Museum. T. G. Pinches and J. N. Strassmaier, *Late Babylonian astronomical and related texts*, (ed. A.J. Sachs), Brown University Press, Providence, Rhode Island, 1955, no. 49; studied by O. Neugeubauer, *Astronomical cuneiform texts*, 3 vols., reprint edition, Springer, Berlin, 1955, no. 53, pp. 115–16.)

often decimal) equivalent to the left of the metrological units instead of the right.[56] Very few multiplication tables survive; instead the small but growing corpus consists primarily of tables of many-place regular reciprocals between 1 and 3, and many-place squares and square roots of integers and half-integers.[57] It has yet to be shown whether these tables would be any use in calculating Ephemerides using the methods of the Procedure Texts.

Ephemerides, always tabular, were drawn up from at least 300 BCE until about 50 CE. They fall into two distinct groups: those for the Moon, and those for the planets. A typical lunar Ephemeris tabulates twelve to eighteen functions of the Moon over a whole year, at new and full moon; about a third of known examples record the data for the new moon on the front of the tablet and the full moon on the reverse. Two different calculational methods are used: System A, which is based on step functions for calculating longitudes (in which a variable jumps discontinuously between constant values at fixed intervals); and System B, which uses linear zigzag functions of the sort used in *Enuma Anu Ellil* tablet 14.[68] Despite the complexity of the data they record, the columns of Ephemerides are never headed. Planetary Ephemerides were much less sophisticated, predicting just five or six key moments in the planets' journeys across the night skies. The latest datable Ephemeris is from Babylon, predicting lunar and solar eclipse possibilities for the period 12 BCE–43 CE. The latest datable cuneiform record of them all is also from Babylon, an astronomical almanac predicting the main celestial events for the year 75 CE.

Conclusions

Only rarely in Mesopotamian history were tables a mainstream document format. They took a long time to catch hold, first appearing over half a millennium after the invention of writing and only establishing themselves in the nineteenth century BCE. At that point their potential as powerful tools for the management of quantitative data was fulfilled remarkably rapidly, over a period of a few decades in the early eighteenth century, and this potential continued to be exploited (allowing for huge gaps in the evidential record) for fully five hundred years, at least in and around the city of Nippur. Perhaps it is no coincidence that tables took off only after the invention of the sexagesimal place value system and the concomitant conceptual separation of quantifier and quantified, for Mesopotamian tables made manifest this distinction between quantitative and qualitative, by drawing physical dividing lines between them. In its turn, the new format enabled numerical data and relationships to be seen and explored in ways hitherto unimaginable. The material objects themselves facilitated conceptual advances in quantitative thinking. But even in the heyday of Mesopotamian tables in the first half of the eighteenth century BCE, they account for only 1 or 2 per cent of all administrative documents and scribes continued to prefer simpler linear or prosaic methods of managing information.

Paradoxically, even as tabular formatting began to infiltrate other textual types, cuneiform scribes became less ambitious in using tables to manipulate and check numerical data as the writing board took over many of the functions of the tablet. It was only with the development of mathematical astronomy in the latter half of the first millennium BCE that tables came into their own again to record and calculate the multiple, complex, and lengthy variables of lunar and planetary theory.

Even more surprising is that Mesopotamian scholastic mathematics employed tables very rarely, preferring to express arithmetical and metrological equivalences as lists. Where we do see truly tabular mathematical documents, their debt to administrative or astronomical practice is transparent. While the inner workings of Mesopotamian mathematical lists are generally well understood there is still much we do not grasp about their functioning within the larger educational system. We understand even less of the history of truly tabular tables in Mesopotamia, and their place in the history of ideas.

Further reading

H. Nissen, P. Damerow, and R. K. Englund, *Archaic bookkeeping*, University of Chicago Press, London and Chicago, 1993, is an excellent overview of the development of writing and accounting in third-millennium Mesopotamia. It also has lots of good photos, including the table of squares from Shuruppag. The classic account of early second millennium mathematical tables is O. Neugebauer, *The exact sciences in antiquity*, 2nd edn, Dover, New York, 1969, pp. 3–52. A more recent summary is K. Nemet-Nejat, 'Systems for learning mathematics in Mesopotamian scribal schools', *Journal of Near Eastern Studies* 54 (1995), 241–60. Almost nothing has been written on administrative tables; see for the moment E. Robson, 'Accounting for change: problematising the development of tabular accounts in early Mesopotamia', in *The social and economic implications of accounting in the ancient world: a colloquium held at the British Museum, November 2000*, ed. M. Hudson and M. Van De Mieroop, to appear, and E. Robson, 'Words and pictures: new light on Plimpton 322', *American Mathematical Monthly* 109 (2002), 105–20. A survey of mathematical tables from later Mesopotamia is given in J. Friberg, 'On the structure of cuneiform mathematical table texts from the −1st millennium', in *Die Rolle der Astronomie in den Kulturen Mesopotamiens* (ed. H. D. Galter), RM- Druck- & Verlagsgesellschaft, Graz, 1993, pp. 383–405.

For more general introductions to Mesopotamian mathematics, see J. Høyrup, 'Mesopotamian mathematics', in *Companion encyclopedia of the history and philosophy of the mathematical sciences* (ed. I. Grattan-Guiness), Routledge, London, 1994, pp. 21–9 and E. Robson, 'The uses of mathematics in ancient Iraq, 6000–600 BCE', in *Mathematics across cultures: the history of non-Western mathematics* (ed. H. Selin), Kluwer, Dordrecht, 2000, pp. 93–113. Astronomy is discussed by J. Britton and C. B. F. Walker, 'Astronomy and astrology in Mesopotamia', in *Astronomy before the telescope* (ed. C. B. F. Walker), British Museum Press, London, 1996, pp. 42–67 and by A. Aaboe, 'Babylonian arithmetical astronomy', in *Episodes in the early history of astronomy*, Springer Verlag, Heidelberg, 2001, pp. 24–65. The first chapter of the same book, 'What every young person ought to know about naked-eye astronomy' (pp. 1–23) sets out the observational basics. M. Roaf's *Cultural atlas of Mesopotamia and the ancient Near East*, Facts on File, New York, 1990, is an excellent overview of Mesopotamian history, while C. B. F. Walker's *Cuneiform*, British Museum Press, London, 1987, is a clear explanation of the writing and numeration systems of ancient Mesopotamia.

Notes

1. Discussed in detail by R. K. Englund, 'Texts from the Uruk period', in *Späturuk-Zeit und Frühdynastische Zeit* (ed. P. Attinger and C. Uelinger), Freiburg and Göttingen, 1998, pp. 15–233; quote from p. 215.
2. R. K. Englund, 'Texts from the Uruk period', in *Späturuk-Zeit und Frühdynastische Zeit* (ed. P. Attinger and C. Uelinger), Freiburg and Göttingen, 1998, pp. 56–64, esp. fig. 17 on pp. 58–9; H. Nissen, P. Damerow, and R. K. Englund, *Archaic bookkeeping*, University of Chicago Press, London and Chigago, 1993, p. 30.
3. See, for instance the accounts of the 'Kushim' brewing archive described in H. Nissen, P. Damerow, and R. K. Englund, *Archaic bookkeeping*, University of Chicago Press, London and Chigago, 1993, pp. 36–46.
4. R. K. Englund, 'CDLI proposal', *The cuneiform digital library initiative*, <http://www.cdli.ucla.edu/proposal.pdf>, UCLA, 2001.
5. R. K. Englund, M. Fitzgerald, *et al.*, 'Ur III catalogue', *The cuneiform digital library initiative*, <http://www.cdli.ucla.edu/progress.html>, UCLA, 2001.
6. AUAM 73.0693, probably from Puzrish-Dagan, now in the Horn Archaeological Museum, Michigan. M. Sigrist, *Neo-Sumerian account texts in the Horn Archaeological Museum*, Andrews University Press, Berrien Springs, 1984, no. 48.
7. AUAM 73.0400, probably from Puzrish-Dagan, now in the Horn Archaeological Museum, Michigan. M. Sigrist, *Neo-Sumerian account texts in the Horn Archaeological Museum*, Andrews University Press, Berrien Springs, 1984, no. 56.
8. M. Sigrist, *Drehem*, CDL Press, Bethesda, 1992.
9. M. Civil, 'Ur III bureaucracy: quantitative aspects', in *The organization of power: aspects of bureaucracy in the ancient Near East*, 2nd edn (ed. McG. Gibson and R. D. Biggs), University of Chicago Press, Chicago, 1992, pp. 35–44, especially p. 44.

10. A more detailed presentation of this discussion is in E. Robson, 'Accounting for change: problematising the development of tabular accounts in early Mesopotamia', in *The social and economic implications of accounting in the ancient world: a colloquium held at the British Museum, November 2000* (ed. M. Hudson and M. Van De Mieroop), to appear.

11. R. M. Sigrist, *Les* sattukku *dans l'Eshumesha durant la période d'Isin et Larsa*, Undena, Malibu, 1984.

12. E. Robson, 'Accounting for change: problematising the development of tabular accounts in early Mesopotamia', in *The social and economic implications of accounting in the ancient world: a colloquium held at the British Museum, November 2000* (ed. M. Hudson and M. Van De Mieroop), to appear.

13. A. T. Clay, *Documents from the temple archives of Nippur dated in the reigns of the Cassite rulers*, 3 vols., The University Museum, Philadelphia, 1906–1912; I. Bernhardt, *Sozial-ökonomische Texte und Rechtsurkunden aus Nippur zur Kassitenzeit*, Akademie Verlag, Berlin, 1976.

14. A. T. Clay, *Documents from the temple archives of Nippur dated in the reigns of the Cassite rulers*, vol. 1, The University Museum, Philadelphia, 1906, p. 5.

15. E.g. F. M. Fales and J. N. Postgate, *Imperial administrative records*, 2 vols., Helsinki University Press, Helsinki, 1992–95.

16. D. J. Wiseman, 'Assyrian writing boards', *Iraq* 17 (1955), 3–13.

17. H. Martin, *Fara*, Chris Martin, Birmingham, 1988, pp. 82–103.

18. H. Martin, *Fara*, Chris Martin, Birmingham, 1988, Appx. III Fiche 2, p. 202.

19. A 681, from Adab, now in the Oriental Institute, Chicago. D. D. Luckenbill, *Inscriptions from Adab*, University of Chicago Press, Chicago, 1930, no. 70; D. O. Edzard, 'Eine altsumerische Rechentafel (OIP 14, 70)', in *Lishan mithurti, Festschrift Wolfram Freiherr von Soden* (ed. W. Röllig), Verlag Butzon, Neukirchen-Vluyn, 1969, pp. 101–4.

20. Yang Zhi, *Sargonic inscriptions from Adab*, IHAC, Changchun, 1989, p. 14.

21. M. A. Powell, 'The antecedents of Old Babylonian place notation and the early history of Babylonian mathematics', *Historia Mathematica* 3 (1976), 417–39.

22. Ist T 7375, from Girsu, now in the Istanbul Archaeological Museum: O. Neugebauer, *Mathematische Keilschrift-texte*, vol. 1, Springer, Berlin, 1935, p. 10. HS 201, from Nippur, now in the Hilprecht Sammlung, University of Jena: J. Oelsner, 'HS 201—Eine Reziproken-tabelle der Ur III-Zeit', in *Changing views on ancient Near Eastern mathematics* (ed. J. Høyrup and P. Damerow), Dietrich Reimer Verlag, Berlin, 2001, pp. 53–60; BM 106425 and BM 106444 from Umma, now in the British Museum; and other tablets from Umma Nippur, and Girsu (unpublished, pers. comm. N. Koslova, B. Lafont, and C. Proust).

23. O. Neugebauer, *Mathematische Keilschrift-texte*, vol. 1, Springer, Berlin, 1935, pp. 8–14 (reciprocals), pp. 36–42 (multiplications), pp. 68–72 (metrology); vol. 2, p. 36, vol. 3, pp. 49–50 (reciprocals and multiplications); O. Neugebauer and A. Sachs, *Mathematical cuneiform texts*, American Oriental Society, New Haven,

1945, pp. 4–6 (metrology), pp. 11–12 (reciprocals), pp. 19–33 (multiplications); J. Friberg, 'Mathematik', in *Reallexikon der Assyriologie*, vol. 7 (ed. D. O. Edzard), Walther de Gruyter, Berlin, 1990, pp. 542–6 [in English].

24. J. Friberg, 'Mathematik', in *Reallexikon der Assyriologie*, vol. 7 (ed. D. O. Edzard), Walther de Gruyter, Berlin, 1990, pp. 549–50 [in English].

25. O. Neugebauer, *Mathematische Keilschrift-texte*, vol. 1, Springer, Berlin, 1935, pp. 68–72; O. Neugebauer and A. Sachs, *Mathematical cuneiform texts*, American Oriental Society, New Haven, 1945, pp. 33–5; J. Friberg, 'Mathematik', in *Reallexikon der Assyriologie*, vol. 7 (ed. D. O. Edzard), Walther de Gruyter, Berlin, 1990, pp. 546–7 [in English].

26. O. Neugebauer, *Mathematische Keilschrift-texte*, vol. 1, Springer, Berlin, 1935, pp. 73–5; O. Neugebauer and A. Sachs, *Mathematical cuneiform texts*, American Oriental Society, New Haven, 1945, p. 34; J. Friberg, 'Mathematik', in *Reallexikon der Assyriologie*, vol. 7 (ed. D. O. Edzard), Walther de Gruyter, Berlin, 1990, pp. 546–7 [in English].

27. E. Robson, *Mesopotamian mathematics, 2100–1600 BC*, Clarendon Press, Oxford, 1999.

28. E. Robson, *Mesopotamian mathematics, 2100–1600 BC*, Clarendon Press, Oxford, 1999, pp. 11–13 and 245–77.

29. E. Robson, 'Neither Sherlock Holmes nor Babylon: a reassessment of Plimpton 322', *Historia Mathematica*, 28 (2001), 1–40; E. Robson, 'Words and pictures: new light on Plimpton 322', *American Mathematical Monthly* 109 (2002), 105–20.

30. For instance H. V. Hilprecht, *Mathematical, metrological and chronological tablets from the temple library of Nippur*, University of Pennsylvania, Philadelphia 1906; followed by O. Neugebauer, *Mathematische Keilschrift-texte*, vol. 1, Springer, Berlin, 1935, pp. 10–13 and *passim*.

31. E. Robson, *Mesopotamian mathematics, 2100–1600 BC*, Clarendon Press, Oxford 1999, List B, p. 26.

32. C. Michel, 'Les marchands et les nombres: l'exemple des Assyriens à Kanish', in *Intellectual life of the ancient Near East* (ed. J. Prosecky), Oriental Institute, Prague, 1998, pp. 249–67.

33. Ist A 20 + VAT 9734, from Ashur, now in the Istanbul Archaeological Museum and the Vorderasiatisches Museum, Berlin. O. Neugebauer, *Mathematische Keilschrift-texte*, vol. 1, Springer, Berlin, 1935, pp. 11, 46–7, 70.

34. VAT 9840, from Ashur, now in the Vorderasiatisches Museum, Berlin. O. Schroeder, *Keilschrift-texte aus Assur: verschiedenen Inhalts*, J. C. Hinrichs, Leipzig, 1920, no. 184; J. Friberg, 'On the structure of cuneiform mathematical table texts from the −1st millennium', in *Die Rolle der Astronomie in den Kulturen Mesopotamiens* (ed. H. D. Galter), RM- Druck- & Verlagsgesellschaft, Graz, 1993, p. 388.

35. Current whereabouts unknown. F. Thureau-Dangin, 'Un petit texte d'Assour', *Revue d'Assyriologie* 23 (1926), 33–4; J. Friberg, 'On the structure of cuneiform mathematical table texts from the −1st millennium', in *Die Rolle der Astronomie in*

den Kulturen Mesopamiens (ed. H. D. Galter), RM- Druck- & Verlagsgesellschaft, Graz, 1993, text 4.

36. K. 2069, from Nineveh, now in the British Museum. O. Neugebauer, *Mathematische Keilschrift-texte*, vol. 1, Springer, Berlin, 1935, pp. 30–3 & pl. 10.
37. K. 56, K. 60, K. 8687, from Nineveh, now in the British Museum. O. Neugebauer, *Mathematische Keilschrift-texte*, vol. 1, Springer, Berlin, 1935, pp. 28–9.
38. P. Gesche, *Schulunterricht in Babylonien im ersten Jahrtausend v. Chr.*, Ugarit-Verlag, Münster, 2001, pp. 37–8.
39. S. W. Cole, *Nippur IV: The early Neo-Babylonian governor's archive from Nippur*, University of Chicago Press, Chicago, 1996, nos. 89, 124.
40. P. R. S. Moorey, *Kish excavations, 1923–1933*, Clarendon Press, Oxford, 1978, p. 50.
41. Now in the Ashmolean Museum, Oxford. P.E. van der Meer, *Syllabaries A, B1 and B: with miscellaneous lexicographical texts from the Herbert Weld Collection*, Clarendon Press, Oxford, 1938, nos. 34, 75, 123, 128; remainder unpublished.
42. Ash 1924.796, in the Ashmolean Museum. O. Neugebauer, *Mathematische Keilschrift-texte*, vol. 2, Springer, Berlin, 1935, pl. 34 [W 1931–38].
43. M. Civil, M.W. Green, and W. G. Lambert, *Ea A* = nâqu, *Aa A* = nâqu, *with their forerunners and related texts*, Biblical Institute Press, Rome, 1979, p. 30.
44. See M. Civil, 'Ancient Mesopotamian lexicography', in *Civilizations of the ancient Near East*, 4 vols. (ed. J. M. Sasson), Scribner's, New York, pp. 2305–14.
45. D. R. Brown, *Mesopotamian planetary astronomy-astrology*, Styx, Groningen, 2000, p. 33. See in general Chapter 1, 'The astronomer-astrologers—the scholars', pp. 30–52; F. Rochberg, 'Scribes and scholars: the *tupshar Enuma Anu Enlil*', in *Assyriologica et Semitica: Festschrift für Joachim Oelsner* (ed. J. Marzahn and H. Neumann), Ugarit-Verlag, Munster, 2000, pp. 359–75.
46. See most recently D. R. Brown, *Mesopotamian planetary astronomy-astrology*, Styx, Groningen, 2000. The letters and reports themselves are published in English translation by H, Hunger, *Astrological reports to Assyrian kings*, State Archives of Assyria, vol. 8, Helsinki University Press, Helsinki, 1992 and S. Parpola, *Letters from Assyrian and Babylonian scholars*, State Archives of Assyria, vol. 10, Helsinki University Press, Helsinki, 1992.
47. See F. Rochberg, 'The cultural locus of astronomy in Late Babylonia', in *Die Rolle der Astronomie in den Kulturen Mesopotamiens* (ed. H. D. Galter), RM- Druck- & Verlagsgesellschaft, Graz, 1993, pp. 31–46; P.-A. Beaulieu, 'The descendants of Sîn-leqi-unninni', in *Assyriologica et Semitica: Festschrift für Joachim Oelsner* (ed. J. Marzahn and H. Neumann), Ugarit-Verlag, Munster, 2000, pp. 1–16.
48. D. R. Brown, *Mesopotamian planetary astronomy-astrology*, Styx, Groningen, 2000, p. 255.
49. See most recently J. and D. Oates, *Nimrud: an Assyrian imperial city revealed*, British School of Archaeology in Iraq, London, 2001, pp. 97–9.

50. F. N. H. Al-Rawi and A. R. George, 'Enuma Anu Enlil XIV and other early astronomical tables', *Archiv für Orientforschung* 38–39 (1991–92), 52–73.
51. 1 mina = 4 hours (weight running through a water-clock), 1 USH = 4 minutes (1 degree of arc).
52. See D. R. Brown, J. Fermor, and C. B. F. Walker, 'The water-clock in Mesopotamia', *Archiv für Orientforschung* 46–47 (1999–2000), 130–48.
53. J. Britton and C. B. F. Walker, 'Astronomy and astrology in Mesopotamia', in *Astronomy before the telescope* (ed. C. B. F. Walker), British Museum Press, London, 1996, p. 50. The Normal Stars were a group of thirty-one stars in the zodiac belt, used as reference points for the movements of the moon and planets. The planets observed were Venus, Jupiter, Mars, Mercury, and Saturn. The Diaries are published by A. Sachs and H. Hunger, *Astronomical diaries and related texts from Babylonia*, 3 vols., Österreichischen Akademie der Wissenschaften, Vienna, 1988–96.
54. H. Hunger and D. Pingree, *Astral sciences in Mesopotamia*, Leiden, Brill, 1999, p. 212. Procedure Texts and Ephemerides are edited and discussed by O. Neugebauer, *Astronomical cuneiform texts*, 3 vols., Springer Verlag, Heidelberg, 1955, and *A history of ancient mathematical astronomy*, vol. 1, Springer Verlag, Heidelberg, 1975, pp. 347–555.
55. I thank John Steele for this information, as well as for detailed comments on the rest of this section. Tablets of this type are published by H. Hunger, A. Sachs, and J. M. Steele, *Astronomical diaries and related texts from Babylonia*, vol. 5, Österreichischen Akademie der Wissenschaften, Vienna, 2001.
56. J. Friberg, 'On the structure of cuneiform metrological texts from the −1st millennium', in *Die Rolle der Astronomie in den Kulturen Mesopotamiens* (ed. H. D. Galter), RM- Druck- & Verlagsgesellschaft, Graz, 1993, pp. 383–405.
57. Lists of published mathematical tables from the later first millennium BCE are given by J. P. Britton, 'A table of 4th powers and related texts from Seleucid Babylon', *Journal of Cuneiform Studies* 43–45 (1991–93), 71–87 and J. Friberg, 'Seed and reeds continued: another metro-mathematical topic text from Late Babylonian Uruk', *Baghdader Mitteilungen* 28 (1997), 251–365. The multiplication tables are published by A. Aaboe, 'A new mathematical text from the astronomical archive in Babylon: BM 36849', in *Ancient astronomy and celestial divination* (ed. N. M. Swerdlow), MIT Press, Cambridge, Mass., 1996, pp. 179–186.
58. J. Britton and C. B. F. Walker, 'Astronomy and astrology in Mesopotamia', in *Astronomy before the telescope* (ed. C. B. F. Walker), British Museum Press, London, 1996, p. 62.

Gr. 2

2 min	Sinus	Logarithmi	Differentiæ	logarithmi	Sinus	
0	348995	33552817	33546723	6094	9993908	60
1	351902	33469860	33463664	6196	9993806	59
2	354809	33387588	33381289	6299	9993703	58
3	357716	33305993	33299590	6403	9993599	57
4	360623	33225056	33218549	6507	9993495	56
5	363530	33144770	33138158	6612	9993390	55
6	366437	33065128	33058410	6718	9993284	54
7	369344	32986107	32979282	6825	9993177	53
8	372251	32907712	32900779	6933	9993069	52
9	375158	32829923	32822881	7042	9992960	51
10	378064	32752740	32745588	7152	9992850	50
11	380971	32676149	32668887	7262	9992740	49
12	383878	32600139	32592866	7373	9992629	48
13	386785	32524706	32517221	7485	9992517	47
14	389692	32449837	32442239	7598	9992404	46
15	392598	32375526	32367814	7712	9992290	45
16	395505	32301761	32293934	7827	9992175	44
17	398412	32228539	32220596	7943	9992060	43
18	401318	32155852	32147793	8059	9991944	42
19	404225	32083692	32075516	8176	9991827	41
20	407131	32012045	32003751	8294	9991709	40
21	410038	31940909	31932496	8413	9991590	39
22	412944	31870276	31861743	8533	9991470	38
23	415851	31800141	31791487	8654	9991349	37
24	418757	31730492	31721716	8776	9991228	36
25	421663	31661332	31652434	8898	9991106	35
26	424570	31592644	31583623	9021	9990983	34
27	427476	31524424	31515279	9145	9990859	33
28	430382	31456671	31447402	9270	9990734	32
29	433288	31389371	31379975	9396	9990608	31
30	436194	31322524	31313001	9523	9990482	30

2

The making of logarithm tables

GRAHAM JAGGER

In 1614 the invention of logarithms burst upon the world with the publication of Napier's *Mirifici logarithmorum canonis descriptio*, or *A description of the admirable table of logarithms*. By the time that Briggs's *Trigonometria Britannica*

Fig. 2.1 Napier's *Mirifici Logarithmorum canonis descriptio*, printed by Andrew Hart in Edinburgh in 1614, contains a dedication to Charles, Prince of Wales, together with some laudatory verses. Book I describes the ideas of arithmetical and geometrical progressions and the definition of the logarithm of a sine. The term *antilogarithm* is used for what is now called the logarithm of the sine, and the *differential* for the logarithm of the tangent. The book continues with a demonstration of the properties of logarithms with regard to proportional numbers and the extraction of square and cube roots. There follow 90 pages of tables. The page shown opposite gives, in the second and third columns, the sines and logarithm of sines respectively of the angles 2° 0′ to 2° 30′ and, in the fifth and sixth columns, the logarithms of sines and cosines respectively of the angles 87° 30′ to 88°, in steps of one minute.

appeared in 1633 logarithms had come of age: Napier's rather clumsy initial formulation had been transformed into one of supreme elegance, the utility of which was to remain unsurpassed until the invention of the electronic calculator. This process of maturation, which we examine here, was marked in particular by the work of two other British mathematicians, Henry Briggs and Edmund Gunter. Briggs's *Logarithmorum chilias prima* (1617) was the first published table of logarithms of numbers to base 10, followed within three years by Gunter's *Canon triangulorum* (1620), the first table of the logarithms to base 10 of trigonometrical functions.

John Napier and his logarithms

John Napier, the eighth Laird of Merchiston, was born in Edinburgh in 1550, the son of Sir Archibald Napier and his wife, Janet Bothwell, daughter of an Edinburgh burgess. It seems that after an initial education at St. Salvator's College, St. Andrews, Napier continued his studies abroad although it is not known where or what he studied. Certainly by 1571 he had returned to Scotland and was married there in 1572. On the death of his father in 1608 Napier moved to Merchiston Castle, Edinburgh, where he lived until his death on 4 April 1617.[1]

Napier first gained a reputation as a scholar and theologian with the publication of his *A plaine discovery of the whole revelation of Saint John* in 1593. This work of Biblical exegesis arose out of the fears entertained in Scotland of an invasion by Philip II of Spain, and in it Napier urged the Scottish king, James VI (in 1603 to become James I of England) to see that 'justice be done against enemies of Gods church' and to 'purge his house, family and court of all Papists, Atheists and Newtrals'. *A plaine discovery* went through several editions in Dutch, French, and German, as well as English.

Napier's interest in aids to calculation was not confined to logarithms. In 1617 he published his *Rabdologiae*, a work which described a number of calculating devices including the method of multiplication and division by numbering rods, known as 'Napier's bones'.[2] These consisted of rectangular rods of square cross-section, each face carrying one of the multiplication tables from 1 to 9. Two additional rods were devised to aid the calculation of square and cube roots. These rods were popular for many years after Napier's death and were sold in a variety of forms made of paper, wood, ivory, and, as *objets de salon*, even precious metals.

THE MAKING OF LOGARITHM TABLES

Fig. 2.2 Statue of John Napier. This is a copy, about 105 cm in height, of a Victorian representation found on the exterior of the Scottish National Portrait Gallery, Edinburgh.

In an appendix to the *Rabdologiae*, headed 'The Promptuary for Lightning Multiplication', Napier describes a more complicated system of engraved rods or strips for mechanical calculation. The appendix begins:[3]

Although this promptuary was discovered by me least of all, it deserves a better place than to be put last in the book. By using it, any multiplication, no matter how difficult or involved, can be done readily and with maximum speed.

Divisions can also be done by its use, but they first have to be converted into multiplications, using either tables of sines, tangents and secants, or the tables in... this work.

The four chapters which follow the introduction describe the construction and use of the promptuary.

But Napier's greatest contribution to mathematics was his logarithms and it is to a description of these that we now turn. There is some evidence to suggest that Napier began his work on logarithms at least as early as 1594. In a letter written in 1624 to his friend Cugerus, Kepler comments on the state of trigonometry and then goes on to say: 'But nothing, in my mind, surpasses the method of Napier, although a certain Scotchman, even in the year 1594, held out some promise of the wonderful Canon in a letter to Tycho'.[4] Kepler had once worked in Tycho Brahe's observatory and so was able to secure information from Brahe directly.

It was not until 1614 that Napier's first work on this subject, *Mirifici logarithmorum canonis descriptio* (known as the *Descriptio*), was published. In addition to tables of logarithms the *Descriptio* also contains an account of the nature of logarithms and a number of examples explaining their use. The East India Company was so impressed by Napier's *Descriptio* that it asked Edward Wright, a Cambridge mathematician and expert in navigation, to translate it into English for the benefit of the Company's seafarers.[5] From the very beginning of logarithms their utility to navigators has been of supreme importance in their development.

Napier's second work on logarithms, *Mirifici logarithmorum canonis constructio* (the *Constructio*), was published posthumously by his son Robert in 1619. The object of this latter work, which comprises material written by Napier many years before, was to explain in detail the way in which the tables in the *Descriptio* had been calculated and the rationale behind them.[6] But whereas the *Constructio* uses the term 'artificial numbers', the *Descriptio* employs 'logarithms' even though this latter term was a later invention.

Napier's Descriptio

Realising that astronomical and navigational calculations involved primarily trigonometrical functions, especially sines, Napier set out to construct a table by which multiplication of these sines could be replaced by addition: the tables in the *Descriptio* are of logarithms of *sines*. They consist of seven columns, and are semi-quadrantally arranged. That is, the first column contains the angle increasing from 0° to 45° (the semi-quadrant) in steps of 1 minute. Only the minute is shown, the degree to which it relates being given at the top left hand corner of the page. Each page covers thirty

The definition of angular functions at the beginning of the seventeenth century

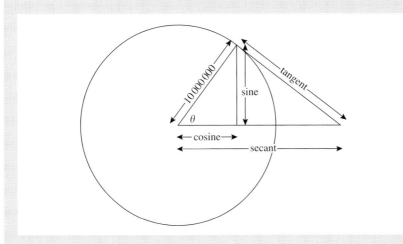

The modern definition of the sine of an angle, the trigonometric function that is equal in a right-angled triangle to the ratio of the side opposite the given angle to the hypotenuse, would have been unknown to Napier. Rather he would have taken it as half the length of the chord subtended by twice the given angle at the centre of a circle of arbitrary radius. This radius, called the *whole sine* because it was the maximum value that the sine could reach, was chosen to be sufficiently large to enable all the quantities to be considered as whole numbers, and avoid fractions, to whatever accuracy was desired. Napier chose a value of 10 000 000.

minutes, the table thus occupying 90 pages. The seventh column contains the complementary angle decreasing from 90° to 45°, also in steps of 1 minute, an arrangement following that introduced by Rheticus in his *Canon doctrinae triangulorum* of 1551.[7] The second and sixth columns, headed *Sinus*, contain the sines of the angles in the first and seventh columns respectively, and the third and fifth columns, headed *Logarithmi*, the logarithms of their respective sines. The fourth column, headed *Differentiae*, contains the differences between these two logarithms. Thus for a given angle in the first column we have essentially, in columns 2 to 6, its sine, the logarithm of its sine, the logarithm of its tangent, the logarithm of its cosine and its cosine, respectively. The same rule hold for angles in the right-most column, this time counting the columns from right to left, save that column 4 now contains the logarithm of the cotangent. In keeping with the definition of angular functions then current, the values in the table are integers in the range 0 to 10 000 000, the value of the whole sine. There are no separators of the digits.

Two features of Napier's logarithms, in their original form, make them difficult to use. The first is that the calculation of the logarithm of a product or a quotient involves not only the addition or subtraction of two numbers, but also the manipulation of the number 161 180 956, the value of log 1. This arises directly from Napier's definition of a logarithm, namely that log ab is log a + log b − log 1 and log a/b is log a − log b + log 1. Note, though, that log bc/a = log b + log c − log a. The second difficulty, which follows directly from the first, is that these logarithms are not well adapted to decimal arithmetic, since the multiplication or division of a number by 10 corresponds to the addition or subtraction of another unwieldy number 23 025 842, log 10.[8] The substitution of a power of ten for this number would resolve this difficulty, but the complete adaptation to the decimal scale also required the choice of the logarithm of unity as zero. It was these two apparent deficiencies which Napier and Briggs later sought to address.

Henry Briggs and the Logarithmorum chilias prima

Napier's invention of logarithms was rapidly and enthusiastically taken up by Henry Briggs. Briggs was born at Warley Wood, in the parish of Halifax, Yorkshire, probably in early 1561.[9] He was educated at the local grammar school and, in 1577, went to St. John's College, Cambridge, where he took his BA in 1581 and MA in 1585. He was made a Fellow in 1588.

In 1596 Briggs became the first professor of geometry at Gresham College in London. In July 1575 Sir Thomas Gresham, the founder of the Royal Exchange, willed his house in Bishopsgate Street to his wife for her lifetime, but thereafter it was to become Gresham College, an institution where free lectures were to be provided for all who cared to hear them. Seven subjects, divinity, astronomy, music, geometry, law, medicine, and rhetoric, were to be offered, one on each day of the week. For each subject an unmarried professor was to be appointed at an annual salary of £50 and with an official residence in the College. Gresham College, which was to play a crucial role in the early development of logarithms, came into existence after the death of Lady Gresham in 1586 and many legal squabbles.[10]

Briggs was later invited by Sir Henry Savile to become professor of geometry at Oxford, and he took up his duties at Merton College in January 1620, a post that he held until his death on 26 January 1630.

The early years of Gresham College were plagued by disputes between the professors and the trustees. It is mainly due to the work and influence of Henry Briggs, who was the professor of geometry at the College from its foundation until his move to Oxford in 1620, that Gresham College has any claim to scientific respectability at all in these years. Edmund Gunter, who was a friend of Briggs, became professor of astronomy in 1620, and Briggs secured the election of Henry Gellibrand to succeed Gunter in 1627

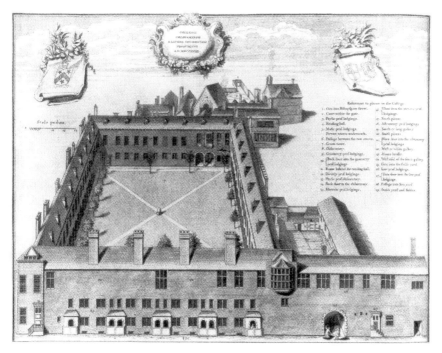

Fig. 2.3 Gresham College, the Royal Society's first home, was inaugurated as a seat of learning in 1597 as part of a bequest made by Sir Thomas Gresham (1519–1579), merchant and founder of the Royal Exchange. The College still exists as an educational institution but the original building, formerly Gresham's mansion in the City of London, was demolished in the eighteenth century. It stood in Bishopsgate, to the south of the modern Liverpool Street Station, on the site now occupied by Tower 42 (the former NatWest Tower).

after the latter's death in 1626. Between them Briggs, Gunter, and Gellibrand ensured that for its first 40 years Gresham College had a local and practical relevance. It was within this group that the developments in theory, computational techniques, and instrumentation took place which were to transform the art of navigation into a science.

During Briggs's early years at Gresham College he worked on problems of navigation, writing *A table to find the height of the pole, the magnetical declination being given*. This table, together with a description by Gilbert of two magnets which he had invented, was published by Blundeville in 1602,[11] and between 1608 and 1612 he collaborated with Edward Wright in compiling observational tables for a new edition of Wright's *Certaine errors in navigation*.[12] Among his contemporaries Briggs was not alone in his interest in navigation. In an age of expanding trade and colonization, the country which could develop new navigational techniques had a competitive advantage over its rivals in the quest for new colonies and their attendant natural resources.[13]

By 1609 Briggs was in correspondence with James Ussher, later to become archbishop of Armagh, and in one of his letters, from 1610, Briggs tells Ussher that he has been studying eclipses. By 1615, however, Briggs was immersed in logarithms. On 10 March 1615 he wrote to Ussher:[14]

> Naper, lord of Markinston, hath set my head and hands a work with his new and admirable logarithms. I hope to see him this summer, if it please God, for I never saw book, which pleased me better, and made me more wonder.

Briggs became an ardent publicist for logarithms and spent the rest of his life in their development. He was, though, aware of the cumbersome features of Napier's canon,[15] which he regarded as defects, and at once set about trying to rectify them. Briggs later recalled that

> ...when expounding [the] doctrine [of Napier's logarithms] publicly in London to my auditors in Gresham College, [I] remarked that it would be much more convenient that 0 should be kept for the logarithm of the whole sine (as in the *Canon mirificus*) but that the logarithm of the tenth part of the same whole sine, that is to say, 5 degrees 44 minutes and 21 seconds,[16] should be 10 000 000 000. And concerning that matter I wrote immediately to the author himself.[17]

In the summer of 1615 Briggs visited Napier in Edinburgh, taking with him some tables that he had prepared along these lines. After some discussion, it emerged that Napier '...was of opinion that the change should be effected

Napier's definition of logarithms

Central to Napier's definition of logarithms was his use of the concept of motion, of points moving along lines with speeds defined in various ways.[†] Napier's definition can be simply stated. Imagine two straight lines, the first, $T'Z$ of infinite length, and the second, TS of fixed length. Two points start from T' and T respectively, at the same instant and with the same velocities, the first point moving uniformly and the second point moving so that its velocity is everywhere proportional to its distance from S. If the points reach L and G respectively at the same instant then the number that measures the line $T'L$ is defined as the logarithm of GS. GS is the sine and TS the whole sine, or radius. It follows from Napier's definition that his logarithms of sines decrease monotonically as the angle, and therefore the sine, increases.

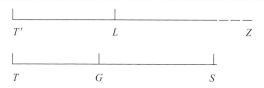

[†]John Napier, *The construction of the wonderful canon of logarithms* (translated by W. R. Macdonald), Blackwood, Edinburgh, 1889, p. 19. As D. T. Whiteside ('Mathematical thought in the later 17th century', *Archive for history of exact sciences* 1 (1960), p. 220) has pointed out, Napier's distance-speed model is medieval rather than modern and its now usual treatment by methods of the calculus is quite foreign to its kinematic nature.

in this manner, that 0 should be the logarithm of unity and 10 000 000 000 that of the whole sine.'[17] Briggs readily agreed that this was the best way of proceeding and began at once to calculate some logarithms on this basis.

It is clear from Briggs's account that he and Napier rapidly arrived at a consensus and that the invention of logarithms to base 10 was a joint accomplishment. By the time of his second visit to Napier in the summer of 1616, Briggs had already composed the 'principal part' of what was to become the *Arithmetica logarithmica* of 1624. Briggs would have visited him again in the summer of 1617 had Napier not died in April of that year.

Shortly after Napier's death Briggs privately printed the 'principal part' of his logarithms, the *Logarithmorum chilias prima*, so called from the first words of the preface to the Latin original. This is an extremely rare book of only 16 pages and is the earliest printed table to base 10.[18] It has no title-page but begins with a Latin preface of which the following is a translation:[19]

Here is the first thousand logarithms which the author has had printed, not with the intention of becoming public property, but partly to satisfy on a private basis the wish of some of his own intimate friends: partly so that with its help he might more conveniently solve not only several following thousands but also the integral table of Logarithms used for the calculation of all triangles. For he has the Table of Sines, accurately drawn up from first elements by himself ten years earlier, through algebraic equations and differences, proportional to the actual sines instead of individual degrees and hundredths of degrees:[20] this he hopes to publish, God willing, together with the appended logarithms, as soon as conveniently may be.

But because these logarithms are different from those which the most renowned inventor [Napier], of ever-to-be-revered memory, published in his *Wonderful Table*,[21] it is to be hoped that his posthumous book will shortly give us abundant satisfaction.[22] He [Napier] did not cease from urging the author [Briggs] (when he twice visited him at his home in Edinburgh and, most kindly welcomed, had gladly stayed with him for several weeks; and had shown him the first part of these Logarithms which he had then solved) to take this work upon himself. He [Briggs] not unwillingly gratified him [Napier].

In a slight volume neither the enjoyment nor the toil has been slight.

Although the preface is not signed, it is clear from its content that the author could only have been Briggs. The reference to Napier's death shows that it could not have been written before April 1617, or after December 1617, since it is undoubtedly the work referred to in a letter from Sir Henry Bourchier to Ussher dated 6 December 1617 in which Bourchier says 'Our kind friend, Mr Briggs, hath lately published a supplement to the most excellent tables of logarithms, which I presume he has sent to you.'[23]

Unlike Napier's logarithms, those of Briggs were the logarithms of the first thousand natural numbers. These, says Briggs, can be used, together with his earlier table of sines, 'for the calculation of all triangles'. The calculations which Briggs had in mind here are those involved in finding all the sides and angles of a triangle given that at least three of the six dimensions (three sides and three angles) of the triangle are known. In this scheme, the determination of the logarithm of a sine would involve first looking up the sine in one table, and the logarithm of that number in another.

THE MAKING OF LOGARITHM TABLES 59

The logarithms are given to 14 figures. The digits are separated into groups of five by commas. The right hand side of each page was originally blank but in one of the British Library copies[24] some of the differences, to four figures, have been written in and the line separating the characteristic (the integral part of the logarithm) from the mantissa (the fractional part) has been ruled in by hand.[25]

The logarithms of Briggs[26] were of integers: a base 10 equivalent of Napier's logarithms of sines was not to appear until 1620, the achievement of Edmund Gunter.

2	Logarithmi.		Logarithmi.
1	00000,00000,00000	34	15314,78917,04226
2	03010,29995,66398	35	15440,68044,35028
3	04771,21254,71966	36	15563,02500,76729
4	06020,59991,32796	37	15682,01724,06700
5	06989,70004,33602	38	15797,83596,61681
6	07781,51250,38364	39	15910,64607,02650
7	08450,98040,01426	40	16020,59991,32796
8	09030,89986,99194	41	16127,83856,71974
9	09542,42509,43932	2	16232,49290,39790
10	10000,00000,00000	43	16334,68455,57959
11	10413,92685,15823	4	16434,52676,48619
12	10791,81246,04762	45	16532,12513,77534
13	11139,43352,30684	6	16627,57831,68157
14	11461,28035,67824	47	16720,97857,93572
15	11760,91259,05568	8	16812,41237,37559
16	12041,19982,65592	49	16901,96080,02851
17	12304,48921,37827	50	16989,70004,33602
18	12552,72505,10331	51	17075,70176,09794
19	12787,53600,95283	2	17160,03343,63480
20	13010,29995,66398	53	17242,75869,60079
21	13222,19294,73392	4	17323,93759,82297
22	13424,22680,82221	55	17403,62689,49424
23	13617,27836,01759	6	17481,88027,00620
24	13802,11241,71161	57	17558,74855,67249
25	13979,40008,67204	8	17634,27993,56294
26	14149,73347,97082	59	17708,52011,64214
27	14313,63764,15899	60	17781,51250,38364
28	14471,58031,34222	61	17853,29835,01077
29	14623,97997,89896	2	17923,91689,49825
30	14771,21254,71966	63	17993,40549,45358
31	14913,61693,83427	4	18061,79973,98389
32	15051,49978,31991	65	18129,13356,64286
33	15185,13939,87789	6	18195,43935,54187
34	15314,78917,04226	67	18260,74802,70083

Fig. 2.4 A page from Briggs's *Logarithmorum chilias prima*, 1617. The right hand side of each page was originally blank but in the British Library copies some of the differences have been written in, and, as in the page shown, the line separating the characteristic from the mantissa has been ruled in by hand. This rare work is a small pamphlet of only 16 pages and is the earliest printed table of logarithms to base 10.

Edmund Gunter and the Canon triangulorum

Edmund Gunter was born in Hertfordshire, of Welsh stock, in 1581. He was educated first at Westminster School and then at Christ Church, Oxford, where he took his BA in 1603 and MA in 1606. In 1615 he became rector of St. George's, Southwark, a position that he retained until his death, and took his BD later the same year. Gunter was not renowned for his preaching: John Aubrey, the inveterate seventeenth-century gossip, recounts that:[27]

> When [Gunter] was a student at Christ Church, it fell to his lot to preach the Good Friday sermon, which some old divines that I knew did hear, but they said that 'twas said of him then in the University that our Saviour never suffered so much since his passion as in that sermon, it was such a lamentable one.

On 6 March 1619 Gunter was elected the third professor of astronomy at Gresham College in succession to Thomas Williams, his tenure of office at Gresham overlapping that of Briggs by some 15 months. Later that year Sir Henry Savile was considering whom to appoint as the first Savilian professor of geometry at Oxford. According to Aubrey

> he first sent for Mr Gunter, from London, (being of Oxford University) ... : so he came and brought with him his sector and quadrant, and fell to resolving of triangles and doing many fine things. Said the grave knight, 'Do you call this reading of Geometry? This is showing of tricks, man!' and so dismissed him with scorn, and sent for Briggs from Cambridge.[28]

Gunter died suddenly in his rooms at Gresham College on 10 December 1626 at the relatively early age of 45. He was buried in an unmarked grave in the church of St. Peter the Poor, Broad Street.

A competent mathematician, Gunter had a gift for devising instruments which simplified calculations in astronomy, navigation, and surveying. In 1623 he published an account, in English, of his sector (which came to be called 'Gunter's Sector') in the *De sectore et radio. The description and use of the sector*.... At the end of this work Gunter wrote that '... of this book, written in Latin, many copies [were] transcribed and dispersed more than sixteen years since.'[29]

It is probable that copies of this earlier version, which was in Latin, fell into the hands of the leading mathematicians of the day, apparently gaining for Gunter 'the friendship of the Earl of Bridgewater, William Oughtred, Henry Briggs, and others.'[30] Some evidence for the connection between

Gunter and Bridgewater is supplied by Oughtred:[31]

In the spring, 1618, I being at London, went to see my honoured friend Master Henry Briggs at Gresham College, who then brought me acquainted with Master Gunter, ..., with whom falling into speech about his quadrant, I showed him my horizontal instrument. He viewed it very heedfully, and questioned about the projecture and use thereof, often saying these words, 'It is a very good one.' And not long after he delivered to Master Briggs, to be sent to me, mine own instrument printed off from one cut in brass, which afterwards I understood he presented to the Right Honourable the Earl of Bridgewater.

In the early 1600s John Egerton, the first Earl of Bridgewater, was a public figure whose star was very much in the ascendant. He was baron of the exchequer of Chester from 1599 until 1605, and was made a knight of the Bath on James I's arrival in England in 1603. He received an honorary MA from Oxford in 1605 and became the first Earl of Bridgewater in 1617. He held many high public offices until his death in 1649. In the climate of the time perhaps Gunter felt that he could do worse than secure the patronage of such an important person as Egerton who was, after all, another 'Oxford man'.

It may be presumed that it was the close relationship between Briggs and Gunter at Gresham College which led to Gunter's *Canon triangulorum* of 1620, a work that did for sines and tangents what Briggs had done for natural numbers.

The title-page gives the full title of Gunter's work: *Canon triangulorum sive tabulæ sinuum et tangentium artificialium ad radium 10000.0000 & ad scrupulum prima quadrantis*. (Trigonometrical canon, that is tables of artificial sines and tangents to a radius of 100 000 000 for each minute of the first quadrant.) The departure here from the convention that Briggs had already adopted of tabulating angular functions in steps of one hundredth of a degree is striking. Gunter uses here the integer 100 000 000 for the whole sine, ten times greater than that used by Napier; the 'decimal point' in Gunter's title serves merely to guide the eye, much as we might use commas.

In addition, the title-page carries an engraving of a spherical triangle *SPZ* indicating clearly the proposed use of the tables in the solution of spherical triangles. Napier uses the same triangle in Chapter 5 of Book II of the *Descriptio* where, in Wright's translation, it is defined as follows:[32]

As for use and exercise sake, let there be a spherical triangle not quadrantal described on the superfices of the *Primum Mobile* PZS representing the pole, the zenith, and the Sunne....

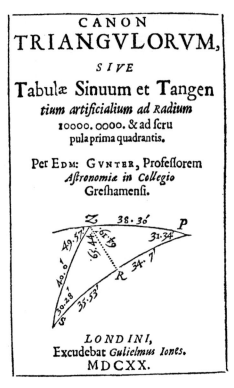

Fig. 2.5 The title-page of Gunter's *Canon triangulorum*, published in 1620 and printed by William Jones. The spherical triangle *SPZ*, previously used by Napier, is reproduced without explanation and it seems likely that the *Canon* was aimed at those already familiar with the solution of such triangles by the methods described in the *Descriptio*, and which had been expounded for several years by Briggs at Gresham College. The *Canon* was the first table to be printed of logarithms of trigonometric functions to base 10.

Since this triangle, previously used by Napier, is reproduced on the title-page of the *Canon triangulorum* without explanation, we may infer that the *Canon* was aimed at those who were already familiar with the solution of spherical triangles by the methods described in the *Descriptio*, and which had been expounded for several years by Briggs at Gresham College.

The *Canon* was printed by William Jones, a noted printer of mathematical material, who was also to print Gunter's *De sectore et radio* in 1623, and Briggs's great work, the *Arithmetica logarithmica*, in 1624.

The *Canon triangulorum* is dedicated to John Egerton, Earl of Bridgewater with whom, as we have seen, Gunter seems to have had previous dealings.

Gunter's table, like those of Napier, presents the logarithms of sines and tangents for every minute of the quadrant, semi-quadrantally arranged. These tables are to eight figures, split into two groups of four by a vertical bar. Leading digits are omitted where these do not change from the line above. This may have been a deliberate device to render the page more readable, but it is not impossible that it is the result of a desire to economize

THE MAKING OF LOGARITHM TABLES 63

on the use of type. Gunter's own description is very brief, taking only 180 words or so in the original Latin. It is of considerable interest and is translated here in full.[33] First, the general layout of the tables is described:

> Our table has six columns, divided from each other by two lines. The first of these is of the angles from 0° to 45° degrees, the sixth, of the angles from 45° to the end of the quadrant. The remaining four in between contain the *Sines* and *Tangents* of these angles.

Gunter then distinguishes between the sines and tangents as usually defined and those in the table. He does not refer explicitly to 'logarithms' but rather to the term Napier used in the *Constructio*, 'artificial numbers':

> But our *Sines* are not half of the chords, and the *Tangents* are not perpendicular to the extremity of the diameter; but in their place their own numbers are substituted, which we call 'artificial' for that reason, with their first discoverer, Lord Merchiston. And thus the second column and the fourth contain the *Sines* and *Tangents* of the angles in the first column; and the third column and the fifth, the *Sines* and *Tangents* of the angles in the sixth column.

There then follows some explanatory material on how to use the tables, using as an example the latitude of London:

> So, if you are looking for the artificial sine of our latitude, which in *London* is 51° 32′, you should find *sine* 51° in the bottom part of the page, and 32 minutes in the sixth column; the common angle gives 9893.7452 for the desired *sine*, to which corresponds on the same line the sine of the complement 9793.8317, which for the sake of brevity we call the *cosine*.[34] Similarly, the *tangent* of 51° 32′ is 10099.9134, and the *cotangent* 9900.0865. When your calculation demands *secants*, they can be supplied by subtracting the *cosine* from double the radius.
>
> Thus, when twice the radius is   20000 | 0000
> Take away from it the *cosine* of 51° 32′   9793 | 8317
> The secant of 51° 32′ is the result   10206 | 1682

The rules used here are the usual trigonometrical identities, $\cos(\theta) = \sin(90 - \theta)$, $\cot(\theta) = \tan(90 - \theta)$ and $\sec(\theta) = 1/\cos(\theta)$. As before, all the numbers are integers, the points and vertical bars being inserted only to guide the eye.

At the end of the *Canon* is a letter to the reader, of which the following is a translation:

> To the reader devoted to practical learning, many greetings. Our table is useful for the solution of spherical triangles, the same as the tables of natural sines and tangents published by others, but it is a little easier to use. For we avoid their multiplication

THE MAKING OF LOGARITHM TABLES

M	Sin. 38			Tan. 38			
0	9789 3419	9896	5321	9892 8098	10107 1901	60
1	9 5036		4334	3 0702	6 9297	59
2	9 6651		3346	3 3305	6 6694	58
3	9 8266		2357	3 5908	6 4091	57
4	9 9880		1368	3 8511	6 1488	56
5	9790 1492		0379	4 1113	5 8886	55
6	0 3104	9895	9388	4 3715	5 6284	54
7	0 4714		8398	4 6317	5 3683	53
8	0 6324		7406	4 8918	5 1081	52
9	0 7933		6414	5 1519	4 8481	51
10	0 9541		5421	5 4119	4 5880	50
11	1 1148		4428	5 6719	4 3280	49
12	1 2753		3434	5 9319	4 0680	48
13	1 4358		2440	6 1918	3 8081	47
14	1 5962		1445	6 4517	3 5482	46
15	1 7565		0449	6 7116	3 2884	45
16	1 9167	9894	9453	6 9714	3 0285	44
17	2 0768		8456	7 2312	2 7687	43
18	2 2368		7459	7 4909	2 5090	42
19	2 3968		6461	7 7507	2 2493	41
20	2 5566		5462	8 0103	1 9896	40
21	2 7163		4463	8 2700	1 7300	39
22	2 8759		3463	8 5296	1 4704	38
23	3 0355		2463	8 7891	1 2108	37
24	3 1949		1462	9 0487	0 9512	36
25	3 3542		0460	9 3082	0 6917	35
26	3 5135	9893	9458	9 5677	0 4323	34
27	3 6726		8455	9 8271	0 1728	33
28	3 8317		7452	9900 0865	10099 9134	32
29	3 9907		6448	0 3458	9 6541	31
30	4 1495		5443	0 6051	9 3948	30
			Sin. 51.		Tan. 51.	M

Fig. 2.6 A page from Gunter's *Canon triangulorum* containing the latitude of London (51° 32′). This table, like Napier's, presents the logarithms of sines and tangents for every minute of the quadrant, semi-quadrantally arranged. This table is to eight figures, split into two groups of four by a vertical bar. Leading digits are omitted where these do not change from the line above. This may have been a deliberate device to render the page more readable, but it is not impossible that it is the result of a desire to economize on the use of type.

by addition, and their division by subtraction. And there is no need of more rules or examples. If you would like the same thing for rectilinear triangles, have bound with ours [i.e. Gunter's tables in the *Canon triangulorum*] the Logarithms of our friend and colleague Henry Briggs [i.e. those in the *Logarithmorum chilias prima*]. For we rest upon that foundation, and we use the same way of working. Farewell, and if this work proves welcome to you, expect more from us of this kind.

Here we have explicitly stated Gunter's view of the utility of his tables in the solution of spherical triangles. His tables are, he says, easier to use than the tables of natural sines and tangents published by others. John Ward, quoting Edmund Wingate, states that Gunter's *Canon* '... will perform that by addition and subtraction only, which heretofore Regiomontanus tables did by multiplication and division...'.[35] It is at least possible that Wingate (or Ward in quoting him) made an error here. Regiomontanus seems only to have published a table of tangents, and that in 1504.[36] It seems more probable that Wingate meant to refer to the *Opus Palatinum* of Rheticus that was published in 1596, or the version of those tables published by Pitiscus in 1613. That Gunter felt that there was no need for more rules or examples indicates that his *Canon* was intended for a readership that was already familiar with Napier's work.

There is some evidence, however, that Gunter did subsequently come to believe that 'more rules or examples' were indeed necessary. There is in the British Library Manuscript Collection a manuscript by Gunter, probably dating from shortly after the publication of his *Canon triangulorum*, which gives many worked examples and refers specifically to Briggs's logarithms (the *Logarithmorum chilias prima*) and 'my canon of artificial sines and tangents'.[37] The content of this manuscript indicates that it formed the basis of the second edition of the *Canon triangulorum* which was published in 1624. This included a table of Briggs's logarithms extended up to 10 000 and gave a rule for extending these up to 100 000. In an extended preface Gunter described the general use of the table illustrated by worked examples, many of which appear in the British Library manuscript.

Briggs's Arithmetica logarithmica *and* Trigonometria Britannica

From Gunter's reference to Briggs in his *Letter to the reader*, and from Briggs's reference to his 'intimate friends', it is likely that he and Briggs had a joint purpose in educating the mathematical community in the advantages of common logarithms and worked together for its fulfilment.

In the years immediately following the publication of the *Descriptio*, these two works, Gunter's *Canon triangulorum* and Briggs's *Logarithmorum chilias prima*, established the techniques and conventions that were to underpin the two great canons of the next decade, the *Arithmetica logarithmica* and the *Trigonometria Britannica*.

In his *Arithmetica logarithmica*, published in 1624, Briggs extended his previous table in the *Logarithmorum chilias prima* to include logarithms, to base 10, of the integers 1 to 20 000 and 90 000 to 100 000, again to fourteen places. In this table the comma is used both to separate the characteristic from the mantissa and to separate groups of five digits for ease of reading. First differences are printed interlinearly.

It was Briggs's intention to fill the gap in these tables but he was overtaken by Vlacq who published the completed canon, to ten digits, in 1628.

In the same year Briggs wrote 'To his very good and much respected frende Mr. John Pell at Trinitie Coll. in Cambridge':[38]

> My desire was to have those Chiliades [thousands] that are wantinge betwixt 20 and 90 calculated and printed, and I have done them all almost by myself, and by some frendes whome my rules had sufficiently informed, and by agreement the business was conveniently parted amongst us; but I am eased of that charge and care by one Adrian Vlacque, an Hollander, who hath done all the whole hundred chiliades and printed them...But he hathe cut of 4 of my figures throughout...

Vlacq's tabular layout differed slightly from Briggs's 1624 edition in that this time the first differences occupy a separate column.

There is no hint in Briggs's letter of any irritation on his part against Vlacq for having anticipated him in the completion of a work that he, Briggs, had so nearly completed. But this is typical of what we know of Briggs whose attitude he himself summed up in the words with which he closed his preface to Wright's translation of the *Descriptio*: 'I ever rest a lover of all them that love the Mathematickes.' The identity of Briggs's 'frendes' who were helping with the calculations can only be guessed at, but in view of their evident close relationship it seems highly probable that Gunter was one of them.

Briggs also calculated tables of natural sines (to fifteen places) and tangents and secants (to ten places), together with log sines (to fourteen places) and tangents (to ten places), at intervals of a hundredth of a degree, with interlinear differences for all the functions. The division of the degree is thus centesimal, but the corresponding arguments in minutes and seconds of arc are also given, the intervals so expressed being 36 seconds. That Briggs tabulated his sines every hundredth of a degree was in itself an innovation: the great canon of Rheticus, the *Opus Palatinum* of 1596, and its derivatives, had all been sexagesimally arranged. Briggs did not live to publish these tables: they were published posthumously in 1633, as the *Trigonometria Britannica*, by Henry Gellibrand, Briggs's successor as professor of astronomy at Gresham College, who also provided some additional material.[39]

Briggs's two great works, the *Arithmetica logarithmica* and the *Trigonometria Britannica*, formed the basis of all logarithm tables published for the next 200 years or so. It was not until the end of the eighteenth century that any further calculation of logarithms took place. As described by Ivor Grattan-Guinness in Chapter 4, in 1785 the French government decided that new tables of angular functions, using the centesimal division of the quadrant, and their logarithms, should be calculated. The production of these tables,

Chilias decimaoctaua.

Num. absolut.	Logarithmi.	Num. absolut.	Logarithmi.	Num. absolut.	Logarithmi.
17501	4,24306,28648,0481	17534	4,24388,10022,1832	17567	4,24469,76012,9672
	2,48147,0057		2,47679,9920		2,47214,7329
17502	4,24308,76795,0538	17535	4,24390,57702,1752	17568	4,24472,23227,7001
	2,48132,8279		2,47665,8675		2,47200,6614
17503	4,24311,24927,8817	17536	4,24393,05368,0427	17569	4,24474,70428,3615
	2,48118,6517		2,47651,7447		2,47186,5916
17504	4,24313,73046,5334	17537	4,24395,53019,7874	17570	4,24477,17614,9531
	2,48104,4771		2,47637,6234		2,47172,5233
17505	4,24316,21151,0105	17538	4,24398,00657,4108	17571	4,24479,64787,4764
	2,48090,3042		2,47623,5037		2,47158,4566
17506	4,24318,69241,3147	17539	4,24400,48280,9145	17572	4,24482,11945,9330
	2,48076,1329		2,47609,3857		2,47144,3915
17507	4,24321,17317,4476	17540	4,24402,95890,3002	17573	4,24484,59090,3245
	2,48061,9631		2,47595,2692		2,47130,3280
17508	4,24323,65379,4107	17541	4,24405,43485,5694	17574	4,24487,06220,6525
	2,48047,7951		2,47581,1544		2,47116,2662
17509	4,24326,23427,2058	17542	4,24407,91066,7238	17575	4,24489,53336,9187
	2,48033,6286		2,47567,0412		2,47102,2059
17510	4,24328,61460,8344	17543	4,24410,38633,7650	17576	4,24492,00439,1246
	2,48019,4638		2,47552,9296		2,47088,1473
17511	4,24331,09480,2982	17544	4,24412,86186,6946	17577	4,24494,47527,2719
	2,48005,3005		2,47538,8196		2,47074,0902
17512	4,24333,57485,5987	17545	4,24415,33725,5142	17578	4,24496,94601,3621
	2,47991,1389		2,47524,7112		2,47060,0347
17513	4,24336,05476,7376	17546	4,24417,81250,2254	17579	4,24499,41661,3968
	2,47976,9789		2,47510,6044		2,47045,9808
17514	4,24338,53453,7165	17547	4,24420,28760,8298	17580	4,24501,88707,3776
	2,47962,8206		2,47496,4992		2,47031,9287
17515	4,24341,01416,5371	17548	4,24422,76257,3290	17581	4,24504,35739,3063
	2,47948,6638		2,47482,3957		2,47017,8780
17516	4,24343,49365,2009	17549	4,24425,23739,7247	17582	4,24506,82757,1843
	2,47934,5086		2,47468,2937		2,47003,8288
17517	4,24345,97299,7095	17550	4,24427,71208,0184	17583	4,24509,29761,0131
	2,47920,3551		2,47454,1933		2,46989,7812
17518	4,24348,45220,0646	17551	4,24430,18662,2117	17584	4,24511,76750,7943
	2,47906,2032		2,47440,0946		2,46975,7354
17519	4,24350,93126,2678	17552	4,24432,66102,3063	17585	4,24514,23726,5297
	2,47892,0528		2,47425,9974		2,46961,6912
17520	4,24353,41018,3206	17553	4,24435,13528,3037	17586	4,24516,70688,2209
	2,47877,9042		2,47411,9019		2,46947,6485
17521	4,24355,88896,2248	17554	4,24437,60940,2056	17587	4,24519,17635,8694
	2,47863,7571		2,47397,8080		2,46933,6074
17522	4,24358,35759,9819	17555	4,24440,08338,0136	17588	4,24521,64569,4768
	2,47849,6116		2,47383,7157		2,46919,5679
17523	4,24360,84609,5935	17556	4,24442,55721,7293	17589	4,24524,11489,0447
	2,47835,4678		2,47369,6249		2,46905,5300
17524	4,24363,32445,0613	17557	4,24445,03091,3542	17590	4,24526,58394,5747
	2,47821,3256		2,47355,5358		2,46891,4937
17525	4,24365,80266,3869	17558	4,24447,50446,8900	17591	4,24529,05286,0684
	2,47807,1850		2,47341,4484		2,46877,4590
17526	4,24368,28073,5719	17559	4,24449,97788,3384	17592	4,24531,52163,5274
	2,47793,0460		2,47327,3624		2,46863,4259
17527	4,24370,75866,6179	17560	4,24452,45115,7008	17593	4,24533,99026,9533
	2,47778,9086		2,47313,2781		2,46849,3944
17528	4,24373,23645,5265	17561	4,24454,92428,9789	17594	4,24536,45876,3477
	2,47764,7728		2,47299,1954		2,46835,3645
17529	4,24375,71410,2993	17562	4,24457,39728,1743	17595	4,24538,92711,7122
	2,47750,6386		2,47285,1143		2,46821,3361
17530	4,24378,19160,9379	17563	4,24459,87013,2886	17596	4,24541,39533,0483
	2,47736,5061		2,47271,0349		2,46807,3094
17531	4,24380,66897,4440	17564	4,24462,34284,3235	17597	4,24543,86340,3577
	2,47722,3752		2,47256,9570		2,46793,1843
17532	4,24383,14619,8192	17565	4,24464,81541,2805	17598	4,24546,33133,6420
	2,47708,2459		2,47242,8807		2,46779,2607
17533	4,24385,62328,0651	17566	4,24467,28784,1612	17599	4,24548,79912,9027
	2,47694,1181		2,47228,8060		2,46765,2388
17534	4,24388,10022,1832	17567	4,24469,76012,9672	17600	4,24551,26678,1415
	2,47679,9920		2,47214,7329		2,46751,2185

17601

Fig. 2.7 A page from Briggs's *Arithmetica logarithmica*, published in 1624 and printed by William Jones. Briggs's table is of the logarithms to base 10 of the integers 1 to 20 000 and 90 000 to 100 000, to fourteen places. The comma is used both to separate the characteristic from the mantissa and to separate groups of five digits for ease of reading. First differences are printed interlinearly.

44. GRAD.

Cente- sima	Sinus.	Tangentes.	Secantes.	Logarithmi Sinuū.	Log. Tangent.	M: S
67	70302,24326,73471 12,41114,66007	99854,66668 34,51503	140613,81555 24,26698	9,84696,91831,0067 7,66635,1036	9,99499,71761 15,16071	40:12
68	70314,65441,39478 12,40900,46938	98889,18171 34,52695	140638,08253 24,27965	9,84704,58466,1103 7,66367,5275	9,99514,87832 15,16065	40:48
69	70327,06341,86416 12,40686,24089	98923,70866 34,53887	140662,36218 24,29232	9,84712,24833,6378 7,66100,0464	9,99530,03897 15,16060	41:24
70	70339,47028,10505 12,40471,97461	98958,24753 34,55081	140686,65450 24,30500	9,84719,90933,6842 7,65832,6583	9,99545,19957 15,16054	42: 0
71	70351,87500,07966 12,40257,67055	98992,79834 34,56275	140710,95950 24,31768	9,84727,56766,3425 7,65565,3653	9,99560,36011 15,16049	42:36
72	70364,27757,75021 12,40043,32870	99027,36100 34,57469	140735,27718 24,33038	9,84735,22331,7078 7,65298,1666	9,99575,52060 15,16043	43:12
73	70376,67801,07891 12,39828,94908	99061,93578 34,58666	140759,60756 24,34309	9,84742,87629,8744 7,65031,0621	9,99590,68103 15,16039	43:48
74	70389,07630,02799 12,39614,53170	99096,52244 34,59862	140783,95065 24,35579	9,84750,52660,0365 7,64764,0514	9,99605,84142 15,16034	44:24
75	70401,47244,55969 12,39400,07654	99131,12106 34,61059	140808,30644 24,36851	9,84758,17424,9879 7,64497,1347	9,99621,00176 15,16029	45: 0
76	70413,86644,63623 12,39185,58364	99165,73165 34,62258	140832,67495 24,38124	9,84765,81922,1226 7,64230,3121	9,99636,16205 15,16025	45:36
77	70426,25830,21987 12,38971,05299	99200,35423 34,63457	140857,05619 24,39398	9,84773,46152,4347 7,63963,5832	9,99651,32230 15,16020	46:12
78	70438,64801,27286 12,38756,48461	99234,98880 34,64657	140881,45017 24,40672	9,84781,10116,0179 7,63696,9484	9,99666,48250 15,16017	46:48
79	70451,03557,75747 12,38541,87847	99269,63537 34,65857	140905,85689 24,41947	9,84788,73812,9663 7,63430,4071	9,99681,64267 15,16012	47:24
80	70463,42099,63594 12,38327,23462	99304,29394 34,67060	140930,27636 24,43224	9,84796,37243,3734 7,63163,9596	9,99696,80279 15,16009	48: 0
81	70475,80426,87056 12,38112,55305	99338,96454 34,68261	140954,70860 24,44500	9,84804,00407,3330 7,62897,0059	9,99711,96288 15,16005	48:36
82	70488,18539,42361 12,37897,83375	99373,64715 34,69465	140979,15360 24,45777	9,84811,63304,9389 7,62631,3456	9,99727,12293 15,16002	49:12
83	70500,56437,25736 12,37683,07676	99408,34180 34,70669	141003,61137 24,47056	9,84819,25936,2845 7,62365,1789	9,99742,28295 15,15999	49:48
84	70512,94120,33412 12,37468,28205	99443,04849 34,71874	141028,08193 24,48336	9,84826,88301,4634 7,62099,1055	9,99757,44294 15,15996	50:24
85	70525,31588,61617 12,37253,44966	99477,76723 34,73080	141052,56529 24,49615	9,84834,50400,5689 7,61833,1252	9,99772,60290 15,15993	51: 0
86	70537,68842,06583 12,37038,57957	99512,49803 34,74287	141077,06144 24,50896	9,84842,12233,6941 7,61567,2387	9,99787,76283 15,15991	51:36
87	70550,05880,64540 12,36823,67181	99547,24090 34,75493	141101,57040 24,52178	9,84849,73800,9328 7,61301,4453	9,99802,92274 15,15988	52:12
88	70562,42704,31721 12,36608,72636	99581,99583 34,76703	141126,09218 24,53461	9,84857,35102,3781 7,61035,7450	9,99818,08262 15,15986	52:48
89	70574,79313,04357 12,36393,74324	99616,76286 34,77911	141150,62679 24,54743	9,84864,96138,1231 7,60770,1378	9,99833,24248 15,15984	53:24
90	70587,15706,78681 12,36178,72247	99651,54197 34,79121	141175,17422 24,56028	9,84872,56908,2609 7,60504,6236	9,99848,40232 15,15982	54: 0
91	70599,51885,50928 12,35963,66404	99686,33318 34,80333	141199,73450 24,57313	9,84880,17412,8845 7,60239,2024	9,99863,56214 15,15980	54:36
92	70611,87849,17332 12,35748,56795	99721,13651 34,81543	141224,30763 24,58599	9,84887,77652,0869 7,59973,8744	9,99878,72194 15,15979	55:12
93	70624,23597,74127 12,35533,43423	99755,95194 34,82757	141248,89362 24,59885	9,84895,37625,9613 7,59708,6389	9,99893,88173 15,15978	55:48
94	70636,59131,17550 12,35318,26287	99790,77951 34,83970	141273,49247 24,61173	9,84902,97334,6002 7,59443,4962	9,99909,04151 15,15977	56:24
95	70648,94449,43837 12,35103,05388	99825,61921 34,85184	141298,10420 24,62461	9,84910,56778,0964 7,59178,4461	9,99924,20128 15,15975	57: 0
96	70661,29552,49225 12,34887,80726	99860,47105 34,86399	141322,72881 24,63750	9,84918,15956,5425 7,58913,4889	9,99939,36103 15,15975	57:36
97	70673,64440,29951 12,34672,52303	99895,33504 34,87615	141347,36631 24,65040	9,84925,74870,0314 7,58648,6242	9,99954,52078 15,15974	58:12
98	70685,99112,82254 12,34457,20119	99930,21119 34,88832	141372,01671 24,66330	9,84933,33518,6556 7,58383,8521	9,99969,68052 15,15974	58:48
99	70698,33570,02373 12,34241,84175	99965,09951 34,90049	141396,68001 24,67623	9,84940,91902,5077 7,58119,1723	9,99984,84026 15,15974	59:24
100	70710,67811,86548 12,34026,44470	100000,00000 34,91268	141421,35624 24,68914	9,84948,50021,6800 7,57854,5852	10,00000,00000 15,15974	60: 0

Fig. 2.8 A page from Briggs's *Trigonometrica Britannica*, published posthumously in 1633 by Henry Gellibrand, Briggs's successor as professor of astronomy at Gresham College. This table is of natural sines (to fifteen places) and tangents and secants (to ten places), together with log sines (to fourteen places) and tangents (to ten places). There are interlineal differences for all the functions. Briggs retains the division of the quadrant into 90 degrees, but divides each degree centesimally. The equivalent angle in minutes and seconds is shown in the right-most column.

the *Tables du cadastre*, was entrusted to Gaspard de Prony with express instructions 'not only to compose tables which leave nothing to be desired in terms of exactitude but to make the greatest and most imposing monument to calculation which shall ever be executed or conceived'.[40] Several attempts were made to publish these tables, none of them successful. A few attempts at recalculating logarithm tables were made in the nineteenth century, those by Charles Babbage (1827)[41] and Edward Sang (1871)[42] being the most notable.

The popularization of logarithms

Almost from the beginning, the utility of logarithms demanded that details of their use be published in the vernacular as well as in Latin—the then usual language of scholarship. Within two years of the publication of the *Descriptio* an English translation was published by Edward Wright. This translation had Napier's blessing for in his preface to this edition he wrote that:[43]

> I thought good heretofore to set forth in Latine for the publique use of Mathematicians [the *Descriptio*]. But now some of our Countrymen in this Island well affected to these studies, and the more publique good, procured a most learned Mathematician to translate the same into our vulgar English tongue...

In 1619 John Speidell, a London teacher of mathematics, published his *New logarithmes*, a book of tables derived directly from those in the *Descriptio*. This work, despite containing no explanation of the tables nor any worked examples, had gone through nine editions by 1627.[44] The omission was rectified that year when Speidell published his *A breefe treatise of sphæricall triangles*, comprising some 28 examples together with the tables which he had first published in his *New logarithmes*.[45]

But logarithms to base 10, as set forth by Briggs in the *Arithmetica logarithmica* and *Trigonometrica Britannica*, were not at once readily accessible to the student and practitioner. These two works were large, relatively expensive and written in Latin, and were, perhaps, destined more for the coffee table than the working library. There was a need for smaller and cheaper volumes, written in English. One of the earliest writers who sought to fill this gap was John Newton (1621–1678). During the Interregnum Newton, a staunch Royalist, taught mathematics in Oxford and in the years 1654 to 1660 wrote eight books, all of them in English, on arithmetic, geometry, trigonometry, and astronomy. Five of these volumes have a significant tabular content: tables of logarithms of the natural numbers and of angular functions and their logarithms, all of them explicitly derived from those of Briggs.

Newton's aim was twofold: to provide mathematical books in English, and to encourage, following Briggs, the use of the decimal division of the degree. With his first book, the *Institutio mathematica* of 1654, Newton made no claim to originality. In its preface he wrote[46]

> We... acknowledge that in this [book] we have presented thee with nothing new, nothing that is our own.... This only we have endeavoured, that the first principles and foundations of these studies (which until now were not to be known, but by being acquainted with many Books) might in a due method and a perspicuous manner, be as it were at once, presented to thy view...

Newton's second volume, the *Tabulæ mathematicæ* (1654),[47] which is usually found bound with the *Institutio mathematica*, contains tables of angular functions and their logarithms, together with a table of the logarithms of the first 10 000 natural numbers, all to six places.

Nowhere is Newton's espousal of the decimal division of a degree more clearly shown than in his statement in the preface to his third work, the *Astronomia Britannica* (1657)[48] where he wrote

> It is therefore our chief and principal aim to shew how much of trouble may be avoyded in computing the motions of the heavenly bodies, if only the form of our Tables were changed from *Sexagenary* into *Decimal*....

After the Restoration Newton became vicar, and later rector, of Ross, Herefordshire. He continued to be a prolific author, but his work in this period was devoted to the education of children.

The year 1657 saw the publication of Oughtred's *Trigonometria*,[49] a volume of about 290 pages of which no less than 240 comprise trigonometric and logarithm tables, derived from those of Briggs and printed by Joseph Moxon, an important printer of tables in the third quarter of the seventeenth century who was, in 1662, to become Hydrographer to Charles II. The first 36 pages of the book, printed by R. and W. Leybourn, contain Oughtred's *Trigonometry*, a highly condensed work treating of both plane and spherical triangles. It is one of the earliest works to adopt a condensed symbolism so that equations involving trigonometric functions could easily be taken in by the eye.

THE MAKING OF LOGARITHM TABLES

On 6 February 1655 Richard Stokes, a Fellow of King's College, Cambridge, wrote to Oughtred as follows:[50]

... Sir, I have procured your Trigonometry to be written over in a fair hand which, when finished I will send to you, to know if it be according to your mind; for I intend (since you were pleased to give your assent) to endeavour to print it with Mr. Briggs his tables...

The tables are clearly an integral part of the *Trigonometria*. Oughtred comments that trigonometrical equations, particularly those involving the use of half-angle formulae, can be easily solved with the aid of logarithms and he gives a number of examples.

Following the tables there is an appendix of some ten pages which gives rules of interpolation in the trigonometrical tables for angles where linear interpolation is inappropriate; that is, near 0° for the logarithm of sines and tangents, and near 90° for the logarithm of tangents. In these regions the rates of change of the tabulated functions are large and highly non-linear.

Undoubtedly the typesetting of the tabular portion of the *Trigonometria* was Moxon's *pièce de résistance*, but something had gone badly wrong in the process of its production, and a second edition of the book, this time in English, was rushed out within months of the first. In the preface to this edition, Stokes wrote:

The Book and Author are still the same, nor will you finde the least altetation [*sic*] except in the Rules for correcting the Canon which was occasioned by a mistake, the number of Figures in the Tables falling short of that required and used in the rules, sprung from the intention of Printing it in octavo, for which volume the number of Figures was resolved on, and upon the changing the Volume forgot to be altered, The reverend Author has both discovered and amended the error, in the appendix, as farre as could be, as you may there perceive.

It is clear that it was Oughtred's intention that the tables of logarithms of sines and tangents should consist of seven figures after the decimal point: in fact only six were printed. Because of this error the rules given in the appendix for using the tables would not work.

Another passage in the preface sheds an interesting light on the attitude of publishers to the publication of scientific books in the vernacular. Stokes felt compelled to apologize for writing in English, being 'much of their opinion who think translations the greatest enemy of Learning' although he did not think 'that censure reaches things of this nature which are sometimes of use to persons who it cannot be expected should understand any Language other than what their Nurses taught them'.

The new appendix to the English edition begins with the statement:

> The Canon of Logarithmes of Sines and Tangents was by the Author intended to consist of seven Figures after the Indices: but in this edition the seventh figure was unhappily left out; so that now the Rules following will not hold in some of those tables, namely of Collaterals, and first and second differences. But they may hold indifferently wel, if unto the numbers in the Canon you joyne a circle or nul.

With this addition, the new appendix was essentially a translation of the previous, Latin, version, although different numerical examples were given.

Although Stokes states that Oughtred was the one who discovered the error it is certain that John Pell (1611–1685) knew about it soon after the first edition was published. Pell's papers contain many pages of numerical experimentation with methods of interpolation for use with Moxon's six-figure tables.[51] By the end of 1657 Pell had devised a method now known as the Lagrange three-point interpolation formula.

Within just a few decades of their invention, logarithms had become ubiquitous among the mathematical community: tables were easily and cheaply available and many texts describing their use had been printed. What is striking is that their development from Napier's initial formulation was the work not of a few isolated individuals but rather the result of an almost explicit collaborative effort by the members of a closely connected group, a group centred on Gresham College with Henry Briggs as its driving force.

Further reading

The articles on Napier, Briggs, and Gunter in the *Dictionary of national biography* and the *Dictionary of scientific biography* provide useful information. The starting point for the history of logarithms must be Charles Hutton's *Mathematical tables; containing the common, hyperbolic, and logistic logarithms, ... to which is prefixed, a large and original history of the discoveries and writings relating to those subjects;....* This work went through eleven editions between 1785 and 1894. A more recent treatment of the subject can be found in James Henderson's *Bibliotheca tabularum mathematicarum. Being a descriptive catalogue of mathematical tables*, London, 1926. Detailed references to the work of Briggs, Gunter, and their contemporaries can be found in David W. Waters' *The art of navigation in Elizabethan and early Stuart times*, London, 1958. A detailed discussion of the Gresham College circle can be found in Christopher Hill's *Intellectual origins of the English revolution*, Oxford, 1965.

Notes

1. I have relied on the *DSB* article for biographical information on Napier. Margaret E. Baron, 'Napier, John', *Dictionary of Scientific Biography*, vol. 9 (ed. C. C. Gillespie), New York, 1974, pp. 609–13.
2. J. Napier, *Rabdology*, tr. William Frank Richardson, intro. Robin E. Rider, Cambridge, Mass., 1990.
3. This is Hawkins's English translation of the original Latin. Hawkins describes in detail the construction and use of the promptuary and gives an almost complete translation of the appendix. W. F. Hawkins, 'The first calculating machine (John Napier, 1617)', *Annals of the History of Computing*, 10 (1989), 39.
4. Quoted in: F. Cajori, 'Algebra in Napier's day and alleged prior invention of logarithms' in *Napier tercentenary memorial volume* (ed. C. G. Knott), Longmans Green, London, 1915, p. 104. M. Napier (*Memoirs of John Napier of Merchiston, his lineage, life, and times, with a history of the invention of logarithms*, Edinburgh and London, 1834, pp. 364ff.) argues that this 'certain Scotchman' was Dr John Craig, whose discussions on the invention of logarithms with John Napier are described by Wood (*Athenæ Oxoniensis* (ed. P. Bliss), London, 1817, cols. 491–2) whose account provides some corroboration for Kepler's assertions.
5. J. Napier, *A description of the admirable table of logarithms* (trans. E. Wright), London, 1616.
6. It was in the *Descriptio* that Napier, following the invention of Stevin, used the decimal point as a separator of units and tenths. In *Art 5* he wrote that 'in numbers distinguished... by a period in their midst, whatever is written after the period is a fraction...'. See: S. Stevin, *De thiende*, Leyden, 1585.
7. J. W. L. Glaisher, 'Tables, mathematical', in *Encyclopaedia Britannica*, 11th edn, 1910–11, vol. 26.
8. This is Napier's value, but log 10 is more nearly 23 025 850.93. See J. Napier, *The construction of the wonderful canon of logarithms* (trans. W. R. Macdonald), Blackwood, Edinburgh, 1889, pp. 90–6.
9. There is some doubt about the year of Briggs's birth. Joseph Mede of Christ's College, Cambridge, wrote on 6 February 1630 that 'Mr Henry Brigges of Oxford, the great Mathematician, is lately dead, at 74 years of age' (quoted in Ward (note 23), p. 120), implying that Briggs was born about 1556. The Halifax parish registers give the baptism of 'Henricus filius Thome Bridge de

Warley' on 23 February 1561 (new style), a date which may be misleading in that then, as now, baptism did not always occur within a few days or weeks of birth. Wood's statement that at the time of his death in 1630 Briggs was 'aged 70 or more' is not wholly inconsistent with either date. It may be remarked, though, that the identification of Henry Briggs with the Henricus Bridge of the parish registers is far from certain. The family name Briggs is totally absent from the registers but there are many occurrences of its variants Brigge, Brege, Bridg(e), Brige, Brydge, Bryge, Brygg(e), Brygg, Brygghe and Bryghe; and Henry was a common first name. See *The parish registers of Halifax, co. York,* vol. 1, (ed. E. W. Crossley), The Publications of the Yorkshire Parish Register Society, 37, 1910, *passim.*

10. For a detailed account of the foundation and early years of Gresham College see I. R. Adamson, 'The administration of Gresham College and its fluctuating fortunes as a scientific institution in the seventeenth century', *History of Education* 9 (1980), 13–25.
11. T. Blundeville, *The theoriques of the seven planets*, London, 1602.
12. E. Wright, *Certaine errors in navigation . . .*, London, 1610.
13. The contributions of Briggs and his contemporaries to the science of navigation are treated extensively by Waters (D. W. Waters, *The art of navigation in England in Elizabethan and early Stuart times*, London, 1958).
14. Quoted in Ward (note 23) p. 121.
15. In the Latin languages *canones* was the name given to introductory instructions as to how the tables were to be used, but by the beginning of the Middle Ages the word was applied to a complete set of tables, and this soon became its standard meaning.
16. That is, the angle whose sine is 1/10.
17. G. A. Gibson, 'Napier's logarithms and the change to Briggs's logarithms', in *Napier tercentenary memorial volume* (ed. C. G. Knott), Longmans Green, London, 1915, p. 26. This quotation is from Gibson's translation of Briggs's introduction to the *Arithmetica logarithmica* of 1624.
18. There are two copies of this work in the British Library. That at shelf mark C.54.e.10.(1), bound with Gunter's *Canon triangulorum* (C.54.e.10.(2.)), is copiously annotated and contains extensive manuscript tables of differences. The second copy, in the Manuscript Collection, *Sloane 917*, is also extensively annotated, by a different person, and contains a number of worked examples.
19. No previous translation of this preface has been published. Sampson's *in extenso* quotation of the Latin original is incomplete. R. A. Sampson, 'Bibliography of books exhibited at the Napier tercentenary celebration, July 1914', in *Napier tercentenary memorial volume* (ed. C. G. Knott), Longmans Green, London, 1915, pp. 194–5.

20. '... per œquationes Algebraicas, & differentias, ipsis Sinubus proportionales, pro singulis Gradibus & gradum centisimis, a primis fundamentis accurate extructum: . . .'.
21. Napier's *Mirifici logarithmorum canonis descriptio* of 1614.
22. It seems clear from the context that this is a reference to Napier's *Constructio* which appeared in 1619, edited by his son, a view shared by Glaisher (note 36), p. 49. The assertion in the *DSB* (note 1) that it refers to the *Rabdologiae*, which was published in 1617, is probably mistaken.
23. Quoted in: J. Ward, *The lives of the professors of Gresham College*, London, 1740, p. 122.
24. C.54.e.10.(2.). (See note 18.)
25. The terms 'characteristic' and 'mantissa' seem to have been coined by Briggs and first appear in his *Arithmetic logarithmica* of 1624.
26. These are sometimes called Briggian or Briggsian logarithms. The first example of the term 'Briggian' seems to have been in 1774 (*Philosophical Transactions*, 64, p. 223). The variant 'Briggsian' seems to have been first used in 1892 (*Science*, 25 November 1892, p. 307).
27. J. Aubrey, *Brief lives* (ed. R. Barber), Boydell, Woodbridge, 1982, p. 121.
28. J. Aubrey (note 27), p. 279.
29. This Latin work, the *New projection of the sphere*, must have been written in about 1607. The date of 1603 given in the *DNB* is unlikely to be correct: Gunter's statement is unambiguous, and in any case, in 1603 he would still have been an undergraduate.
30. G. J. Gray, 'Gunter, Edmund', *Dictionary of National Biography*, vol. 23, London, Smith, Elder, 1890, pp. 350–1.
31. W. Oughtred, *To the English gentrie, and all others studious of the mathematicks . . . The just apologie of Wil: Oughtred, against the slaunderous insimulations of Richard Delamain, in a pamphlet called Grammelogia, or The mathematicall ring, or Mirifica logarithmorum projectio circularis*, London, 1632.
32. J. Napier, *A description of the admirable table of logarithms* (trans. E. Wright), London, 1618, p. 59.
33. I am grateful to Carol Bostock-Smith for her ready help with translations of both Briggs's and Gunter's (see note 19) Latin.
34. The term 'cosine' seems to have been Gunter's invention.
35. Ward (note 23) p. 77.
36. J. W. L. Glaisher, *Report of the British Association on mathematical tables*, London, 1873, p. 41.
37. E. Gunter, *The general use of the former canon and table of logarithmes*, British Library Manuscript Collection, *Sloane 910*, f. 33.

38. British Library Manuscript Collection, *Sloane 3498*. Quoted in: J. W. L. Glaisher, 'On early logarithmic tables, and their calculators', *Philosophical Magazine*, 4th Ser., 45 (1893), 376–82.
39. H. Briggs, *Trigonometria Britannica...excudebat Petrus Rammasenius*, Gouda, 1633.
40. My translation of the original French: '...non seulement à composer des Tables qui ne laissassent rien a désirer quant à l'exactitude, mais à en faire le monument de calcul le plus vaste et le plus imposant qui eùt jamais été exécuté ou même conçu...'; quoted in: Glaisher (note 36), p. 56.
41. C. Babbage, *Tables of logarithms of the natural numbers from 1 to 108 000...*, London, 1827.
42. E. Sang, *A new table of seven-place logarithms of all numbers from 20,000 to 200,000, etc.*, London, 1871.
43. J. Napier (note 6), Preface.
44. J. Speidell, *New logarithms. The first invention whereof, was, by the Honourable Lo: John Napair Baron of Marchiston... These being extracted from and out of them, they being first over seene, corrected, and amended, require not at all any skill in Algebra or Cossike numbers...*, London, 1619. It is often, incorrectly, stated that the table in this work was the first published example of what are now called Naperian, or hyperbolic, logarithms (i.e., logarithms to base *e*). Speidell's table of such logarithms, and then of natural numbers rather than trigonometric functions, did not in fact appear until the second edition of *New logarithms* in 1620.
45. J. Speidell, *A breefe treatise of sphericall triangles, wherein is handled the sixteen cases of a right angled triangle, being all extracted out of one Diagram, and reduced into Theorems, with the total sine in the first place, so that by addition only, they may be effected. As also, the twelve cases of an oblique sphœricall triangle...*, London, 1627.
46. J. Newton, *Institutio mathematical. A Mathematical Institution Shewing the Construction and Use of the Naturall and Artificial Sines, Tangents, and Secants, in Decimal Numbers and also of the Table of Logarithms. In the general solution of ant Triangle whether Plain or Spherical, with Their more particular application in Astronomie, Dialling and Navigation*, London, 1654.
47. J. Newton, *Tabulæ mathematicæ. Tables of the natural Sines, Tangents and Secants, and the Logarithms of the Sines and Tangents to every degree and hundred part of a degree in the Quadrant. Their common Radius being 10,000,000. With a Table of Logarithms of all absolute numbers increasing by naturall succession from an Unite to 10,000...*, London, 1654.
48. J. Newton, *Astronomia Britannica, exhibiting the doctrine of the sphere, and theory of the planets decimally by trigonometry, and by tables. Fitted for the meridian of London, according to the Copernican systeme as it is illustrated by Bullualdus, and the easie way of calculation, lately published by Doctor Ward*, London, 1657.

49. W. Oughtred, *Trigonometria*, London, 1657. The English edition, *Trigonometry*, was published in the same year.
50. *Correspondence of scientific men of the seventeenth century,* vol. 1 (ed. S. J. Rigaud), Oxford, 1841, p. 82.
51. British Library Manuscript Collection, *Add MS 4415*, ff. 126v–131.

The Breuiat of the Table of 10. per Cent.

Yeares.		Yeares.	
1	11000000	16	45949729
2	12100000	17	50544702
3	13310000	18	55599173
4	14641000	19	61159090
5	16105100	20	67274999
6	17715610	21	74002499
7	19487171	22	81402749
8	21435888	23	89543024
9	23579476	24	98497326
10	25937424	25	108347059
11	28531167	26	119181765
12	31384283	27	131099941
13	34522712	28	144209936
14	37974983	29	158630929
15	41772481	30	174494022

These Numbers in this Breuiat must also be esteemed Numerators, each of them hauing for Denominator 10000000.

Now follow the Questions: In working whereof, we will vse this Breuiat.

The

3

History of actuarial tables

CHRISTOPHER LEWIN AND
MARGARET DE VALOIS

Actuaries make financial sense of the future by combining techniques of risk and finance. The actuarial profession was formed in 1848 but long before then mathematicians were producing tables in these two areas, starting with the relatively simple concepts of compound interest and

Fig. 3.1 Table of compound interest from Richard Witt's *Arithmetical questions*, 1613.

probabilities of survival. This chapter shows how actuarial tables have developed over the past four centuries.

Compound interest was in use among ancient civilisations,[1] though as far as we know they did not have any tables relating to it. It was probably not until the Middle Ages that compound interest tables first came into being, for use in banking transactions. Other tables were developed later, for different purposes. In 1653 Poul Klingenberg (1615–1690) moved to Copenhagen and persuaded the King to establish the world's first tontine (an idea which had been suggested by Lorenzo Tonti, an Italian banker who lived in Paris).[2] The subscribers to a tontine paid the State a capital sum in order to receive annual payments for the rest of their lives. Once the State's total annual payout was established, this remained fixed despite the death of some subscribers, so each of the surviving subscribers received a sum which gradually increased each year. In order to market the scheme, Klingenberg published a life table which showed the expected number of survivors for every year in future. It is believed that this was not based on scientific data but was purely illustrative as an advertising aid. The table suggested that of 2000 children aged up to eight, only 900 would survive for 20 years, 250 for 40 years and 35 for 60 years.

Proper life tables based on real data started to emerge later in the seventeenth century and became a fundamental tool. These life tables were then combined with compound interest tables to facilitate evaluation of the sums of money to be paid for life assurance and annuity contracts. Life tables show the number of survivors in a population which is reduced with the passage of time by one event, namely death, but around 1900 the concept was extended to cover multiple events, such as the resignations, deaths, and retirements in a workforce.

Compound interest tables

A surviving fifteenth-century Italian manuscript includes compound interest tables showing how much an invested sum will accumulate to after various numbers of years.[3] However, it was not until the second half of the sixteenth century that any serious attempt was made to publish compound interest tables. In 1558, in an arithmetic textbook published at Lyons, Jean Trenchant included a table on simple and compound interest.[4] Only a few tables were given. These included tables calculated at 4% interest per period, showing the accumulation over a number of periods of an initial investment of one unit of currency, i.e. a table of 1.04 raised successively to the power of 1, 2, 3, etc. There was also a table showing the total accumulated result of investing a *series* of payments of the same amount each period. As well as these tables at 4% interest, a table based on interest at 10% per year showed the accumulation of an invested sum for complete months and also for periods of years up to 6.

Trenchant discusses whether it is better to receive 4% interest per quarter on a loan or regular payments of 5% per quarter for 41 quarters with no return of the sum invested at the end of that period. The latter is shown to be marginally worse, because, if 100 units of currency is invested, the extra unit from the annuity, each quarter for 41 quarters, accumulates at 4% per quarter to slightly less than the 100 of principal which would be returned at the end of that time if the loan were selected instead. Trenchant also points out that to receive 4% interest per quarter is better than 16% per year. He quotes a case where the rent of a farm is 500 per year (in arrears) for 3 years and it is desired to pay it in advance by a single payment. He shows how to use the 4% table in reverse to find the sum which must be invested now in order to have 500 at the end of the first year. This is called the 'present value' equivalent to 500 in a year's time. He then adds to this the present values corresponding to payments of 500 in 2 and 3 years' time respectively.

In 1585 a more detailed work on compound interest was published by the famous mathematician, Simon Stevin of Bruges, as part of his own arithmetic textbook.[5] A number of worked examples were included for problems of both simple and compound interest. There was also a set of tables calculated for interest rates from 1% to 16% and also for rates of interest corresponding to 15, 16, 17, 18, 19, 21, and 22 years' purchase of freehold properties. The tables were 'the other way round' from those which had been published by Trenchant 27 years previously, i.e. they showed the present values of a sum due to be received at the end of each period, rather than the sum to which a sum invested now would accumulate by the end of the period. In other words, whereas Trenchant's table showed 1.04 raised to successive powers, Stevin's table showed 1 divided by 1.04 raised to successive powers. Stevin also provided tables corresponding to the total present value of a *series* of payments to be received at the end of each period. This is called an annuity certain. For all problems of the type presented in Trenchant's book, Stevin's tables were in a more convenient form, because they avoided the need for division.

Accumulation table and present value table for a single payment

Period of years	Sum to which 1 invested now will accumulate in period	Present value of 1 at end of period
1	1.04	$1/1.04 = 0.96154$
2	$1.04^2 = 1.0816$	$1/1.04^2 = 0.92456$
3	$1.04^3 = 1.12486$	$1/1.04^3 = 0.88900$

Stevin also gives a table of the accumulation of a *series* of payments, calculated at an interest rate of six and two thirds per cent; this is in the same form as the accumulation tables given by Trenchant. However, he does not give similar tables for other rates of interest, but instead demonstrates how these accumulated values can be derived from his present value tables for a series of payments by dividing the present value of an annuity of 1 per year for the number of years concerned, by the present value of a payment of 1 at the end of that period. In an example where someone pays a lump sum instead of an annual payment for 22 years, he shows how to find the rate of interest underlying the transaction by inspection of the tables for various rates of interest.

In England it had for a long time been illegal to charge interest on loans (though this did not prevent it happening underground!) until the passing of an Act of Parliament in 1571, which allowed interest to be paid up to 10% per year. Even after the Act became law, it was found to be ambiguous. By about 1600, however, it was becoming generally accepted that charging interest was in order, and books on compound interest started to appear.

The principal compound interest book of this period[6] was written by Richard Witt, a 44-year old mathematical practitioner, and published in 1613. It delved deeply into the subject in a very practical way, and it is evident from the clarity of expression and the care which was taken that the author thought in much the same way as modern actuaries. However, all Witt's payments were certainties, to be received on definite dates, and there is no evidence that risk entered into his thoughts at all.

The book begins with a table (or 'breviat') calculated at a rate of interest of 10% per year (see Fig. 1.1 at the opening of this chapter). It is, quite simply, an accumulation table showing how much £1 invested now will accumulate to after 1, 2, 3, ... years, i.e. a table of 1.1 raised to the power of the number of years concerned. This table is used in various worked examples to obtain other compound interest functions as necessary. For example, suppose that you wish to find the present value of a series of payments of £1 per year for the next 5 years, starting with a payment in one year's time. Witt's table shows that after 5 years 1 will have accumulated to 1.61051. His method would be to subtract 1 from this to give 0.61051, divide this by 1.61051 and multiply the result by 10 to give the correct result of 3.7908. His use of processes such as this shows that the mathematics of compound interest were fully understood. However, having demonstrated these methods, he goes on to give tables of these other compound interest functions, to avoid the user having to make tedious calculations. The book contained 124 interesting worked examples, of which the following is an example:

Q70. One oweth £900 to be paid all at the end of 2 years: he agreeth with his creditor to pay it in 5 years, viz. every year a like sum. They demand what each of these 5 payments shall be, reckoning 10 per cent. per annum interest, and interest upon interest.

Numerous other writers followed Witt in producing compound interest tables. Perhaps the most interesting[7] is the *Table of leasses and interest* printed by William Jones dwelling in Red-cross street (London). The author's name is not given, though it may have been a Mr Aecroid. The book was clearly intended as a practical ready reckoner for everyday use by valuers and others who were concerned in the renewal of leases. The main point of interest lies in the fact that the book contains a description of the rather odd method of construction of the tables.[7a]

Life tables

The first scientifically based life table was, curiously enough, constructed by a London draper. John Graunt, who had a shop in Birchin Lane, decided to investigate the London Bills of Mortality for a long series of years, with a view to drawing conclusions about London's population and other matters. His conclusions were published in a pioneering book in 1662.[8]

The Bills of Mortality were printed statements of the numbers of people who died each week, classified according to the apparent cause of death. The original purpose was to warn of the rise of epidemic diseases, such as the plague, so that the better off could flee in time to the perceived safety of the countryside. The data underlying the Bills were collected by the parish clerks. When anyone died, the church bell was rung and then the official searchers (whom Graunt describes as 'ancient matrons, sworn to their office') went to view the corpse and make enquiries, to establish what the person had died from. They reported their findings to the local parish clerk and he, every Tuesday night, took a weekly return of the deaths and christenings to a central office, where the week's Bill for London as a whole was made up and printed on the Wednesday. It was published on the Thursday and taken round to the families who were willing to subscribe four shillings per annum.

Upon analysing his data, Graunt estimated that of 100 children conceived, 36 die before the age of 6. (The data were not subdivided by age, so Graunt had to guess which causes of death were most likely to relate to children under 6.) He also asserted that perhaps one of these 100 individuals survives to age 76. To get the numbers of survivors at intermediate ages at ten-yearly intervals, he then sought 'six mean proportional numbers between 64, the remainder living at 6 years, and the one which survives 76'. Graunt's table shows that, of 100 people conceived, 36 die before age 6, leaving 64 surviving to age 6. Of these 64, 24 are assumed to die before age 16, leaving 40 surviving to age 16, and so on.

Life table (based on Graunt)

Age	Number alive	Deaths before next listed age
0	100	36
6	64	24
16	40	15
26	25	9
36	16	6
46	10	4
56	6	3
66	3	2
76	1	1
86	0	—

It is apparent that, although the number surviving to age 6 (i.e. 64) is based on data from the Bills of Mortality, the other figures in the table are purely speculative, and are designed to secure a smooth progression. Each 'number alive' after age 6 is exactly or approximately equal to five-eighths of the

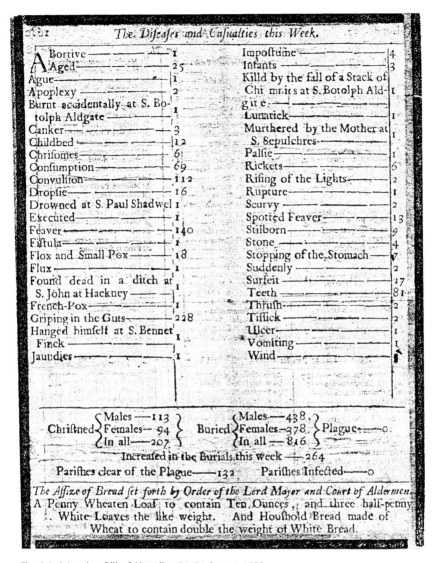

Fig. 3.2 A London Bill of Mortality, 24–31 August 1680.

previous one. This is equivalent to an annual survival rate of about 95.4%, independent of age. However, Graunt's mathematical concept was very far from the truth. He does not seem to have noticed the serious discrepancy between the overall annual mortality rate implied by his life table (if the population is assumed to remain the same size), which is about 1 in 18, and the overall annual mortality rate of 1 in 27 which was shown by his data from the deaths in the Bills of Mortality and his estimates (which were probably reasonably correct) of the size of London's population. If Graunt had realised the existence of this discrepancy, he might well have concluded that the obvious way to make the figures consistent would have been to assume for his life table a mortality rate for adults which in some way increased with age, in which case the table would have been much more correct.

We now come to Edmund Halley's life table, which formed the basis for his great work on the valuation of life annuities—the creation of actuarial science. The underlying data were collected by Caspar Neumann, pastor of Breslau, from the church registers there, for the period 1687–91, and sent to the Royal Society in London. The important point about these data was that they were classified by age at death, unlike the London Bills of Mortality.

Halley's paper[9] was published in the widely-circulated *Philosophical transactions* of the Royal Society in January 1693, and made an impact throughout Europe. His life table is shown in the accompanying illustration. There has been much confusion about the meaning of the 1000 persons shown at age 1 'curt'. in the table, with many writers mistakenly assuming that it meant 1000 births. In fact it means 1000 children living that will be age 1 at their next birthday. This corresponds to about 1174 children conceived, or 1110 live births.[10] Halley's life table later proved to be a sufficiently accurate estimate of mortality for many practical purposes. Although Halley had pointed out how to use the table in conjunction with compound interest functions in order to obtain the purchase price which should be paid for a

Age. Curt.	Per- fons.	Age. Curt.	Per- fons	Age. Curt.	Per- fons	Age. Curt.	Per- fons	Age. Curt.	Per- fons	Age. Curt.	Per- fons
1	1000	8	680	15	628	22	586	29	539	36	481
2	855	9	670	16	622	23	579	30	531	37	472
3	798	10	661	17	616	24	573	31	523	38	463
4	760	11	653	18	610	25	567	32	515	39	454
5	732	12	646	19	604	26	560	33	507	40	445
6	710	13	640	20	598	27	553	34	499	41	436
7	692	14	634	21	592	28	546	35	490	42	427
43	417	50	346	57	272	64	202	71	131	78	58
44	407	51	335	58	262	65	192	72	120	79	49
45	397	52	324	59	252	66	182	73	109	80	41
46	387	53	313	60	242	67	172	74	98	81	34
47	377	54	302	61	232	68	162	75	88	82	28
48	367	55	292	62	222	69	152	76	78	83	23
49	357	56	282	63	212	70	142	77	68	84	20

Fig. 3.3 Halley's life table, 1693.

stream of regular payments throughout life, i.e. a 'life annuity', the calculations were very laborious in practice. Consequently, the mathematician Abraham de Moivre invented an approximate method which enabled life annuities to be valued without using a life table. [11] This method assumed that, in the case of a group of lives who were all the same age (over 12), an equal number of them would die each year until age 85 inclusive, so that none would be left alive at age 86. This led to a very simple calculation indeed, with an error of less than 5% at all ages. The method did not work so well with later life tables, however.

From 1728 onwards the London Bills of Mortality recorded deaths within broad age groups, and after a few years this enabled writers such as John Smart[12] and Thomas Simpson[13] to construct their own life tables from the London Bills (which tables generally showed higher mortality than Halley's table—perhaps because of the widespread social evil of excessive consumption of gin at that time).

However, these life tables suffered from the fact that they were constructed only from the numbers of deaths at each age. No information was available about the corresponding numbers of people alive at each age, because there were no population censuses. This could lead to severe distortions if, for some reason, the population had not yet reached a stationary state. Such reasons might include immigration, for example, where the immigrants had an age distribution different from that of the native population. Alternatively, there might have been a plague epidemic which carried off much of the population at all ages some years previously, following which the city population would probably rapidly have been replenished by young adults coming in from the surrounding countryside, leading to a population bulge at the ages to which those incomers had now attained. Even if neither of these causes was applicable in a particular case, the population age structure might be very far from stationary due to past fluctuations in the birth rate.

Because of such problems, it was difficult to be confident of the numerical results obtained from the analysis of Bills of Mortality. Several Continental investigators decided, therefore, to try an entirely different approach. Unlike Britain, some Continental countries had sold life annuities to the public for much of the seventeenth century. The investigators took the records of Government annuities which had been sold during a particular period and followed the annuities through over a period of many years, until all the annuitants had died. The records were kept accurately, because the payments depended upon them, and distortions in the results obtained were therefore kept to a minimum. Life tables of this kind were produced in the 1740s by Struyck[14] and Kersseboom[15] in Holland and by Deparcieux[16] in France. This method became known as the 'cohort' method, because it followed through a cohort of lives until they had all died. In later years it was seen to produce out-of-date figures, which could not keep up with the significant improvements in mortality which started to emerge due to medical advances and improving sanitation in towns.

In due course population censuses became available and it became possible to relate the deaths occurring in a year to an accurate figure for the number of people alive in a year at the same age. Mortality rates seem first to have been calculated in this way by Pehr Wargentin[17] in Sweden in 1766. The mortality rates can then be converted into survival rates which are

38]   T A B L E S.

TABLE VII.

Shewing the EXPECTATIONS of Human Life at every Age, deduced from the *Northampton* Table of Observations.

Ages	Expectat.	Ages	Expectat.	Ages	Expectat.	Ages	Expectat.
0	25.18	25	30.85	50	17.99	75	6.54
1	32.74	26	30.33	51	17.50	76	6.18
2	37.79	27	29.82	52	17.02	77	5.83
3	39.55	28	29.30	53	16.54	78	5.48
4	40.58	29	28.79	54	16.06	79	5.11
5	40.84	30	28.27	55	15.58	80	4.75
6	41.07	31	27.76	56	15.10	81	4.41
7	41.03	32	27.24	57	14.63	82	4.09
8	40.79	33	26.72	58	14.15	83	3.80
9	40.36	34	26.20	59	13.68	84	3.58
10	39.78	35	25.68	60	13.21	85	3.37
11	39.14	36	25.16	61	12.75	86	3.19
12	38.49	37	24.64	62	12.28	87	3.01
13	37.83	38	24.12	63	11.81	88	2.86
14	37.17	39	23.60	64	11.35	89	2.66
15	36.51	40	23.08	65	10.88	90	2.41
16	35.85	41	22.56	66	10.42	91	2.09
17	35.20	42	22.04	67	9.96	92	1.75
18	34.58	43	21.54	68	9.50	93	1.37
19	33.99	44	21.03	69	9.05	94	1.05
20	33.43	45	20.52	70	8.60	95	0.75
21	32.90	46	20.02	71	8.17	96	0.50
22	32.39	47	19.51	72	7.74		
23	31.88	48	19.00	73	7.33		
24	31.36	49	18.49	74	6.92		

Fig. 3.4 The Northampton table (by Richard Price).

combined to build up an accurate and up-to-date life table. This is the 'census' method usually employed today.

James Dodson, a London mathematician, constructed his own life table as a basis for calculating the initial tables of premium rates for the Equitable Life Assurance Society, which was founded in 1762 as the first life assurance company offering long-term business financed on proper scientific principles.[18] Prior to its formation the Society had applied for a Government charter but this was refused on the ground that the premiums might prove to be too low. However, the Society actually experienced lower death claims than had been expected, and in 1782 adopted a new lower scale of premiums, based on the Northampton life table produced by Richard Price.[18a] Even that table proved too conservative and further surpluses arose.

Other life companies were now starting up and they, too, adopted the Northampton table. This gave all the offices a built-in margin for adverse contingencies and profit, which may not have been a bad thing in those early days, since it helped to ensure the establishment of the life assurance industry on firm financial foundations. Apart from possible defects in the construction of the Northampton table, the reasons for the relatively light mortality experienced by the offices may have included the fact that life assurance was mainly used by the better-off professional classes, and that only people in good health were accepted. However, many people felt that it was unsatisfactory to use a table which differed so much from the actual experience of the offices, and considerable interest was therefore awakened when in 1815 Joshua Milne[19] published a table showing lighter mortality. This was his famous Carlisle table, derived from the Bills of Mortality for the City of Carlisle between 1779 and 1787, which were produced by Dr John Heysham. Although the table was based on relatively scanty data, was poorly graduated from one age to another, and perhaps showed somewhat *too* light mortality, even for life assurance, it was in widespread use by 1853, no doubt through competitive pressures to quote lower premiums in order to attract new business.

In 1829 John Finlaison,[20] the Government's Actuary (who was known as the Actuary of the National Debt) reported on a number of painstaking mortality investigations which he had carried out on a 'cohort' basis. The first showed the experience of 1002 nominees of the English tontine of 1693, the last of whom died in 1783. The other investigations related to the participants in other Government annuity schemes of the eighteenth century. Some of the resulting tables were produced separately for men and women. The report demonstrated conclusively that the terms on which the

Government was still selling annuities to the public were too generous. The Govenment, which until then had ignored Finlaison's warnings, now adopted new tables (which were used until 1884) and saved considerable sums. Finlaison's report also raised some fundamental philosophical questions, which were doubtless discussed in actuarial circles:

> *Is there a law of mortality?*
> *Does mortality vary by district, climate, and social class?*
> *Does mortality improve over time?*
> *Has male mortality changed at the same rate as female mortality?*
> *With the decline of smallpox due to vaccination, have other diseases risen?*
> *Is there an age beyond which humans cannot live, even if more individuals approach that age?*

Any table of mortality rates produced from crude data contains irregularities due to chance fluctuations, data errors, etc. These need to be smoothed out, in other words the table needs to be 'graduated'. Finlaison used a moving-average method for this purpose.

Technical improvements in life tables

Towards the end of the eighteenth century new methods were starting to be adopted to ease the burden of calculation involved in an accurate application of Halley's method of valuing life annuities. Halley had looked at each future payment of the annuity and worked out from the life table the probability of survival to that age. This probability was then multiplied by the present value (obtained from compound interest tables) of £1 paid at the time the annuity payment in question fell due. While this might not sound too difficult for just one payment, it was very time consuming and liable to error for a whole series of future payments, the results of which had then to be summed to get the total value of the annuity. (For accurate results all the multiplications involved had to be done with the aid of tables of logarithms). It was realised that there would be considerable efficiencies if tables could be provided which contained the summations and some of the multiplications already done within them. The first step was taken in 1772 by William Dale,[21] who laid out his calculations in columns with a summation backwards from the end of life. This enabled the value of a life annuity at any age to be derived quite simply by one multiplication and division. The method was then improved in 1785 by

Tetens[22] and eventually came to be known as the method of 'commutation columns', which still continues in use for some purposes.

Throughout the nineteenth century there was much interest in the search for a mathematical law of mortality. Apart from the philosophical interest which would result from the discovery of such a formula, there was the practical consideration that the probability of death could then be found at any age from specified parameters, and the time consuming collection of data would be less necessary.

The first well-known attempt to devise a law of mortality was due to Benjamin Gompertz,[23] and his law was subsequently refined by Thomas Young,[24] Thomas Rowe Edmonds,[25] and William Makeham.[26] Despite the fact that the Gompertz formula is no longer considered to be a universal law of mortality, and indeed the search for such a law has been discontinued, formulae of this type are nevertheless still used occasionally today as an approximation for some practical purposes.

Because of the emerging need for life tables to be up to date, in line with the latest advances in medicine and sanitation, they increasingly came to be calculated from mortality rates derived from the experience of a short period of recent years. In principle the mortality rate is the number of deaths at a given age in the period, divided by the number of people who were exposed to the risk of death at that age. Attention now started to be paid to the best method of determining how many people were in fact exposed to the risk of death. For example, in 1839 Woolhouse[27] assumed that new entrants joining part way through a year had half as much exposure as individuals who were there at both the start and end of the year. As will be seen below, it was partly a lack of a proper understanding of how to calculate exposures to risk that led to the relatively short life of the Highland Society of Scotland's sickness tables, which were produced in 1824.

Later in the nineteenth century there were developments in the methods used to graduate tables produced from crude data. Woolhouse[28] was the first to explain the method of graduation using differences in 1866. He considered this method to be superior to the previous method of using moving averages (such as that which had been used by Finlaison in 1829). He wanted to find simple methods of adjusting the crude rates so that they ran smoothly from age to age. One of the methods he proposed was for the correction of an isolated error in a run of figures. The initial step was to calculate the first differences between the successive crude figures. The second differences (i.e. the differences between the first differences) were also

> ### The search for a law of mortality
>
> Benjamin Gompertz outlined the first sketches of such a law in 1820. In 1825 he formalized his law and checked it against the life tables of the time. He postulated that the mortality rate (or, strictly speaking, a very similar function called the 'force of mortality') at a given age, x, was given by the formula
>
> $$Bc^x$$
>
> where B and c were constants.
>
> In 1826 Thomas Young examined graphical curves of mortality to see whether they conformed to a formula. He produced a high-powered polynomial that would probably have resulted in spurious results if used in practice.
>
> In 1832 Thomas Rowe Edmonds was the first to use Gompertz's formula to produce mortality tables.
>
> In 1859 William Makeham suggested that the Gompertz law was not accurate and started to develop an improved formula for the force of mortality:
>
> $$Bc^x + A$$
>
> where B, c and A are constants. The formula proved particularly useful in practice for calculating functions for multiple lives.

calculated. Irregularities in the pattern of the differences would then identify the crude figure which was in error. The correction to be made to that crude figure would be obtained by the following rule:

From the middle second difference subtract separately the preceding and following differences; add together the two remainders and divide by six. The result, with its proper algebraic sign, will be the correction to be made...

Woolhouse then went on to consider difference methods of adjustment to make a whole table proceed more smoothly, rather than just eliminate isolated errors. Difference methods of graduation are still widely used today, alongside graphical and other methods.

Life office mortality tables

In 1838 a Committee was formed in order to conduct an investigation into the combined mortality experience of 17 life offices, 'to afford the means of determining the Law of Mortality which prevails among assured lives.' Fifty-eight offices funded the project, in return for access to the results. Details of nearly 84 000 policies were collected.

The resulting life tables[29] were published by the Committee in 1843. The tables showed survivorship rates but these were not at that stage combined with compound interest factors to derive monetary functions which would make the tables easier to use. Before the Committee had come to any conclusion over the calculation and publication of tables of monetary functions, Jenkin Jones, the actuary to the National Mercantile Life Assurance Society, seized the initiative and republished the main results[30] with comments and monetary functions. The tables were not used very much in Britain but were widely used in mainland Europe, even into the early twentieth century. Their main significance was that they promoted the use of life office mortality data rather than population data when setting premiums, and pioneered a spirit of co-operation between the offices in this area which has lasted to the present day.

Jones pointed out that the continued use of the Northampton table by the Equitable meant that it had only 1489 deaths in the 12 years preceding 1829, compared with 2248 expected by the table. The Carlisle table was better in this respect but it was so irregular from one age to another that it was unsuitable as a basis for calculating the premiums for temporary assurances. Overall the new tables agreed very closely with the Carlisle table, though generally giving a lower expectation of life than that table. Jones also concluded that the Irish had higher mortality than the British, and that women had higher mortality than men between ages twenty to fifty, with lower mortality than men thereafter.

Mortality tables based on life office data continued to be produced in later years. Each new set of tables that was produced came to be known as an *investigation*. At first a new committee was set up and new data were collected for each investigation. However, in 1912 Elderton and Fippard[31] showed how the census method of mortality table construction could be applied to life office data, thus enabling a continuous mortality investigation to take place. The method was in theory less accurate than that which had been used in previous investigations and proceeded by taking an annual census of the population exposed to risk and calculating the rate of mortality as a/b, where a was the number of deaths in the year and b was the mean population plus one-half of the deaths. The old methods had involved considerable trouble and delay, with the need to prepare data retrospectively, and it was felt that the new method would be more satisfactory in practice. In 1924 the continuous collection of data began and a permanent Continuous Mortality Investigation Bureau was formed. Modern mortality tables[32] are now produced regularly by the Bureau under the guidance of a Committee appointed by the actuarial profession.

The original data for the continuous investigation were provided by 60 offices. Female lives were not included and each policy was assumed to be held by a different individual. In 1937 two sets of tables were produced, one for heavy mortality and one for light. In 1954 the Committee investigated the effect of individuals holding more than one policy. From 1964 an investigation according to cause of death was often included. Female lives were included in the collection of data from 1973 onwards. The mortality of individuals taking out life annuities is usually lighter than that of assured lives, and using tables constructed from assurance data would therefore cause the cost of annuities to be set too low. In 1948 the collection of annuitants' data began, and the resulting life tables included projection factors to allow for future improvements in mortality. In 1990, to the delight of many actuaries, the Committee commissioned the production of computer software to generate most of the existing basic tables. The Standard Tables Program is updated regularly and is still used today. The 1992 set of tables produced by the Bureau was based on over 15 million lives exposed to risk—a far cry from the 84 000 policies covered in 1843!

Sickness tables and Friendly Societies

By 1835 at least a million people were members of Friendly Societies.[33] The societies were clubs, usually for working men, though some women's clubs also existed.[34] The societies usually restricted themselves to a particular locality: they had a strong social element and typically had between fifty and a hundred members. The provision of death and sickness benefits was common and was regarded as important, but was usually subsidiary to the society's social function and the meetings were often held at public houses. Originally the weekly contributions were usually fixed rather arbitrarily and often proved inadequate for the purpose, leading to the dissolution of the society as the members grew older, but eventually actuarial certification of the contribution tables was required. This led to a need to have actuarial tables for the purpose of valuing the benefits to be provided.

Although Richard Price[35] had done a little work in this area in 1789, this was not a firm enough foundation for the purpose.

In 1824 the Highland Society of Scotland collected data from 73 friendly societies and computed sickness rates.[36] This was the first investigation into sickness rates using real life data. Unfortunately the resulting tables were found to understate the actual rates of sickness experienced by many societies and the tables were withdrawn from use in 1851. John Finlaison suggested that one reason for this problem was a mistreatment of withdrawals, which were incorrectly regarded as exposed to risk for a whole year in the year of withdrawal.

There were a few very large friendly societies with branches in many towns and villages. The largest was the Manchester Unity Friendly Society. In 1903 Alfred Watson (a distinguished actuary who later became Government Actuary and was knighted in 1915) published the results[37] of a comprehensive investigation into the Society's sickness experience during 1893–7. Investigations of the same Society's experience had previously been carried out using data for 1846–8, 1856–60, and 1866–79. Watson began his 1903 report by explaining that the experience of the Society had become somewhat higher than the rates determined from the last investigation. The Board of Directors had agreed to look into this and a new method of compiling the rates was developed. Each of the Society's branches had to complete a set of cards, providing details of their members and the sickness they had experienced. Over 3500 sets of cards were painstakingly collected, covering some 800 000 members.

The report contained sickness tables divided by town, city, and rural areas, showing a higher rate of sickness in the city. Rates were also subdivided by age and occupation and included the average length of claim and the probability of becoming sick again after recovery. The picture revealed by the investigation was that sickness rates were rising and mortality rates falling. All in all, bad news for friendly societies. There has been no subsequent major investigation into sickness absence among members of friendly societies and the result is that the Manchester Unity tables are still used, with appropriate adjustments, today.

Multiple decrement tables and pension funds

A life table shows a hypothetical population of births which is thereafter diminished by one decrement only, namely death. In real life, however, populations are often diminished by other causes as well. For example, a population of spinsters would be reduced by marriage as well as death. A population of employees might be reduced by resignations, ill-health retirements, and age retirements, as well as death. Such situations can be represented by a multiple decrement table, which shows the number of survivors in the population at each age, after allowing for the various decrements at earlier ages.

Multiple decrement tables were first developed to investigate the effect that smallpox was having on mortality. Bernoulli[38] adjusted Halley's table for this purpose in 1760. In 1862 Wittstein published a paper[39] mentioning independent and dependent decrement rates. Using given probabilities of marriage and death at each age, he derived a table starting with 10 000 unmarried women aged 16 and showed how many of them would survive at each age in the future, subdivided between married and unmarried, and how many would marry at each age.

However, multiple decrement tables really came into their own following the development of funded occupational pension schemes in the second half of the nineteenth century. The prospective liabilities had to be revalued every five years, so that a check could be made that there was enough money to cover future pension payments. Multiple decrement tables were ideal for the purpose of carrying out these valuations. The methods used to estimate how many people die at a certain age can also be used to estimate the number of members leaving a pension scheme at a certain age. Appropriate assumptions need to be made about the probabilities of death, resignation, and retirement at each age: these can be checked against the scheme's actual experience after it has been in operation for a few years.

A further development was the preparation of monetary functions based on the multiple decrement table. These monetary functions fulfilled the same purpose as their counterparts in life tables, i.e. to ease the burden of calculation. An additional feature was that the monetary functions for pension schemes for salaried staff often incorporated an increasing scale to take account of expected salary increases in future on promotion. The first recorded use of a scale relating salary to age was in 1821, in relation to a superannuation fund for civil servants.[40]

Two actuaries called Manly and Thomas wrote the first comprehensive paper[41] on the valuation of staff pension funds in 1901. They gave a multiple decrement table which they called a *combined mortality, withdrawal, and retirement table*. This started with 20 000 active employees aged 15 and showed how many of them would die, withdraw (for example, through resignation) or retire each year in future, and hence how many of them would survive in service at each future age. The table was based on assumed probabilities, based to some extent on the experience of existing pension funds. Monetary functions were derived, based on a pension age of 65 and an interest rate of 4% per year (the rate paid or guaranteed by the railway companies, which ran many of the pension funds then existing in Britain). The authors allowed for mortality being higher among early retirements, who are 'men utterly broken down in the service, who are generally suffering from brain trouble or some chronic incurable disease'.

Meeting users' needs

Users of actuarial tables are, of course, looking for accurate, relevant and up-to-date figures, in a form which they can easily use to carry out the calculations required. The history of actuarial tables shows that their makers have, by and large, tried to respond to these needs. Particular attention has been paid to accuracy, since even quite a small error could cost someone quite a lot of money. Errors could be caused either by the author or the printer. Richard Witt in 1613, in an endeavour to reassure purchasers of his book on leases and annuities, stated on the title page:[6] 'Examined also and corrected at the press by the Author himself.' An edition of Robert Recorde's *The ground of arts* in 1632 contained a very erroneous table of the value of an annuity certain, no doubt puzzling the schoolboys it was meant to educate.[42] John Richards[43] had to admit in 1739 that there were errors in the tables of annuity values on three lives contained in his book *The gentleman's steward*, which he had published in 1730. What made it worse for him was that the errors had been pointed out by Weyman Lee, a writer whose work was generally of poor quality! Richards corrected the worst errors but could hardly have inspired confidence in his readers when he went on to say: 'I am aware that a great many small mistakes of this kind are crept into the Tables; but hope that few of them amount to so much as an unit in the first place of decimals, and that affects it to but 1/10 of a year's value, which is not so considerable as to need any apology.'

On the whole, though, users of actuarial tables have had good reason to be satisfied. As we have seen, many improvements were introduced over the years to make things easier for users—and in Victorian times it even became usual to give the logarithms of some of the functions tabulated, in order to avoid having to look them up in logarithm tables. Brilliant minds devoted themselves to the search for approximations, short cuts, and laws of mortality. However, with the spread of mechanical desk-calculators in the first half of the twentieth century, tables no longer needed to incorporate logarithmic equivalents. Following the advent of desktop electronic calculators and the introduction of suitable calculation software in the second half of the twentieth century, it became unnecessary to tabulate monetary functions. Actuarial tables are therefore nowadays much simpler than they were a hundred years ago. Nevertheless they still perform a vital role in the finances of life assurance companies and pension funds, and new tables are still needed every few years to cope with the latest mortality changes. Finally, the number of centenarians in the population is increasing rapidly

Fig. 3.5 The Institute of Actuaries, High Holborn, London.
(By Kind permission of the Institute of Actuaries.)

and Finlaison's question of whether there is an inextensible upper limit to human age has yet to be answered.

Given the long history of actuarial science, it is perhaps appropriate that the Institute of Actuaries is located behind one of the few surviving Tudor frontages in London. However, the profession is not standing still. It is gradually expanding into new areas of work, such as capital project appraisal, general insurance, healthcare systems, banking, risk management, alternative risk transfer, and credit derivatives—any field, in fact, which requires knowledge of both risk and finance.

Further reading

Perhaps the best book for further technical reading would be the *History of actuarial science*, edited by Steven Haberman and Trevor Sibbett, and published by Pickering & Chatto in 1995 in 10 volumes. This contains reprints of many historic actuarial tables and accompanying text. The useful preface by Professor Haberman gives more detail on some of the topics covered here, including the origin and graduation of life tables. Another very worthwhile technical book is *A history of probability & statistics and their applications before 1750*, by Anders Hald, published by Wiley in 1990. Some additional books of tables are described in the article entitled 'Table' (written by Augustus de Morgan?) in *The English cyclopaedia, arts and sciences division*, volume VII, columns 976–1016, London, *c.* 1873. On a somewhat lighter note, one of the present authors (Christopher Lewin) has written *Pensions and insurance before 1800: a social history*, which will be published by Tuckwell.

Notes

In the notes below 'H & S' refers to *History of Actuarial Science*, edited by Steven Haberman and Trevor Sibbett (10 vols, Pickering & Chatto, London, 1995).

1. D. J. Struik, *A concise history of mathematics*, Dover, New York, 1965 (originally published 1954).
2. A. Hald, *A history of probability and statistics and their applications before 1750*, Wiley, New York, 1990.
3. F. B. Pegolotti, *La pratica della mercatura*, edited by Allan Evans, 1936 (publications of the Mediaeval Academy of America, Cambridge, Mass., no.24).
4. J. Trenchant, *L'arithmetique*, Lyons, 1558, 1566, 1571 and later editions.
5. S. Stevin (of Bruges). *La pratique d'arithmetique*, Leyden, 1585 (also in a collected edition of 1634).

6. R. Witt, *Arithmeticall questions, touching the buying or exchange of annuities*, London, 1613.
7. Anon. *Tables of leasses and interest, with their grounds expressed in foure tables of fractions*, London, 1628. By Mr Aecroid?
7a. One of the tables, for example, was based on an interest rate of 11.18%, which can also be expressed as 19/170. (The reason for using such a peculiar rate of interest was that it corresponded approximately to the rate which would have been obtained if a 21-year lease was bought for a capital sum equal to 8 times the annual rent.) The working method used to calculate this table was to tabulate three columns: the first was 170 raised to the successive powers $1,2,3,4,\ldots$; the second was 189 raised to the same powers; the third column started with 170 and each subsequent number was probably obtained by multiplying the number immediately above it in the column by 189 and adding the number from the first column in the current row. The main compound interest functions were then obtained by dividing a number in one of the columns by the number in another column in the same row.
8. J. Graunt, *Natural and political observations made upon the bills of mortality*, London, 1662. H & S, I, 23–127.
9. E. Halley, 'An estimate of the degrees of the mortality of mankind, drawn from curious tables of the births and funerals at the city of Breslaw; with an attempt to ascertain the price of annuities upon lives', *Philosophical transactions* 17(1693), 596–610, table and 654–6. H & S, I, 165–84.
10. J. Graetzer, *Edmund Halley und Caspar Neumann*, Breslau, 1883. Translation into English in Institute of Actuaries Library. Halley rather misleadingly mentions 1238 births in a supplementary paper but this figure should not be taken as the number which is consistent with the numbers in his life table.
11. A. De Moivre, *Annuities on lives*, London, 1725. H & S, III, 1–125.
12. J. Smart, *A letter from Mr. Smart to George Heathcote Esq., re inclosing tables extracted from ye bills of mortality for the last ten years, and from them shewing the probabilities of life, in order to estimate annuities &c*, February 1738. H & S, I, 197–205.
13. T. Simpson, *The doctrine of annuities and reversions*, London, 1742.
14. N. Struyck, *Inleiding tot de Algemeene geographie...*, Amsterdam, 1740. An English translation of the Appendix, entitled *Appendix to the conjectures on the state of the human race and the calculation of life annuities*, is in H & S, I, 207–41.
15. W. Kersseboom, *Essays in political arithmetic contained in three treatises*, The Hague, 1748. This is an English translation, made by E. F. Rickman in 1886, and existing in manuscript in the library of the Institute of Actuaries, of works originally published in Dutch. The First Treatise is designated the Third Edition and is dated 1748; the other two treatises are dated 1742.

16. A. Deparcieux, *Essai sur les probabilités de la durée de la vie humaine*, Paris, 1746. An English translation is in the Library of the Institute of Actuaries. H & S, I, 243–9.
17. P. Wargentin, *Mortality in Sweden according to the 'Tabell-Verket' (General Register Office)*, 1766. Translated into English for the ninth international congress of actuaries by the Thule Insurance Company in 1930. H & S, II, 12–69.
18. J. Dodson, *First lecture on insurances*, 1756 (H & S, V, 79–143). A typescript of a translation of the manuscript made in 1984, with a commentary, is held in the Library of the Institute of Actuaries. Extracts from the manuscript were printed in an article by M. E. Ogborn, 'The actuary in the eighteenth century' in *Proceedings of the centenary assembly of the Institute of Actuaries*, 1950, vol. 3, pp. 357–86. Ogborn also explains how he thinks a modified version of Dodson's life table was calculated.
18a. R. Price, *Observations on reversionary payments*, 4th edn, London, 1782. H & S, IX, 1–24.
19. J. Milne, *A treatise on the valuation of annuities and assurances*, 2 vols. London, 1815. H & S, II, 79–118.
20. J. Finlaison, *Life annuities—Report of John Finlaison, actuary of the national debt, on the evidence and elementary facts on which the tables of life annuities are founded*, House of Commons, London, 1829. H & S, II, 216–85.
21. W. Dale, *Calculations deduced from first principles...for the use of the societies instituted for the benefit of old age, by a member of one of the societies*, London, 1772. H & S, IV, 1–17.
22. J. N. Tetens, *Introduction to the calculation of life annuities for life and reversions which depend upon the life and death of one or more persons*, Vol. I, Leipzig, Weidemanns Erben und Reich, Leipzig, 1785, pp. 88–103, 136, 216–22. An English translation appears in H & S, IV, 30–43.
23. B. Gompertz, 'A sketch of an analysis and notation applicable to the estimation of the value of life contingencies', *Philosophical transactions* 110 (1820), 230. Also Gompertz, Benjamin. 'On the nature of the function expressive of the law of human mortality, and on a new mode of determining the value of life contingencies. In a Letter to Francis Baily', *Philosophical Transactions*, London, 1825, pp. 513–83. H & S, II, 119–91.
24. T. Young, 'A formula for expressing the decrement of human life. In a letter to Edward Hyde East, Bart., M.P., F.R.S.', *Philosophical transactions* 116 (1826). H & S, II, 192–215.
25. T. R. Edmonds, *Life tables founded upon the discovery of a numerical law regulating the existence of every human being: illustrated by a new theory of the causes producing health and longevity*, London, 1832. H & S, II, 286–92.

26. W. Makeham, 'On the law of mortality and the construction of annuity tables', *Journal of the Institute of Actuaries* 8 (1859), 301–10. H & S, II, 324–34.
27. W. S. B. Woolhouse, *Investigation of mortality in the indian army*, London, 1839. H & S, X, 19–29.
28. W. S. B. Woolhouse, 'On interpolation, summation, and the adjustment of numerical tables (Part III)', *Journal of the Institute of Actuaries* 12 (1866), 136–76. H & S, X, 166–208.
29. *Tables exhibiting the law of mortality, deduced from the combined experience of seventeen life assurance offices*, London, 1843. H & S, X, 30–78.
30. J. Jones, *A series of tables of annuities and assurances calculated from a new rate of mortality amongst assured lives*, London, 1843.
31. W. P. Elderton, and Fippard, Richard C. 'Notes on the construction of mortality tables', *Journal of the Institute of Actuaries* 56 (1912), 260–72. H & S, X, 132–46.
32. *Continuous mortality investigation reports, numbers 1–19*, Institute of Actuaries and Faculty of Actuaries, London, 1973–2000.
33. C. Ansell, *A treatise on friendly societies*, London, 1835.
34. An example is 'Mrs. Cooksey's Womens Club', which was located at Rowley Regis, near Dudley, Worcestershire—a lodge book from the 1860s survives in the collection of one of the authors. The Club paid benefits on death and maternity and also hired a surgeon to attend members in their confinements. A great deal of ale was consumed at the meetings, which appear to have taken place at Cooksey's hotel, run by Mrs Ruth Cooksey.
35. R. Price, *Observations on reversionary payments*, 5th edn, London, 1792, pp. 409–30 deal with the contributions required to provide sickness benefits. H & S, IX, 1–24.
36. Highland Society of Scotland. *Report on friendly or benefit societies, exhibiting the law of sickness, as deduced from returns by friendly societies in different parts of Scotland*, Edinburgh, 1824. H & S, IX, 25–178.
37. A. W. Watson, *Sickness and mortality experience of the Independent Order of Oddfellows Manchester Unity Friendly Society, 1893 to 1897*, Manchester, 1903. H & S, IX, 219–300.
38. D. Bernoulli, 'An attempt at a new analysis of the mortality caused by smallpox', read 1760. *Histoire de l'Academie Royale des Sciences*, Paris, 1766. A translation into English from L. Bradley, *Smallpox inoculation: an eighteenth century mathematical controversy*, Nottingham, 1971, is in H & S, VIII, 1–37.
39. T. Wittstein, 'The mortality in companies with successive entering and exiting members', *Archiv for Mathematics and Physics with special attention to the needs of Teachers in higher educational Institutions*, 39th volume, Griefswald, C.A. Koch, 1862. English translation by Trevor Sibbett in H & S, X, 91–113.

40. *Observations on a superannuation Fund proposed to be established in the several public departments by the Treasury minute of the 10th August 1821.* H & S, VI, 101–68.
41. H. W. Manly, and Thomas, Ernest C. 'On the valuation of staff pension funds: with tables and examples', *Journal of the Institute of Actuaries* 36 (1902), 209–87. H & S, VI, 169–249.
42. R. Recorde, *The ground of arts, teaching the perfect worke and practice of arithmeticke*, London 1632, revised by Robert Hartwell.
43. J. Richards, of Exon. *The gentleman's steward and tenants of manors instructed*, London, 1730, and *Annuities on lives*, London, 1739.

de 0.00 à 0.25

Log Sin.

c	00"	10"	20"	30"	40"	50"	60"	70"	80"	90"	00"	c
00	−∞	5̄.1961198̄8	5̄.4971498̄7	5̄.6732411̄3	5̄.7981798̄7	5̄.8950898̄8	5̄.9742711̄3	4̄.0412179̄2	4̄.0992098̄6	4̄.1503663̄9	4̄.1961198̄8	99
01	4̄.1961198̄8	4̄.2375125̄6	4̄.2753011̄2	4̄.3106632̄3	4̄.3424279̄1	4̄.3722111̄3	4̄.4002398̄6	4̄.4267587̄9	4̄.4513923̄8	4̄.4748734̄7	4̄.4971498̄7	98
02	4̄.4971498̄7	4̄.5183391̄6	4̄.5384255̄.	4̄.5578477̄0	4̄.5763111̄.	4̄.5940598̄7	4̄.6110932̄1	4̄.6274835̄3	4̄.6432778̄9	4̄.6585178̄6	4̄.6732411̄2	97
03	4̄.6732411̄2	4̄.6874815̄5	4̄.7012638̄4	4̄.7146338̄0	4̄.7275987̄7	4̄.7401879̄0	4̄.7524423̄5	4̄.7643315̄8.	4̄.7759034̄5	4̄.7871844̄6	4̄.7981798̄4	96
04	4̄.7981798̄4	4̄.8089037̄0.	4̄.8193691̄4	4̄.8295383̄0.	4̄.8395725̄2.	4̄.8493523̄5	4̄.8588776̄7	4̄.8682177̄0.	4̄.8773610̄7	4̄.8863159̄1	4̄.8950898̄4	95
05	4̄.8950898̄4	4̄.9036900̄1	4̄.9121231̄7	4̄.9203957̄0.	4̄.9285135̄8	4̄.9364825̄1	4̄.9443078̄5	4̄.9519946̄7	4̄.9595478̄1	4̄.9669718̄3	4̄.9742710̄6	94
06	4̄.9742710̄6	4̄.9814496̄5	4̄.9885115̄0.	3̄.0002299̄78	3̄.0095403̄36	3̄.0090331̄6	3̄.0156637̄3	3̄.0221946̄0.	3̄.0286287̄1.	3̄.0349688̄8	3̄.0412178̄3	93
07	3̄.0412178̄3	3̄.0473781̄4	3̄.0534512̄8	3̄.0594426̄4	3̄.0653515̄0.	3̄.0711810̄4	3̄.0763337̄.	3̄.0826105̄0.	3̄.0882143̄7	3̄.0937468̄6	3̄.0992097̄5	92
08	3̄.0992097̄5	3̄.1046047̄8.	3̄.1099336̄1	3̄.1151978̄5.	3̄.1203990̄4.	3̄.1255336̄7	3̄.1306182̄0.	3̄.1356389̄9	3̄.1406024̄1	3̄.1455097̄4	3̄.1503622̄4	91
09	3̄.1503622̄4	3̄.1551611̄2	3̄.1599075̄5	3̄.1646602̄5	3̄.1692475̄7	3̄.1738433̄2	3̄.1783957̄8	3̄.1828914̄4	3̄.1873457̄8	3̄.1917534̄9̄8	3̄.1961197̄0.	90
10	3̄.1961197̄0.	3̄.2004410̄7	3̄.2047198̄6	3̄.2089569̄1	3̄.2131530̄2	3̄.2173089̄8	3̄.2214255̄4	3̄.2225034̄5	3̄.2295434̄2	3̄.2335461̄6	3̄.2375123̄5.	89
11	3̄.2375123̄5.	3̄.2414426̄4	3̄.2453376̄8.	3̄.2491980̄9	3̄.2530245̄0.	3̄.2568174̄8	3̄.2605776̄3	3̄.2643054̄9	3̄.2680016̄4.	3̄.2716665̄9	3̄.2753008̄7.	88
12	3̄.2753008̄7.	3̄.2789049̄9.	3̄.2824794̄4	3̄.2862247̄2	3̄.2895412̄9.	3̄.2930296̄1	3̄.2964901̄4	3̄.2999233̄1.	3̄.3033295̄5	3̄.3067092̄9	3̄.3100629̄3	87
13	3̄.3100629̄3	3̄.3133908̄7.	3̄.3166935̄0.	3̄.3199712̄0	3̄.3232224̄3.	3̄.3264353̄2	3̄.3296584̄6.	3̄.3328401̄1	3̄.3359986̄2	3̄.3391343̄3	3̄.3424475̄6	86
14	3̄.3424475̄6	3̄.3453386̄3	3̄.3484078̄6	3̄.3514555̄5	3̄.3544820̄0.	3̄.3574875̄0.	3̄.3604723̄5	3̄.3634368̄3	3̄.3663812̄0	3̄.3693057̄5	3̄.3722107̄3	85
15	3̄.3722107̄3	3̄.3750964̄2.	3̄.3779763̄05	3̄.3808108̄9.	3̄.3836401̄7	3̄.3864511̄5.	3̄.3892440̄4	3̄.3920190̄9.	3̄.3947765̄2.	3̄.3975165̄5	3̄.4002394̄0	84
16	3̄.4002394̄0	3̄.4029453̄9	3̄.4056344̄2	3̄.4083070̄1.	3̄.4109632̄4	3̄.4136033̄4.	3̄.4162274̄7	3̄.4188358̄5	3̄.4214286̄5	3̄.4240060̄7	3̄.4265682̄8	83
17	3̄.4265682̄8	3̄.4291154̄7.	3̄.4316478̄0.	3̄.4341654̄5.	3̄.4366685̄8	3̄.4391573̄8.	3̄.4416319̄9	3̄.4440925̄8	3̄.4465393̄1	3̄.4489723̄4.	3̄.4513918̄0	82
18	3̄.4513918̄0	3̄.4537978̄7.	3̄.4561906̄7	3̄.4585703̄7.	3̄.4609371̄0.	3̄.4632909̄9	3̄.4656322̄0	3̄.4679608̄6.	3̄.4702771̄0.	3̄.4725810̄4	3̄.4748728̄3	81
19	3̄.4748728̄3	3̄.4771525̄9	3̄.4794204̄5.	3̄.4816765̄2	3̄.4839209̄3.	3̄.4861538̄1.	3̄.4883752̄6	3̄.4905854̄1	3̄.4927843̄7.	3̄.4949722̄5	3̄.4971491̄6	80
20	3̄.4971491̄6	3̄.4993152̄1	3̄.5014705̄2.	3̄.5036152̄8.	3̄.5057493̄0.	3̄.5078729̄9.	3̄.5099863̄4	3̄.5120894̄6.	3̄.5141824̄4.	3̄.5162653̄8	3̄.5183383̄8	79
21	3̄.5183383̄8	3̄.5204015̄4.	3̄.5224549̄4.	3̄.5244987̄0	3̄.5265328̄3.	3̄.5285575̄1	3̄.5305727̄9	3̄.5325787̄7.	3̄.5345755̄2.	3̄.5365631̄4	3̄.5385416̄9	78
22	3̄.5385416̄9	3̄.5405112̄8.	3̄.5424719̄7.	3̄.5444238̄5.	3̄.5463670̄0.	3̄.5483014̄9.	3̄.5502274̄0	3̄.5521448̄1	3̄.5540538̄0.	3̄.5559544̄2	3̄.5578467̄7.	77
23	3̄.5578467̄7.	3̄.5597309̄9	3̄.5616069̄0	3̄.5634748̄3.	3̄.5653347̄6.	3̄.5671867̄5	3̄.5690308̄9.	3̄.5708672̄2.	3̄.5726958̄2	3̄.5745167̄6.	3̄.5763500̄9	76
24	3̄.5763500̄9	3̄.5781358̄8.	3̄.5799354̄0.	3̄.5817251̄0.	3̄.5835086̄4.	3̄.5852848̄9.	3̄.5870533̄0.	3̄.5888157̄4	3̄.5905704̄6.	3̄.5923181̄2.	3̄.5940587̄7.	75
c	90"	80"	70"	60"	50"	40"	30"	20"	10"	00"	c

Log Cos.

4

The computation factory: de Prony's project for making tables in the 1790s

IVOR GRATTAN-GUINNESS

A large set of logarithmic and trigonometric tables was produced at the end of the eighteenth century under the direction of Gaspard Riche de Prony.[1] Although they were completed in 1801, their size made publication a costly task and it was never done, despite the fact that printing was started more than once and various efforts were made over the years to find finance.

Fig. 4.1 De Prony's ambitious tables proved too voluminous and expensive for commercial publication. They eventually saw the light of day in a much reduced form over fifty years after his death. The title page does not even mention his name. (Courtesy of the Tomash collection.)

Fig. 4.2 De Prony was sketched in one of his student's notebooks at the Ecole Polytechnique in 1802 or shortly afterwards. (Courtesy of Ivor Grattan-Guinness.)

Eventually a reduced edition of some tables appeared in 1891. This project also turned out to be influential to Charles Babbage's work.

de Prony's career: education and mathematics for the engineer

Gaspard Clair François Marie Riche de Prony (1755–1839) was born in the Beaujolais region of Southern France to the Riche family. They owned land at a location nearby called Prony, and he added this name to his own. I shall refer to him as 'de Prony', in conformity with his own preferred form in his many publications; but his younger brother Claude used only the original surname during his career as a distinguished naturalist.

After studying at the École des Ponts et Chaussées in Paris in the late 1770s, de Prony joined the associated engineering corps but maintained connections with the school. In the reforms of educational institutions following the Revolution of 1789 he was appointed professor of analysis and mechanics with Joseph-Louis Lagrange (1736–1813), at the newly founded École Polytechnique from 1794, transferring to a post of graduation examiner in 1816. From 1798 he was also made Director of the École des Ponts et Chaussées, another post which he retained until the end of his life.

He was elected in 1795 to the new classe des sciences mathématiques et physiques of the Institut de France, the learned body which replaced the old Académie des Sciences, and he retained this position in 1816 when that institution was restored after the fall of Napoleon. In 1801 he was appointed to a supplementary position at the Bureau des Longitudes. He held all these posts throughout his life, and also some subsidiary ones at various times, such as the directorship of a small École de Géographes which he ran from 1795 until its closure in 1803. Such was the style of professional life in the Paris of his time: the so-called 'cumul' system, where a scientist or administrator held several posts at once.

de Prony is remembered today only for his tables, and for a simple dynamometer (the 'de Prony brake') which he introduced in the late 1810s; but in his lifetime he was very well known, for his numerous activities in engineering practice and education, and in many related areas of science. He wrote a few treatises, a large number of papers, some short pamphlets, and published several editions of his lectures on mechanics at the École Polytechnique. He brought to much of his work a philosophy which bore strongly upon his tables project: namely the desire to render theories mathematically in ways which were both amenable to numerical calculation and accessible to observational data. For example, at the École Polytechnique he included an optional but substantial course on difference equations and related topics, publishing it at length in the school's *Journal*.[2] He made a stark contrast to his colleague professor Lagrange, who preferred a theoretical algebraic approach to the calculus of his own devising (although the 'Lagrange interpolation formula' belongs to this time also).

The tables project

In 1790, when he was in his 36th year, citizen de Prony was appointed chief engineer to the département of Oriental Pyrenees, but he announced his desire to stay in Paris. This predicament may have inspired him to launch the tables project, which was partly born out of the reform of weights and measures then being undertaken. Perhaps opportunely, therefore, early on in the new decade de Prony set up a Bureau de Cadastre in Paris, to prepare a detailed map of France to facilitate the accurate measurement of property as a basis of taxation.[3] In connection with this plan, it was decided that a very large set of logarithmic and trigonometric tables would be prepared.

From the mid-1790s the Bureau seems to have been housed with de Prony's Ecole de Géographes, and had responsibility for the calculations of the navigational tables required by the (newly formed) Bureau des Longitudes for publication in their annual journal *Connaissance des temps*. The mathematical details of the tables project will be surveyed in the next section: the chronology is described here, as well as it can be traced.

This qualifier needs some explanation. Despite the fame (or notoriety) of the project, connected information is hard to find, and some sources contradict details given in others.[4] I have not been able to find even the address of the Bureau de Cadastre, nor have I traced its archive (if it still exists). No useful information exists on the organisation of the work-room (which must have been one of the largest in Paris at that time) or the full personnel or the budgets for the project; and apparently the waste sheets used for producing or checking the tables were discarded after use. Given these provisos, here is brief outline of the history of the tables.

de Prony devised the project explicitly following the principles of the division of labour which had been laid down by Adam Smith in *A treatise on the wealth of nations* of 1776. As he later recalled,

[...] I came across the chapter where the author treats of the division of work [*travail*]; citing, as an example of the great advantages of this method, the manufacture of pins. I conceived all of a sudden the idea of applying the same method to the immense work with which I had been burdened, and to manufacture logarithms as one manufactures pins. I have reason to believe that I had already been prepared for this conception by certain parts of mathematical analysis, on which I was then lecturing at the Ecole Polytechnique.[5]

de Prony gave some details of the project in a 'Notice' read to the classe des sciences mathématiques et physiques in 1801, soon after it was finished.[6] The personnel were divided into three sections according to the work they did. The first section contained a handful of mathematicians, including A. M. Legendre, C. A. Prieur de la Côte d'Or, and Lazare Carnot; the former two were also involved with the reform of weights and measures, and latter two also acted as influential political figures. They chose the mathematical formulae to be used for calculation and checking, and also considered the choice of initial values of the numbers or angles, the number of decimal places to be adopted in each table, and so on. The second section comprised several 'Calculators', including the mathematicians A. M. Parseval (of the well-known formula in infinite series) and the textbook writer J. G. Garnier, who determined the values, and the differences of various orders, that needed to be calculated. They also prepared a page of tables for the numerical work by laying out the columns of the chosen values and the first row of entries, and preparing the instructions on the preparation of the remaining entries on the page. These calculations were done by the third *section*, a large team of between 60 and 80 assistants.[7] Many of these workers were unemployed hairdressers: one of the most hated symbols of the ancien régime had been the hair-styles of the aristocracy, and the obligatory reduction of coiffure 'as the geometers say, to its most simplest expression' left the hairdressing trade in a severe state of recession. Thus these artists were converted into elementary arithmeticians, executing only additions and subtractions.[8]

When a page was completed, it was returned to the second section, to check the figures using formulae chosen by the first section. The project was run twice, in that two sets of each table were produced from different equations, so that each set could be checked against the other. By 1794 seven hundred results were being produced each day.[9]

After the project was completed in 1801 several of the calculators were transferred to the Bureau des Longitudes to carry on similar work there.[10] In addition, some of the equipment that had been used (presumably on the

Fig. 4.3 Hairdressers to the aristocracy were among the economic victims of the French Revolution. De Prony trained them in elementary arithmetic and set them to work as production line labourers in his table-making factory.

de Prony's influence upon Babbage

Charles Babbage (1791–1871) gives in his autobiography two occasions for the origins of his interest in mechanical calculation, one from 1812 and the other from 1819.[1] The latter date coincides with Blagden's involvement in the possible printing of the tables: Babbage does not refer to de Prony here, but he must surely have heard of the project, and he may have realised that mechanization was the better way to produce tables of this size. Moreover, his policy of using large numbers of decimal places, working with sequences of differences of various orders is reminiscent of de Prony (although of course it was not novel with either man). Indeed, soon afterwards, in a pamphlet of 1822 on 'the application of machinery to the purpose of calculating and printing mathematical tables'—his main publication for securing governmental support for production of the Difference Engine—he gave some account of de Prony's project and noted that mechanical methods would speed up the process of calculation.[2] Clearly he was struck by de Prony's production of the tables following an industrial process, and was hoping to imitate the process by mechanical means. In his book on manufactures, he rehearsed some of the same material on de Prony's project and on mechanical calculation.[3]

The nature and extent of de Prony's influence on Babbage needs some exploration. It seems that Babbage had already envisioned his idea of mechanical calculation before learning of de Prony's tables; but he may well have been helped to crystallize some of his ideas, recognise difficulties, and so on, by acquaintance with that project. Babbage does not seem to have met de Prony or corresponded with him, but he examined the Observatoire set of tables during his visit to Paris perhaps in 1826 and had copies made of some of de Prony's writings on the project.[4] Thus his contribution was to substitute manual labour by engineered automation in the construction of tables.

Notes

1. C. Babbage, *Passages from the life of a philosopher*, Longman, Green, London, 1864, reprinted Gregg, Farnborough, 1969, pp. 42–3.
2. C. Babbage, *A letter... on the application of machinery to the purpose of calculating and printing mathematical tables*, Booth, London, 1822.
3. C. Babbage, *On the economy of manufactures and machinery*, 1st edn, Knight, London, 1832; 4th edn, 1835, arts. 241–50.
4. C. Babbage, *Tables of logarithms of the natural numbers from 1 to 108,000*, Mawman, London, 1827, p. vi. Pertinent materials are in the Babbage archive at the Science Museum (London), file BAB U21, and in his correspondence held at the British Library, Add. Mss. 37182, 37183 and 37201.

preparation of the map) went to the Dépôt Générale de la Guerre, a military organization concerned with geodesy and cartography. At that time de Prony read his 'Notice' to the *classe*, and the task was set to find money to finance the printing.

One of de Prony's supporters was Jean Baptiste Joseph Delambre (1749–1822), a leading practical astronomer of his time. Delambre himself was much concerned with the compilation of tables: by 1801 he had completed a much smaller set of logarithmic and trigonometric tables initiated by the mathematician and instrument designer J. C. Borda, who had died in 1799[11] and as part of his responsibility at the Bureau des Longitudes he was involved in producing solar and lunar tables for navigators. He was also a leader of the expedition in the mid-1790s to triangulate the longitude between Paris and Barcelona, writing up the details in three large volumes. Delambre was to be an influential aid to de Prony, for in 1803 he was appointed founder secrétaire perpétuel of the classe for the (so-called) 'mathematical' sciences, having served previously in the rotating post of secrétaire. He gave the project some publicity in one of his regular reports of the activities of the classe. His pieces and de Prony's 'Notice' were reprinted in the *Mémoires* of the *classe* in 1804, whereupon de Prony added an historical note on the errors found in previous tables.

The publishers Firmin Didot were charged with the task of printing, and estimated that a volume of 1200 large pages would result, together with the introduction, at a cost of 80 000 francs;[12] apparently about 500 pages were in proof by 1802. But financial crises in France began to intervene, and printing stopped. Nevertheless, hopes remained alive for many years to complete the work. Firmin Didot announced the volume(s) in some of their catalogues of the 1810s. Legendre published some entries from the logarithmic tables in his treatises on transcendental and elliptic integrals.[13] At the end of that decade, and following the fall of Napoleon, collaboration was sought with the British government for financial support, especially through the good offices of the physician Sir Charles Blagden, who was then a representative of the British government in France; but Blagden's death in 1820, and the British desire to have the tables converted from the original centesimal system of angular division to the sexagesimal system, a massive task in itself, scuppered these hopes. In pamphlets and notices of the 1820s de Prony announced the possible collaboration to the Académie des Sciences (as it had reverted to) and he hoped that the 'unique monument' could be kept in its original form. He revived interest to the extent that printing seems to have begun again; but once more it ceased, and Firmin Didot formally returned their set to the Bureau des Longitudes in 1833. Only a few printed pages seem to have survived.[14] After de Prony's death in 1839, hopes still remained for printing;[15] but they were not realized.

Firmin Didot must have lost much money as a result of the failure to publish. Seemingly they were able to take the loss; and this capability reflects the astonishing scale of publishing in science and technology in France at that time. In the new phase of school and higher education the Ecole Polytechnique and the Ecole des Ponts et Chaussées were only two (important) institutions among a national network which no other country then matched. Several other national organizations, especially military ones, supported research both by offering employment and establishing journals. Many treatises appeared from commercial houses such as Firmin Didot; they also supported a large number of journals. Thus someone in applied mathematics and mechanics might have up to 40 outlets for his papers by the 1820s.[16]

Towards publication

Both sets of the manuscript tables are still preserved. One is held in the library of the Paris Observatoire. de Prony himself retained the other set; but his niece and heir did not include it in his manuscripts that she sent to the École des Ponts et Chaussées. It was found only in 1858, in the possession of the niece's daughter and son-in-law, by Pierre Alexandre Francisque Lefort(-Latour) (1809–1878), in the course of making an extensive study of the other set, in the Observatoire. Lefort was a graduate of the Ecole Polytechnique and the École des Ponts et Chaussées (during the time that de Prony had been respectively graduation examiner and Director), and then a career engineer in the Corps des Ponts et Chaussées. His grandfather-in-law J. B. Biot reported his finding to the Académie des Sciences, and that set was placed in the library of the Bibliothèque de l'Institut.

Each set comprises 19 volumes, the last one of which consists of an account of the methods of calculation (and a few basic tables). The others are large-format tomes, often of 251 folios each (containing the tables for a run of 25 100 values of the numbers or angles). The first eight volumes contain logarithms of numbers from 1 to 200 000 calculated to 14 decimal places; there then follow (not necessarily in this order) a volume of sines of parts of the radius, three volumes each of sines and of tangents (including their

THE COMPUTATION FACTORY

Fig. 4.4 De Prony produced two independent manuscripts of his tables in order to check one against the other. One of these nineteen folio volume sets is now in the Bibliotheque de l'Institut in Paris. In the 1980s John Fauvel showed them in an Open University video for a course on the history of mathematics. (Courtesy of the Open University.)

logarithms for some ranges of the argument), one volume each of the ratios of arcs to sines and of arcs to tangents, and the product of sines with cosines. In fact, only the Institut set has this last volume: presumably the other set had had a copy also, but it is now lost. Further, the last volume in the Institut was originally with its brother in de Prony's family: it was transferred from the Observatoire some time after 1858, and gives a less complete account.

In the mid-1870s the Scottish mathematician Edward Sang (1805–1890), who had recently produced some extensive logarithmic tables of his own, initiated a debate with Lefort about some of de Prony's procedures as Lefort had recently described them.[17]

As de Prony had stated, his design of the tables was like an inverse to normal methods. Instead of building up the final values from differences of various orders and by interpolative formulae, the exact values of $f(x)$ were calculated for some initial values x from a known formula (usually a series expansion), and the differences pertaining to succeeding values $(x + h)$ were calculated and inserted into an adopted formula to produce the required values of $f(x + h)$. A main motive for this strategy was to avoid multiplication and division and to reduce the calculations to sums and (especially)

Fig. 4.5 In contrast to de Prony's computation factory the Scottish mathematician Edward Sang's table-making project was a cottage industry with his daughters as the only other workers. But like de Prony Sang died with 47 volumes of his 28 and 15 place tables unpublished. His daughters fought a long and hard but ultimately successful battle to deposit the manuscript with the Royal Society of Edinburgh.

differences, which the hairdressers could fairly be expected to handle. De Prony used two different sets of formulae for the two sets of tables for checking purposes. His repertoire of techniques was also determined by the perceived need to calculate the entries to so many decimal places (the numbers varied between 14 and 29).

In the design some obvious moves were missed. For example, as Sang pointed out, de Prony did not employ the obvious relation

$$\log_{10}(10x) = \log_{10} 10 + \log_{10} x = 1 + \log_{10} x$$

either as a means of reducing or simplifying interpolations or as a check on the decimal parts of the logarithmic tables. (Maybe he had used it but regarded it as too obvious to need stating.) He also criticised the strategy of rounding or truncating the differences of the highest orders (which he called a method of 'vitiated differences'), on the ground that rounding or truncating errors were prematurely introduced already in the values found at the higher orders of difference and therefore cumulation could occur upon transmission back to the lower orders. Lefort showed that the possible errors obtainable were not as large as Sang feared; but he had his own criticisms of the project, such as deploring the apparent discard of the loose

On de Prony's mathematical methods

So that the hairdressers could be called upon simply to add and subtract, avoiding multiplication and division, de Prony devised formulae expressing expansions in terms of differences of various orders. One of these was a theorem due to Euler,

$$u(x+nh) - u(x) = \sum_{r=1}^{n} nC_r \Delta^r u(x), \quad (1)$$

where the forward difference was defined as usual:

$$\Delta u(x) := u(x+h) - u(x). \quad (2)$$

An iterative formula on orders of differences on neighbouring values of function, due to the seventeenth-century astronomer G. Mouton, played an important role:

$$\Delta^{n-1} u(x) = \Delta^{n-1} u(x-h) + \Delta^n u(x-h) \quad (3)$$

(in terms of the definition of $\Delta u(x)$ in (2)). The names of the required differences were laid out in columns on the page by the second section for the hairdressers to calculate as per instructions in the body of the page. These computations proceeded up to the column of differences (prepared by the second section) which contained (almost) equal figures in each row, when of course the process stopped.

Among special formulae used, the logarithms of prime numbers were calculated from an expansion which can be written as:

$$2\log x = \log(x+1) + \log(x-1) + (2\log e) \sum_{r=0}^{\infty} [z^{2r+1}/2r+1)] \quad (4)$$

$$\text{where } z := (2x^2 - 1)^{-1}. \quad (5)$$

Further terms were deliverable via expansions such as

$$\log(y+x) + \log(y-x) = (2\log e) \sum_{r=0}^{\infty} [(x/y)^{2r+1}/(2r+1)]. \quad (6)$$

Logarithms of compound numbers were produced from those of their prime factors by addition.

For the trigonometric tables the standard power-series expansions were used for the initial entries over $10°$ intervals, and addition formulae for sines and cosines were deployed for intermediate values, for example $\sin(a+b)$, where a took values in steps of $0.1°$ and b in steps of $0.01°$. (The quadrant was

On de Prony's mathematical methods *cont.*

divided centesimally. For convenience the usual symbol for degrees indicates these angles here.)

For higher-order differences de Prony adopted 'the extremely elegant and simple formulae proposed by the citizen Legendre', rather similar in form to Mouton's (3):

$$\Delta^n \sin x = -(2 \sin \Delta x/2)^2 (\Delta^{n-1} \sin x + \Delta^{n-2} \sin x), \quad n \geqslant 1 \qquad (7)$$

for extreme values of x in the range, and for intermediate values

$$\Delta^n \sin x = \cos x \, \Delta^n \sin 0 + \sin x \, \Delta^n \sin 1. \qquad (8)$$

The tangent tables were devised by various formulae. Up to 50° difference formulae were deployed; then up to 94° recourse was made to

$$\tan(50° + a) = 2 \tan 2a + \tan(50° - a); \qquad (9)$$

and for the last few a power series in $\cot x$ (which would then be taking very small values) was used.

For log sine tables, results such as

$$\log(ax/\sin ax) = \log ax - \log \sin ax, \quad \text{with } a > 0, \qquad (10)$$

were used, with $(ax/\sin ax)$ expanded in power series of ax. For differences of various orders, some use was made of Lagrange's symbolic generalized version of Taylor's theorem:

$$\Delta^n u(0) = [\exp(\Delta x f'(x) - 1)]^n, \quad \text{where } u(m) := f(x + m\Delta x). \qquad (11)$$

This formula required knowledge of the derivatives of $f(x)$: for $\log \sin x$, these were expressed as polynomials in $\cot x$, which was small when x was close to 100°. Log tan tables were produced from differences such as

$$\Delta^n \log \tan x = \Delta^n \log \sin x - \Delta^n \log \cos x, \qquad (12)$$

and various similar formulae derivable from it.

Some check formulae were used. A favoured one for the sine formulae was Euler's:

$$\sin x + \sin(40° - x) \sin(80° + x) = \sin(40° + x) + \sin(80° - x). \qquad (13)$$

sheets upon which calculations had been made before transcription of the results onto the pages, and disliking de Prony's method of producing the final values for the log sine table. By and large, Lefort's reservations and Sang's criticisms carry weight; even though the Service Géographique de l'Armée found few errors in the tables when preparing their version for publication, the manner of compilation seems to have been corrigible.

One aspect of the tables which is not too clearly described is the effect of errors of calculation. One would presume that the errors would (or could) iterate onto entries in the vicinity of their occurrence; but they seem to be occasional, even if fairly frequent in places (and written in red ink above the black originals). In a passage which excites suspicion, Lefort reports detecting 'arbitrary corrections' in places in the tables.[18]

What remains unexplained is the reason that de Prony chose to calculate these tables to such extraordinary numbers of decimal places in the first place. Did the taxation system really require the cadastral survey to be such an exact science, or did de Prony want to keep the threat of the Pyrenees appointment of 1790 well at bay? Delambre knew the project well, and expressed himself with careful vagueness to Blagden in 1819 when the collaboration with the British Government was in the offing. In contrast to his own completion of the Borda tables, he opined that 'These tables will not serve in the usual cases, but only in extraordinary cases.'[19]

When de Prony's set had been reported to the Académie des Sciences in 1858, hopes had again been expressed that printing could be achieved. The collation of the two sets was completed in 1862 by officers of the Dépôt Général de la Guerre (the last volume of the set is so annotated and signed). In his discussion of the tables, Lefort was doubtful of the value of publication, but thought that a rounded-off version to eight places was desirable.[20] Either by accident or influence, thirty years later the Service Géographique de l'Armée of the Dépôt published just such an edition, as a replacement to the Borda/Delambre tables, which went only to six or seven places. This edition comprised 640 (printed) large quarto pages, of the (truncated) logarithmic, and log sin, cos, tan and cot tables, with the latter still rendered in centesimal angular division.[21]

The year of publication, 1891, was close to the centenary of the project; but this detail was not mentioned in the preface written by the Director of the Service Géographique de l'Armée, and may not have been noticed. In fact, this volume did not even mention de Prony in its title (so that its existence has understandably been overlooked by historians); and in the short preface the Director reported that the Borda/Delambre tables were practically sold out, and mentioned the need for tables which go beyond the seven places given there. He did not refer to the fine set of tables prepared for hydrographers by Valentin Bagay (1772–1851) and published by Firmin Didot, where several of the same functions had been calculated to seven (or a few more) decimal places.[22]

The rounded-off tables were transcribed from the *Observatoire* set and the proofs checked twice against the original. It seems that few errors were found. So congratulations are due to de Prony for his design of the project—and hats off to the hairdressers, too, at least for the early places of the long computations.

Further reading

The most substantial survey of de Prony's career is M. Bradley, *A career biography of Gaspard-Clair-François-Marie Riche de Prony: bridge-builder, educator and scientist*, Edwin Mellen Press, Lewiston, Queenston and Lampeter, 1998. See also X. Walckenaer (editor), 'Commémoration en souvenir de Prony (1755–1839)', *Bulletin de la Société d'Encouragement pour l'Industrie Nationale* 139 (1940), 68–98; and I. Grattan-Guinness, *Convolutions in French mathematics, 1800–1840: from the calculus and mechanics to mathematical analysis and mathematical physics*, 3 vols., Birkhäuser, Basel, and Deutscher Verlag der Wissenschaften, Berlin (DDR), 1990, especially Chapters 2, 3, 5, 6, 8, and 16. His manuscripts and library—both massive collections—are held in the École des Ponts et Chaussées in Paris.

On the context of the project in French cartography of the time, see J. Konvitz, *Cartography in France, 1660–1848*, Chicago University Press, Chicago, 1987, pp. 47–62. R. Herbin and A. Pebereau, *Le cadastre français*, Lefebvre, Paris, 1953, is a moderately useful history of French cadastral surveys. For the work of Borda and others on tables and on related projects of this time, see J. M. Mascart, *La vie et les travaux du Chevalier Jean-Charles Borda (1733–1799)*, Lyon, 1919 (as *Annales de l'Université de Lyon, n.s.*, sec. 2 (droit, lettres), fasc. 33), especially pp. 580–93. On the historical context, see L. Daston, 'Enlightenment calculations', *Critical enquiry* 21 (1994), 183–202.

Notes

1. The author's archival researches carried out in Paris over the years were aided by grants from the Royal Society and the British Academy. An earlier version of this

chapter appeared as 'Work for the hairdressers: the production of de Prony's logarithmic and trigonometric tables', *Annals of the History of Computing* 12 (1990), 177–85.

2. G. F. C. M. Riche de Prony, 'Suite des leçons d'analyse', *Journal de l'École Polytechnique*, (1)1, cah. 2 (1796). 1–23; cah. 3, 209–73; cah. 4, 459–569. Also in his *Méthode directe et inverse des différences*, Imprimerie de la République, Paris, 1796, pp. 31–53, 107–71, 172–282.

3. For the map de Prony used a 'central' projection, where the supposedly spherical Earth was projected onto a tangent plane; for a country of the shape and size of France, good representations of geodesics and areas were obtained.

4. For example, de Prony gives various dates for the date of commencement of the project: 'an 2' (1793–4) in G. F. C. M. Riche de Prony, *Notice sur les grandes tables*, Baudouin, Paris, 1801 (also in *Mémoires de la classe des sciences mathématiques et physiques de l'Institut de France*, 1803–4, 49–55), p. 49; but 1791 in *Notice sur les grandes tables*, Didot, Paris, 1824, p. 4, and also a manuscript career document held at the École des Ponts et Chaussées, ms. 1786. See footnote 7 for another example of inconsistency.

Various other files in the *École des Ponts et Chaussées* are relevant (see J. Konvitz, *Cartography in France, 1660–1848*, Chicago University Press, Chicago, 1987, pp. 165–6), as are the minute-books of the Bureau des Longitudes. There are also files in the Archives Nationales, F^{17} 1393 and 13571, and very probably other files there (for example F^{14} 2146 is quite useful): the very poor state of cataloguing has doubtless hidden other files.

5. G. F. C. M. Riche de Prony, *Notice sur les grandes tables*, Didot, Paris, 1824, p. 4; compare his friend the educationist and novelist Maria Edgeworth on him in *Maria Edgeworth in France and Switzerland*, (ed. C. Colvin), Clarendon Press, Oxford, 1979, p. 151.

6. G. F. C. M. Riche de Prony, *Notice sur les grandes tables*, Baudouin, Paris, 1801 (also in *Mémoires de la classe des sciences mathématiques et physiques de l'Institut de France*, 1803–4, pp. 49–55).

7. These are the numbers indicated in G. F. C. M. Riche de Prony, *Notice sur les grandes tables*, Baudouin, Paris, 1801, p. 53; but according to his *Instruction élémentaire sur les moyens de calculer les intervalles musicaux*, Didot, Paris, 1832, pp. 67–8, between 150 and 200 assistants were involved.

8. The quotation here, and the attached information, comes from C. Dupin, 'Introduction d'un nouveau cours [at the *Conservatoire des Arts et Métiers*]', *Annales des sciences industrielles* 16 (1824), 193–216, 225–52 (also published in his *Discours et leçons sur l'industrie*, vol. 2, Bachelier, Paris, 1825, pp. 149–209), p. 23.

9. See J.-J. Lefrançois de Lalande, *Bibliothèque astronomique avec l'histoire d'astronomie depuis 1781 jusqu'à 1802*, Imprimerie de la Republique, Paris, 1803, pp. 743–4.

10. See C. G. Bigourdan, 'Le Bureau des Longitudes', *Annuaire du Bureau des Longitudes* (1928), A1–A72, especially A25–A28.
11. J. C. Borda, completed by J. B. J. Delambre, *Tables trigonométriques décimales, ... précédées des tables des logarithms*, Imprimerie République, Paris, 1801.
12. This was a huge sum of money, about five times a typical annual academic income of 13 000–18 000 francs.
13. A. M. Legendre, *Exercices de calcul intégral*, vol. 3, Courcier, Paris, 1816, table V (repeated in his *Traité des fonctions elliptiques et des intégrales Eulériennes*, vol. 2, Huzard-Courcier, Paris, 1826), pp. 260–8. He took the logarithms for all odd numbers from 1163 to 1501, and the prime numbers thereafter up to 10 000.
14. These pages are in the file in the *Archives Nationales* cited in footnote 4. P. A. F. Lefort, 'Description des grandes tables', *Annales de l'Observatoire Impérial de Paris*, 4 (1858), Additions, [123]–[150], [138] reports having seen six incomplete printed copies of the sine tables, but he gives no provenance.
15. This information comes from the obituary C. Dupin, 'Eloge de M. le Baron de Prony', *Moniteur universel* (1840), 624–9 (also as *Chambres des pairs, impressions diverses*, no. 48; and in *Journal du génie civil* 11/2 (1846), 326–51, 338).
16. See the table in I. Grattan-Guinness, 'The *ingénieur savant*, 1800–1830: a neglected figure in the history of French mathematics and science', *Science in context*, 6 (1993), 405–33.
17. E. Sang, 'Remarks on the... tables computed... under the direction of M. Prony', *Proceedings of the Royal Society of Edinburgh*, 8 (1872–5), 421–36; P. A. F. Lefort, 'Observations relatives aux remarques...', *Proceedings of the Royal Society of Edinburgh*, 8 (1872–5), 563–81; E. Sang, E. 'Reply to M. Lefort's observations...', *Proceedings of the Royal Society of Edinburgh*, 8 (1872–5), 581–7. For a recent survey of Sang's logarithmic work see A. Craik, 'Edward Sang (1805–1890), calculator extraordinary', *BSHM Newsletter*, 45 (Summer 2002), 32–43.
18. P. A. F. Lefort, 'Description des grandes tables', *Annales de l'Observatoire Impérial de Paris*, 4 (1858), Additions, [123]–[150], [146].
19. This letter is quoted in P. A. F. Lefort, 'Note sur les deux exemplaires manuscrits des grandes tables', *Comptes rendus de l'Académie des Sciences*, 46 (1858), 994–9, 999, and also in his 'Description des grandes tables', *Annales de l'Observatoire Impérial de Paris*, 4 (1858), Additions, [123]–[150], [146]. The letter is not among the Blagden papers in the archives of the Royal Society, but there are Blagden materials elsewhere.
20. 'Description des grandes tables', *Annales de l'Observatoire Impérial de Paris*, 4 (1858), Additions, [123]–[150], [146]. In fact, de Prony himself reported that during the run of the project, in response to a ministerial request, he had his Bureau prepare a set of some trigonometric tables to nine decimal places (G. F. C. M. Riche de Prony, *Notice sur les grandes tables*, Didot, Paris, 1824, p. 7).

This set is in the archives of the École des Ponts et Chaussées along with various other manuscript tables: see the *Catalogue des manuscrits de la Bibliothèque de l'cole des Ponts et Chaussées*, Plan and Nourrit, Paris, 1886, pp. 4–5.

21. Service Géographique de l'Armée, *Tables des logarithmes... et de sinus et tangentes...*, Imprimerie Nationale, Paris, 1891.
22. V. Bagay, *Nouvelles tables astronomiques et hydrographiques*, Firmin Didot, Paris, 1829.

No. 2.
$u = x^4 - 6x^3 + 11x^2 - 6x$

x	u
0	0
1	0
2	0
3	0
4	24
5	120
6	360
7	840
8	1680
9	3024
10	5040
11	7920
12	11880
13	17160
14	24024
15	32760
16	43680
17	57120
18	73440
19	93024
20	116280
21	143640
22	175560
23	212520
24	255024
25	303600
26	358800
27	421200
28	491400
29	570024
30	657720
31	755160
32	863040
33	982080
34	1113024
35	1256640
36	1413720
37	1585080
38	1771560
39	1974024
40	2193360
41	2430480
42	2686320
43	2961840
44	3258024
45	3575880
46	3916440
47	4280760
48	4669920
49	5085024
50	5527200

No. 3.
$u = x^4 - 72x^3 + 1798x^2 - 18072x$

x	u
0	0
1	16345
2	29512
3	39897
4	47872
5	53785
6	57960
7	60697
8	62272
9	62937
10	62920
11	62425
12	61632
13	60697
14	59752
15	58905
16	58240
17	57817
18	57672
19	57817
20	58240
21	58905
22	59752
23	60697
24	61632
25	62425
26	62920
27	62937
28	62272
29	60697
30	57960
31	53785
32	47872
33	39897
34	29512
35	16345
36	00000

5

Difference engines: from Müller to Comrie

MICHAEL R. WILLIAMS

Most people in the computing disciplines know of the famous difference engine designed by Charles Babbage, but many are unaware of several others that were produced after Babbage's efforts and even one mention of

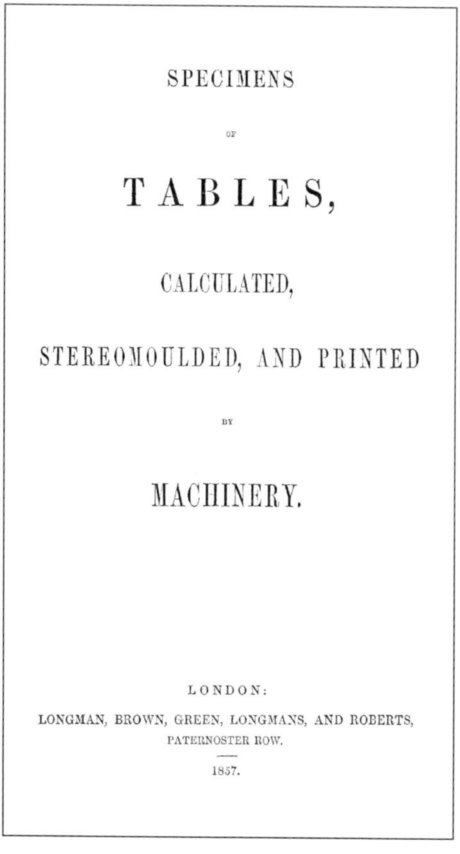
SPECIMENS

OF

TABLES,

CALCULATED,

STEREOMOULDED, AND PRINTED

BY

MACHINERY.

LONDON:

LONGMAN, BROWN, GREEN, LONGMANS, AND ROBERTS,

PATERNOSTER ROW.

1857.

Fig. 5.1 The Swedish difference engine builders Georg and Edvard Scheutz published their *Specimens of tables calculated, stereomoulded, and printed by machinery* in London, 1857. It was perhaps the first set of tables to be published that were entirely machine produced. (Collection of E. Tomash.)

the concept that predates him. No one can claim that difference engines were a mainstay of table-making ventures, but some were of use in the actual production of extensive tables and, if nothing else, they make an interesting story about the efforts to mechanize the drudgery of the task.

Difference methods were once of fundamental importance in the construction of tables but they have fallen into disuse since the invention of the digital computer. Indeed, they were already of lesser importance during the era of the massive electro-mechanical machines developed by, among others, Bell Laboratories and Harvard University in the 1940s. Because of their relative obscurity, it is worth a small digression (see sidebar, p. 125) to explain these methods before examining the machines that used them.

Müller's machine

Many of the early mechanical calculating machines were constructed on a one-off basis and, as such, were prone to problems which were overcome once they began to be produced on a commercial scale in the 1800s. One of the very early machines (perhaps the first that could be said to work even reasonably reliably) was produced by the German clergyman Philipp Matthäus Hahn in 1774. This device was typical in that it had mechanical difficulties that prevented it from being easily used. Hahn and his son made several different versions of this device, but he certainly could not be said to have been in the commercial production of them.

Johann Helfrich Müller (1746–1830) was a military engineer who later was employed by the Governor in Giessen, Germany. When he heard about Hahn's machine and its difficulties, he made his own version. This was a modest four-function calculating machine based on a design first created by G.W. Leibniz over a hundred years earlier. It apparently worked well enough that a friend, Philipp Engel Klipstein, edited and published a user's manual in 1786.[1] This work consists of 47 pages of instruction for the simple machine and three pages, evidently written by Müller himself, giving very sketchy descriptions of his ideas for improved devices, mainly a printing mechanism for a calculating machine. A small portion of Müller's appendix is devoted to an entirely different machine which, he claims, he designed to produce sequences of calculations by the method of differences.

Almost nothing is known of Müller's design (if indeed it was anything other than just the concept) but he indicates that if someone would be

Difference methods and difference engines

If one is faced with the task of producing a table of the values of a polynomial function, say $x^2 + 2x + 3$, then the effort is quite manageable if you are only interested in values of x between 1 and 20. On the other hand it becomes a daunting task if you require all values of x between 1 and 100 000. The method of differences involves tabulating a few values of the function and then noting the differences between these values (Δ^1). For functions more complex than a simple quadratic, it will be necessary to take the differences between these differences (the second differences: Δ^2 and possibly even the third (Δ^3), fourth (Δ^4), or greater orders of differences. For example, a tabulation of the quadratic function noted above would yield:

$x =$	1	2	3	4	5	6	7	8	9
$F(x) = x^2 + 2x + 3 =$	6	11	18	27	38	51	66	83	102
First difference, $\Delta^1 =$		5	7	9	11	13	15	17	19
Second difference, $\Delta^2 =$			2	2	2	2	2	2	2

Obviously the second difference is a constant value 2. To obtain the value $F(10)$ it is not necessary to evaluate the original function: one need only add the constant second difference (2) to the last first difference (19) to get the fact that $F(9)$ differs from $F(10)$ by $2 + 19 = 21$ and thus $F(10)$ must be $F(9) + \Delta^1 = 102 + 21 = 123$. We have thus replaced the two multiplications and two additions required to evaluate the function F by two simple additions.

Every polynomial whose largest power is x^n will have a constant nth difference and thus is amenable to being calculated in this way. The replacement of many multiplications by a simple sequence of additions results in a huge reduction in the labour of table making. This is true even if relatively few values are required but the function is quite complex. It is unfortunate that this method is only applicable to the evaluation of polynomials but non-polynomial functions (such as logarithms and trigonometric functions) may be approximated by polynomials, and these may be evaluated by difference methods for short intervals before a new approximating polynomial must be used.

A 'difference engine' is simply a calculating machine that contains several registers: one to hold the value of $F(x)$, and several others to hold the various orders of differences ($\Delta^1, \Delta^2, \ldots, \Delta^n$). The machine must be capable of adding, in sequence, the contents of each order of difference to the one above it and finally adding the contents of Δ^1 to the register containing $F(x)$ to generate the next function value. In terms of modern technology, the construction of such a machine is trivial. In the very early days of mechanical calculating machine design, it was both a major conceptual and physical task.

willing to finance the construction he was willing to make one. He claimed that the machine would print a single copy of the table being calculated, very likely using the printing mechanism he had in mind for the simple calculator noted above, but that it was certainly possible to create additional copies by employing an ordinary workman to turn the handle of the machine. He estimated that a table of x^3 ($1 \leq x \leq 100\,000$) could be produced in just over 10 days of effort, even if one only worked for 8 hours per day. Other than the fact that the machine would be capable of dealing with three orders of differences (which would have been required for a cubic function), no other details are mentioned.

Today Klipstein's book is quite rare[2] which would indicate that it was not widely circulated in his own day. It is notable, for example, that Charles Babbage (who had an extensive collection of books on all aspects of calculation[3]) was evidently unaware of its existence[4] until his friend John Herschel gave him a copy that he had found during a trip to the Continent. It would appear that no one answered Müller's plea for financial aid and that nothing came of this idea until Babbage independently thought of it about 35 years later. It is not entirely clear whether nobody was actually interested in such a device or the limited circulation of Klipstein's book did not bring it to the attention of the people in need.

The Babbage machines

The machines designed by Charles Babbage (1791–1871) are, of course, the most famous difference engines. It is quite impossible in a short chapter to describe these devices in detail, let alone the full facts that led up to them, so readers are advised to consult several of the books and papers that deal with this subject in depth discussed in the Further Reading. The following will simply touch on a few highlights of the whole story.

The origins of Babbage's ideas are not at all clear. He indicates, in his autobiography, that sometime in 1812 or 1813:

> One evening I was sitting in the rooms of the Analytical Society, at Cambridge, my head leaning forward on the table in a kind of dreamy mood, with a table of logarithms laying open before me. Another member, coming into the room, and seeing me half asleep, called out, 'Well Babbage, what are you dreaming about?' to which I replied, 'I am thinking that all these tables (pointing to the logarithms) might be calculated by machinery'.[5]

This was written 50 years after the reported incident and Babbage admits that he did not remember the occasion but a friend (likely John Herschel) related it to him. It is also clear from Babbage's own writings that he was heavily influenced by the work of de Prony.[6] Whatever the true origin of his interest in table making machines, it was certainly spurred on by the experience he had publishing his famous set of logarithms.[7] He took great pains to ensure these were correct and it must have been obvious to him that other tables could benefit from some form of mechanism that would ensure the accuracy of their calculation. We know that he was working on the preparation of his logarithm tables and simultaneously considering the problems involved in the construction of the difference engine in the early 1820s.

By 1822 he had constructed a working model of a difference engine, capable of evaluating any quadratic to six figures of accuracy. He had shown it to a number of friends and wrote to Sir Humphry Davy, the President of the Royal Society, describing it as: '...producing figures at the rate of 44 per minute, and performing with rapidity and precision all these calculations...'[8]

The purpose of this letter was to get the Royal Society's support for an application to the Government to pay for the construction of a full-sized machine. After asking the Royal Society to comment, the Government agreed and, by July 1823, Babbage had received £1500 with which to begin the work. The Government later added to this amount and kept adding to it, particularly when the members of the Royal Society praised Babbage's efforts.

Babbage had hired a mechanic, Joseph Clement, to help him with the construction. Clement was, perhaps, the best mechanic in London at the time, although he had only recently arrived there from the North of England.[9] This step was both the salvation and ruin of the Engine construction project. Clement was a perfectionist and was capable of the finest workmanship—a talent that was badly needed by Babbage at this stage of his ambitious project. Unfortunately Clement's attitudes also extended to the creation of the most exquisite drawings, creating superb accuracy and finish on pieces that did not actually require it, long careful consideration of the design of each piece, the tools to make it, and the material from which it should be fashioned. As Babbage also possessed some of these same attributes, the two of them were more concerned that it should be done well, rather than it should be done quickly. The Government, of course, had the opposite approach. This led to long periods when the Government would not automatically pay Babbage's accounts, required further reports from the Royal Society, and generally delayed the progress. Finally, after many years, a dispute, over whether Clement should move his workshop into a new fire-proof building on the grounds of Babbage's house, caused the two men to part company and the construction of the machine to come to a standstill. The only parts that were even assembled were, in 1833, as a demonstration piece for the Government and this device now resides in the Science Museum in London.

When Babbage and Clement parted company, Clement exercised his rights under British law and kept all the tools and drawings that had been made for the difference engine. Thus Babbage found himself in the difficult position of being unable to continue with the construction even if he had managed to hire another mechanic. Clement eventually returned the drawings, but by that

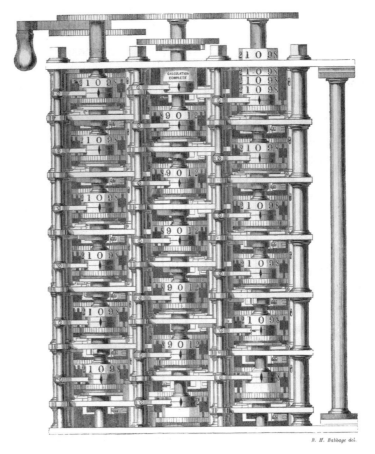

Fig. 5.2 Demonstration piece of the difference engine, 1833.

time Babbage had conceived of the analytical engine and, mainly because of the bad experiences in attempting the construction of the difference engine, vowed never again to venture down that road.[10] However Babbage was mindful of his previous arrangement with the Government and, in 1849, he took time off from his studies of the analytical engine to create a set of drawings for his Second difference engine.[11] He had no intention of ever attempting its construction but produced them to satisfy his own ideas of what he owed to the Government. This was the only fully complete set of plans that Babbage ever produced—all his others were for unfinished versions of his engines.

Had the difference engine ever been finished, it would have been about 10 feet high, 10 feet wide, and 5 feet deep. The front would have shown 7 vertical axles (one for each of the 6 orders of differences and the 7th for the function value) each of which would carry 18 brass wheels engraved with the digits 0–9 to represent the numbers. These axles also contained the carry mechanism. Behind this first row was another set of axles for the adding mechanisms, and behind that was the equipment to engage and disengage the adding devices when required. The Engine was to operate in 4 distinct cycles, each corresponding to a quarter turn of the drive wheel:

1. the numbers stored on $\Delta^1, \Delta^3, \Delta^5$ were added to $F(x), \Delta^2, \Delta^4$;
2. the carries resulting from these additions were dealt with;
3. the numbers stored on $\Delta^2, \Delta^4, \Delta^6$ were added to $\Delta^1, \Delta^3, \Delta^5$;
4. the resulting carries were taken care of.

A printing mechanism would be used to stamp the function value into a soft substance which could then be used to make stereomoulds for printing the tables.

Of course Babbage was justly famous for more than his difference engine and he is especially remembered for his analytical engine and his work in economics. His name appears on several geographic features: a mountain range in Western Australia, a river in the Yukon Territory of northern Canada, and a crater on the Moon.

The Scheutz machines

When Babbage was having difficult times with the Government, Dr Dionysius Lardner, a friend and promoter of science (although not a good scientist himself) wrote an article about the difference engine for the *Edinburgh Review*.[12]

This was mainly done to promote the idea and urge the Government to continue with the financing arrangement, but it also provides us with quite a good view of what Babbage had in mind for the design. Although Babbage supplied the information, Lardner took liberties in the description and changed some of the physical aspects of the machine. One of the major ones was that he described the number storage registers as consisting of a horizontal row of rotating rings. Babbage had actually decided on the vertical arrangement of the numbers. This vertical arrangement resulted in less friction and thus less power being used to effect an addition. Lardner's article happened to be read by a printer from Stockholm, Pehr Georg Scheutz. Scheutz produced a technical journal and thus made a point of reading the latest technical articles from other countries. He was immediately struck with the elegance of the concept and began experimenting to construct his own machine to what he thought was Babbage's design—but wasn't because of the liberties Lardner's had taken in his explanation. His son Edvard, who was a student at the Royal Technological Institute, aided him in this work. Their initial experiments were in wood, but by 1837 they began to make metal parts and could see that a complete machine was feasible. By 1843 they had completed a machine, with only two 5-digit registers and the printing mechanism, and submitted it to the Royal Swedish Academy of Sciences for approval. The Scheutz father and son then asked the Government for a grant to create a full-scale machine and, after some difficulty, they received a small grant enabling them, with professional help, to finish their full-scale difference engine in 1854. It was capable of working to four orders of difference, each number being kept to 15 digits.

When being worked by an experienced operator it could generate 120 lines of a table per hour. In one experiment done by the Scheutz team, it generated the logarithms of numbers from 1 to 10 000 in just 80 hours, including the time taken to reset the differences for the 20 different approximating polynomials.

The Scheutz Engine was brought to London where Babbage gave it his unreserved praise. After staying in London for a time while the Scheutzs looked for a buyer, it was taken to Paris for the Great Exhibition and was awarded their Gold Medal. It was eventually purchased for use in the newly formed Dudley Observatory in New York State. The story of the Dudley Observatory is a classic 'town vs. gown' situation where, in this case, the town won and the research oriented astronomers left for more congenial situations. The Scheutz difference engine, although used to produce a few

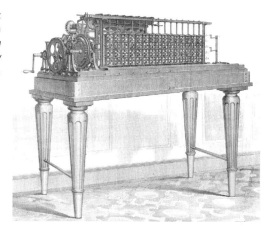

Fig. 5.3 Drawing of the Scheutz engine in Dudley Observatory, from *Report of the Astronomer in Charge of the Dudley Observatory for the year 1863*.

limited calculations (even, at one point, being hooked up to a windmill), was largely ignored and, after 64 years, was purchased by Dorr E. Felt (the inventor of the Comptometer) for his collection of calculating devices. It now resides in the National Museum of American History, Smithsonian Institution, in Washington DC.

In the 1850s the British Registrar General decided that it was time to produce a new set of life tables for use in the growing insurance industry and it was decided that a difference engine should be obtained to aid in their production. This second copy of the Engine was the one machine (other than the work of Comrie to be described later) that can be said to have made a major contribution to the creation of mathematical tables. The story of the use of this machine by William Farr is described by Edward Higgs in Chapter 8.

Wiberg machine

Another difference engine was created as a direct result of the work of the Scheutz team. Martin Wiberg (1826–1905) was a well educated Swede (Ph.D. from Lund University) who, knowing of the Scheutz machine, set about creating his own difference engine. It would appear that his motive was not to produce these machines for sale, but to use them to create tables for a scientific press. He is known to have experimented with several models and, like the Scheutz team before him, asked the Swedish government for money. He appears to have had a champion in Prince (later to be King) Oscar, which may have helped in securing some official financing.

Fig. 5.4 The Wiberg difference engine. (Collection of E. Tomash.)

A Wiberg machine was sent for exhibit at the Great Exhibition in London in 1861 but arrived too late to be shown. It was, however, examined by Babbage and he wrote letters of reference for Wiberg to introduce him to French scientists which resulted in the machine being examined by three members of the French Academy of Sciences who published their report in 1863.[13] They noted that a set of interest tables[14] had already been calculated and proposed that the Academy give its approval to this beautiful and ingenious machine.

Wiberg certainly attempted to produce other tables using this device. He is known to have calculated a set of logarithm tables that could have been published shortly after his Paris visit but were delayed while he worked out problems with the printing. It is not entirely clear if these problems were associated with financing or mechanical difficulties but it resulted in a 13 year delay in their appearance in print.[15] As noted by R. C. Archibald: 'The result, involving an attractive face of type, with top and bottom tails, and of different sizes, is certainly a great advance over the product of the Scheutz machines, some 20 years earlier.'[16]

Wiberg exhibited copies of the logarithm tables in the Centennial Exhibition in Philadelphia in 1876. They do not, however, appear to have been a successful publication and today they are very rare.

Grant's difference engine

George Bernard Grant (1849–1917) was the son of a ship builder. He was born in Maine, attended Dartmouth College and, in 1869, entered the Lawrence Scientific School at Harvard College. Because of his previous experience at Dartmouth, he was able to complete the normal four-year program in only three years and still have time to think about the construction of calculating machines. He was unaware of the work of either Babbage or Scheutz but knew of the existence of simple mechanical calculators. During his first year at Harvard he had to laboriously calculate a 'cut and fill' table and this prompted his first thoughts about building a machine to do the same job. After spending some time contemplating the difficult mechanical movements it would require, he gave up the study of such a machine and returned to manual methods of table construction.

About 1870 he learned of Babbage and his difference engine and this inspired him to return to his own and he 'designed a machine that might possibly have worked, but I could convince nobody that it would do so, and give it up again'.[17] At the urging of some of his professors, mainly Wolcott Gibbs,[18] and John Bachelder,[19] he again returned to the design and 'Though I have built no large machine, the efficiency of the design for its purpose may be considered as having been proved, as through the liberality of the superintendent of the Coast Survey, Prof. Benjamin Peirce, I have been able to build a model of small capacity, which had worked to satisfaction'.[20]

After graduation he continued to experiment with designs of calculating machines and took out patents relating to them in 1872 and 1873. Once again Wolcott Gibbs came to his aid and arranged financing for the construction of a full-size machine, which was exhibited at the 1876 Centennial Exhibition in Philadelphia. The report of the Exhibition Commissioners described it as follows:[21]

This machine for computing and tabulating mathematical tables, such as logarithm, sines and tangents and their logarithms, reciprocals, square and cube roots, etc., possesses great merit for the simplicity and certainty of its action and for the ease with which its parts can be adapted to the requirements of any given function...is arranged to combine and print functions involving one hundred elements. The combination of the several parts is extremely simple; the number of elements can be indefinitely increased, and the machine acts with the greatest certainty.

While the above quotations certainly imply that the machine worked properly, another description indicated that it was not a completely finished product. Perhaps it is not surprising that it was not fully functional considering that it had such a short gestation from its conception and was, evidently, only finished a few days prior to the opening of the Exhibition:

It occupies a space of about five feet in height by eight feet in length, and weighs about two thousand pounds, containing, when in full working order, from twelve thousand to fifteen thousand pieces. The long body contains the calculating mechanism, while at the front end is an apparatus for printing a wax mould of the results... The machine is driven either by hand, by a crank at the front end, or by a power appliance at the rear end... When the machine is worked by hand, a speed may be made of ten to twelve terms per minute, and from twenty to thirty when by power by the attachment at the rear end. All that limits the speed is the imperfection of the mechanism, and in the case of the present machine—the first ever constructed of so complex a character—imperfections are to be expected which will not exist in future machines. Thirty of the elements of this machine were placed in a light wooden frame and worked successfully at a speed of over one hundred terms per minute, and if the whole machine were used and sufficient power applied, this speed would be perfectly predicable, provided that the mechanism of the driving-gear and printing apparatus were in accurate working order and made sufficiently strong to stand the wear and tear resulting from the same.[22]

Grant's Engine was, under the conditions of the financing arrangements, to be donated to the University of Pennsylvania when the Exhibition was over. In fact there appears to be some mystery as to its eventual fate. Most of the exhibits were given to the Smithsonian Institution (and formed the basis of their museum collections) but they have no record of ever having received Grant's machine. Neither can it be traced to the University of Pennsylvania. The most reasonable explanation for its disappearance is likely contained in the 'imperfections' hinted to in the above quotations. It is quite likely that Grant simply took it back with the intention of putting it into proper order. According to one biographer: 'Shortly after graduation, as a result of his calculating-machine work, he started a machine-shop for

Fig. 5.5 Grant's difference engine on exhibit at Philadelphia, from *The masterpieces of the Centennial International Exhibition*, 1876.

gear cutting... and became one of the founders of the gear-cutting industry in the United States.'[23]

His efforts at founding a business likely caused him to put the difference engine aside and he simply never got around to cleaning up the final details. The same biographer indicated that he continued his interest in the area and 'during the last years of his life he conducted considerable experimental work in connection with the development of such machines'. There is, however, no hint as to whether this work was concerned with difference engines or simply with the four-function mechanical calculators he was also known to have designed and sold.

Hamann machine

By the beginning of the twentieth century the technology of mechanical calculating machine was already well advanced with commercial versions of several different types readily available. The German engineer Christel Hamann (1870–1948) was in the forefront of this technology and had invented a number of devices, particularly the so-called 'proportional lever' which was used in the Mercedes Euklid automatic multiplication machines.

At about the same time, Julius Bauschinger (1860–1934) and Johann Theodor Peters (1889–1941) began a project to construct a new set of 8-place logarithm tables of both the natural numbers and the trigonometric functions. Most of the well-regarded tables of the day only contained 7 places of decimals and these were found to be inadequate because of the recent advances in astronomic observations and geodesy. At the time of this project Julius Bauschinger had just left his position as Director of the Astronomical Research Institute in Berlin and had become Director of the Imperial Observatory at Strasbourg. He was thus an important figure in astronomy and appreciated the usefulness of logarithms to all branches of science. Johann Peters was a professor in the Royal Astronomical Calculating Institute in Berlin and an expert on methods of table production. Archibald describes Peters as 'perhaps the greatest mathematical table maker of all time'.[24] Simultaneously with the start of hand calculation, they asked Christel Hamann about the possibility of constructing a difference engine to help with the work. The resulting machine is illustrated in the frontispiece of the two-volume set of logarithm tables[25] and a detailed description of its internal workings can be found in the first volume's preface. Unfortunately the machine itself has been lost.

The Hamann difference engine was a great deal simpler than any of the previously described devices. Because of the way Bauschinger and Peters had set up the work, it was only necessary to use the machine to interpolate between short intervals, a task which could be mainly accomplished using only the second differences. The machine consisted of two independent arithmetical mechanisms and a simple printer. In order to ensure 8-digit accuracy in the results, the machine could both manipulate and print 16-digit numbers. It was not automatic in its operation but required the user to enter initial values and then repeatedly turn the two cranks to add the second difference to the first and then the first difference to the function value. When used by an experienced operator, the calculation of 36 table entries would take only 5 minutes, including the time taken to enter the initial values.

Hamann delivered the machine in 1909 and it was immediately put to work. The resulting 2 volumes of tables were published in 1910 and 1911.

Other minor difference engines

A number of other difference engines are known to have been constructed. Among them was one built by Alfred Deacon of London. It appears that Deacon read the same article by Lardner that inspired Georg Scheutz. He began to experiment with various mechanical devices and eventually produced a small difference engine that would operate with three orders of differences, the numbers being each of 20 digits.[26] It is entirely possible that Charles Babbage later acquired this machine (which is now apparently lost) because he offered the Great Exhibition of 1862 '...a small difference engine, made in London, in consequence of its author having read Dr. Lardner's article in the *Edinburgh Review*'.[27]

Many other devices were designed to help people with calculations using the method of differences. Several versions of the famous Brunsviga calculators were produced that were specifically designed for difference methods. In one, the Brunsviga Dupla, there were two result registers with the ability to transfer their numerical contents back on to the setting levers—such a machine could be easily used to calculate tables from their first or second differences. Perhaps the most ambitious of these devices was constructed by A. J. Thompson who mounted four standard calculating machines in a wooden platform and modified them so that the result from one machine could be transferred mechanically to the setting levers of the machine below. He used this aid when calculating a new set of 20-place logarithm tables[28] published in the early 1950s.

Leslie John Comrie (1893–1950) and the National Accounting Machine

Leslie Comrie was one of the great figures in table making of the twentieth century. Born in New Zealand, he attended Auckland University and graduated with a degree in chemistry in 1916. After graduation he joined the army (even though he was very deaf) and was sent, in 1917, as part of the New Zealand Expeditionary Force to fight in the French trenches of the First World War. His military service came to an abrupt end when, in February 1918, he had his left leg blown off by a British shell. Remarkably, by the Fall of that same year he had enrolled in University College London where he was first introduced to machine calculation by using a Brunsviga (a pin-wheel type machine of the original Odhner design). His acquaintanceship with massive calculation problems developed into a life-long passion when he became a graduate student in Cambridge.[29] After a short spell of teaching in the USA, he returned to Britain to work in the Nautical Almanac Office.

At the Nautical Almanac Office Comrie quickly developed a reputation as an expert in all aspects of mechanical calculation and his advice was often sought on the best uses of various new machines. He would obtain copies of each new machine as it appeared, usually take it apart, and then produce a description of which jobs were best suited for its use. It was in 1931 that he examined the National Accounting Machine Class 3000 (produced by the National Cash Register Co. but also sold under the name Ellis) and noted that it contained a number of extra registers which were designed to be used for the standard accounting sub-total applications. The interesting thing about these extra registers was that they were not simply used for temporarily storing results, but they could be used as the main accumulators for the device. Different registers were activated (by metal clips attached to the printing carriage) as the carriage moved various columns of figures (debits, credits, etc.) in the account sheet over the printing mechanism.

Fig. 5.6 L. J. Comrie checking a set of tables.

The different registers could also be brought into a calculation by depressing keys on the left of the machine. Comrie realized that this automatic selection of which register was active at any given moment could be put to good use to make the machine into a difference engine operating with six orders of difference.

While Comrie certainly used this device in making some of the many tables he produced in both the Nautical Almanac Office and in his later private ventures, its main help was in using difference methods to check older tables for accuracy. A great many errors in earlier tables were found from Comrie's use of the National as a difference engine and were reported in *Mathematical tables and other aids to computation* and elsewhere.

It is interesting that Comrie, legendary table maker and champion of the use of mechanical equipment, was highly sceptical of the use of large computing machines for the production of tables. While recognizing that tables of all kinds would still be of use for some years, he thought that the production of tables by Aiken's Harvard Mark I was simply a waste of effort.

In 1936 Comrie left the Nautical Almanac Office over a dispute about the use of the equipment and staff for non-governmental table making, and founded the world's first computing service bureau, Scientific Computing Service, which had a very significant influence on the British computing scene. He suffered a series of strokes in late 1949 and 1950 and died in December 1950 at age 57. In recent times a crater on the far side of the Moon and an asteroid have been named in his honour.

Were the difference engines important?

There is, of course, one other difference engine that must be mentioned. In 1991, on the 200th anniversary of Babbage's birth, the London Science Museum unveiled its construction of Babbage's second difference engine. This was created from the design left by Babbage in 1851 but constructed using modern manufacturing methods. Every attempt was made to use materials that might have been available to Babbage and to ensure that the individual parts were no more accurate that Joseph Clement could have managed in the mid-1800s. Recently a printing unit, again to Babbage's design, has been added. The result certainly proves that Babbage's design works. It has been used to tabulate complex polynomials and, although it must be treated with care, is certainly up to the task.

The fact that several different working difference engines were created in Babbage's lifetime (and shortly after) using technology that was a great deal simpler (and smaller) than that employed by him, certainly helps to show that Babbage's ideas were rather more than what was necessary at the time. Only one of these machines—the second copy of the Scheutz machine employed by Farr—was heavily used, with great difficulty, in practical situations. That, too, gives credence to the idea that Babbage's designs were more advanced than was required by the scientific community of the day. Many of Babbage's contemporaries have been vilified because of their statements that the money spent on his devices was a waste of resources. I think that the marginal use made of difference engines for subsequent tables helps to confirm that these comments were not all misguided. Of course when one takes into account the other manufacturing advances that came from the efforts of Babbage, Clement, and his workmen, then the balance sheet looks a little better. However, as stated in the introduction, no one can claim that difference engines were a major factor in the production of mathematical tables.

Further reading

When attempting to find further information on any difference engine, it is always best to read the original sources where possible. The works by Charles Babbage are often available in large libraries and many of his writings, particularly his autobiography *Passages from the life of a philosopher, 1864 (numerous reprintings)* contain references to his engines. Another interesting source is the book by Babbage's son H. P. Babbage, *Babbage's calculating machines,* London, 1889. Perhaps the best modern work on Babbage's machine is Doron Swade's *The cogwheel brain,* Little, Brown, and Co., 2000. Swade was the curator at London Science Museum which constructed Babbage's difference engine No. 2 from the original plans. This book not only gives lots of detail on Babbage and his machine, but chronicles the interesting story of the Museum's efforts to produce the first of Babbage's large-scale machines.

The complete story of the Scheutz difference engines can be found in Michael Lindgren's *Glory and failure,* MIT Press, Cambridge, Mass., 1990. This highly readable work not only deals with the Scheutz machine, but covers the difference engines created by Müller and Babbage as well. Unfortunately the majority of the creators of difference engines did not feel it was worth documenting them in ways that would be easily available to the modern reader, so one must rely on the work of historians who have had the privilege of researching these topics in depth. There is little material easily available on the difference engines of George Grant other than his original papers noted in the references. The work of Martin Weiberg is similarly scattered throughout short items, often containing no more than a brief mention, in the works noted in the references.

On the other hand, we are fortunate that the best account of the work of Comrie is contained in Mary Croarken's fine book, *Early scientific computing in Britain,* Oxford University Press, 1990. She not only details the life and accomplishments of Comrie (including a description of how the National 3000 Accounting Machine was actually used), but does an excellent job of discussing all forms of computing, both analog and digital, up to the creation of the digital computer in both Manchester and Cambridge.

Notes and references

In the notes below 'WB' refers to *Works of Babbage,* edited by Martin Campbell-Kelly (11 vols, Pickering & Chatto, London, 1989).

1. E. Klepstein, *Beschreiburg seiner neu erfundenen rechenmaschine,* Frankfurt, 1786.
2. In 2003 a copy was offered for sale at a price of $20,000.
3. M. R. Williams, 'The scientific library of Charles Babbage', *Annals of the History of Computing* 3 (1981), 235–40.
4. For a discussion on this point see M. Lindgren, *Glory and failure: the difference engines of Johann Muller, Charles Babbage, and Georg and Edvard Scheutz,* MIT Press, Cambridge, Mass., 1990.

5. C. Babbage, *Passages from the life of a philosopher*, Longman and Co., London, 1864, p. 42 WB, XI, 30–31.
6. See for example, C. Babbage 'On the division of mental labour', *Economy of machinery and manufactures*, London, 1835, 191–202. WB, VIII, 135–43.
7. M. Campbell-Kelly, 'Charles Babbage's table of logarithms (1827)', *Annals of the history of computing* 10 (1988), 159–69.
8. H. P. Babbage, *Babbage's calculating machines*, E. Spon and Co. London, 1889.
9. M. R. Williams, 'Joseph Clement: the first computer engineer', *Annals of the history of computing* 14 (1992), 69–76.
10. M. R. Williams, 'Babbage and Bowditch: a trans-Atlantic connection', *Annals of the history of computing* 9 (1987), 283–90.
11. Actually his third, if you count his initial trial model from 1823 as being the first.
12. D. Lardner, 'Babbage's calculating engines', *The Edinburgh review* 59 (1834), 263–327. WB, II, 118–86.
13. *Comptes rendus*, 56 (1863), 330–9.
14. M. Wiberg, *Med Maskin uträknade och stereotyperade Ränte-tabeller, Jemte en Dagräknings-Tabell*, Stockholm, 1860.
15. M. Wiberg, *Tables de logarithmes calculées et imprimées au moyen de la machine à calculer*, Compagnie anonyme de Forsete, Stockholm, 1876.
16. R. C. Archibald, 'Martin Wiberg, his tables and difference engine', *Mathematical tables and other aids to computation* 2 (1947), 371–73.
17. G. B. Grant, 'On a new difference engine', *American journal of science, third series*, August 1871, 113–17.
18. Wolcott Gibbs, known as the 'greatest chemist in America' during the nineteenth century, was the Rumford Professor and Lecturer on the Application of Science to the Useful Arts from 1863 to 1887.
19. John N. Bachelder had been the man in charge of the Scheutz machine at the Dudley Observatory and was thus one of the few who had practical experience with the operation of such a machine.
20. Grant, 'On a New Difference Engine'.
21. Reports and Awards, US Centennial Exhibition, 1876, Volume 7.
22. J. M. Wilson, *The masterpieces of the Centennial International Exhibition illustrated . . .*, Philadelphia, 1876, vol. 3, pp. 27–32.
23. L. L. Locke, entry for George Bernard Grant, *Dictionary of American Biography*.
24. R. C. Archibald, *Mathematical table makers*, Scripta Mathematica, New York, 1948.
25. J. Bauschinger and J. T. Peters, *Logarithmisch-trigonometrische tafeln mit acht dezimalstellen enthaltend die logarithmen aller zahlen von 1 bis 200000 und die logarithmen der trigonometrischen funktionen für jede sexagesimalsekunde des quadranten*, 2 vols, Wilhelm Engelmann, Leipzig, 1910–11.

26. G. and E. Scheutz, *Specimens of tables calculated, stereomoulded and printed by machinery*, privately published, London, 1857. WB, II, 194–222.
27. H. P. Babbage, *Babbage's calculating machines*, 197.
28. A. J. Thompson, *Logarithmica Britannica*, 2 vols, Cambridge, 1952.
29. Comrie was awarded a Ph.D. in astronomy in 1924.

Fig 6.1 (a) Difference Engine No. 2, experimental printout for a seventh-order polynomial. Argument is left-hand three figures.

Fig. 6.1 (b) Difference Engine No. 2 printing mechanism, showing inking rollers and ink bath (upper); paper drums in raised position taking an inked impression from the printing wheels.

6

The 'unerring certainty of mechanical agency': machines and table making in the nineteenth century

DORON SWADE

When machines are mentioned in the context of table making it is their role as calculators that tends to be emphasized. The mechanization of other processes, specifically transcription, typesetting, proof reading, printing, binding, and distribution, have received less attention perhaps because they are seen to be part of book production and not distinctively belonging to table making. The theme of this chapter is the interplay between machines and the wider processes involved in table making.

The successful transition from manual to machine-produced tables is largely a twentieth-century phenomenon. During the first half of the century electromechanical punched card equipment and electrically driven mechanical calculating devices were developed and pressed into service.[1] In the second half of the century electronic technologies brought new levels of speed, certainty, and convenience. The need for reliable computation and permanent printed record was met by computers and calculators with on-line printers, and portable devices provide calculation at the place and time of need. In the electronic era of the twentieth century the printed table as a computational aid, portable or desk-bound, largely, though not entirely, vanished. Paradoxically the technology that conclusively solved problems of reliability in the production of printed tables all but eliminated their need.

As material artifacts volumes of printed mathematical tables fall into the category of 'books' and as such the history, economics, and technology of their production is shared with the bibliography of literary texts. One clear respect in which printed tables differ from literary texts is that their content (apart from prefaces) is essentially numerical not alphabetical. The distinction is not a pedantic one. The originating agency of literary texts is irreducibly human and, while tables can be viewed simply as numerical texts, they differ from literary texts in at least one fundamental respect in that the numerical content of many tables can be generated by machines that embody mathematical or computational rules. So there is a dimension to the role of machines in the bibliography of table making that is distinctive and unshared by literary text: the mechanization of authorship. It is perhaps because of this distinction that the starting point and ultimate emphasis in discussions about technology and table making tend to centre on calculation at the expense of processes generic to book production.

The mechanical calculating aids devised by Schickard, Pascal, and Leibniz in the seventeenth century feature routinely in histories of computation, as do those of the eighteenth century, notably by Hahn, Stanhope, and Müller. The seventeenth century devices were ornate curiosities, *objets de salon*, unreliable and unsuited to daily use. Those of the eighteenth century, Hahn's and Müller's in particular, are extravagantly ornamented testaments to the instrument makers' art. They worked, but were expensive, and few were made.

The device often billed as the first to be made commercially for general sale is the arithmometer introduced by Thomas de Colmar in 1820. However, it took over half a century for de Colmar's calculator to emerge as a product robust enough for daily use and it was not until the latter decades of the century that they sold in large numbers.[2] Because of their unreliability alone arithmometers had practically no impact on mathematical table making until the last decades of the nineteenth century. By the time the arithmometer had matured as a product, desktop machines, marketed under various names of which Brunsviga is amongst the best known, came on stream.

It was not only the time it took for such devices to evolve from erratic novelties into reliable workhorses that inhibited their use. Even reliable manual calculating devices have limited use in table making. The execution of single arithmetical operations such as the multiplication of two numbers using an arithmometer is not entirely automatic but relies on the continuous informed intervention of the operator. Multiplying two numbers involves entering the digits on sliding dials, rotating a handle the correct number of times for each decade, and correctly lifting and repositioning the moveable carriage for each digit of the multiplier.[3] The correctness of the final result therefore relies not only on the repeated correct mechanical functioning of the device but on the faultless execution by an operator of a sequence of manual procedures which surround and intervene in the process: the correct input of data, the correct physical operation of the machine, and the error-free transcription of results. In this and other late nineteenth century examples the supposed infallibility of mechanism provides at best security against error only for part of the overall process. In the case of the arithmometer this amounts to the calculation of products of pairs of numbers once entered, and for their cumulative summation.

The use of mechanical calculators is further circumscribed by the place of computation in the overall process of table making. Because of the exacting labour of calculation and checking, and anxieties about accuracy, computing tables from scratch was avoided whenever possible. Instead, reputable existing tables were used as the starting point. These were rechecked, reprinted, merged, extended, revised, reduced in the number of significant digits, or otherwise modified for new editions to meet new needs. Recomputation of existing tables was exceptional and computation *ab initio* was only resorted to when new tables were needed. Perhaps more importantly, not all tables originate in the calculation of mathematical functions. Tables of observational data, for example, often owe little or nothing to calculation as an originating process and represent a class of table that invokes many generic aspects of book production. So while the ability to machine-generate calculable tables is a distinctive feature of certain classes of numerical tables, this feature is not a defining one for the genre.

Printing and stereotyping

Until the 1970s the primary medium of transmission and dissemination of the informational content of tables was print and paper. In the nineteenth century table production relied on the same well-established trades and practices as literary book production i.e. typesetting, proof reading, printing, binding, and distribution. The success of early attempts to automate manual typesetting were neither immediate nor complete. The first patented cold-metal composing machine for typesetting was by William Church in 1822. This machine, like all its cold-metal successors, held ordinary type in magazines above the machine, and these are released one at a time by an operator using a keyboard. The machines were fast, but justification (the alignment of text to produce straight margins at the edges of the page) was still manual. Cost savings on labour were not great and few nineteenth-century printers used them. Those that did used them primarily for periodicals and newspapers where speed was more important than the niceties of typography.[4]

The process most directly affected by machinery was printing and paper making. The manufacture and economics of paper production changed radically over the century.[5] The expense of paper as a proportion of overall book production costs dropped significantly. In 1800 all paper was made, expensively, by hand. By 1900 more than 99% of paper was machine made, and prices had dropped tenfold.[6] Major influences here were the sourcing of rags used as the raw material for paper, the abolition of excise duties, and the effect of automated paper making machinery. During this time the efficiency of printing presses increased dramatically as machine presses replaced manual presses. While the fundamental physical process of inked impression on paper as an inscriptional record remained unchanged, throughput increased and labour costs dropped.

Perhaps the most significant development for table makers was the practice of stereotyping which involved making printing plates (stereotypes) from casts taken from the type already set. Plaster-mould stereotyping was used in literary book production from the 1780s and continued in use till the early twentieth century. The use of 'flong' which came into general use in England in the 1850s ran in parallel with plaster-mould stereotyping. Flong is laminated paper made from alternate layers of blotting paper and tissue paper bonded together. Dampened flong was beaten onto the face of the type, dried and hardened by heating *in situ* and the mould so formed was then used to cast metal stereotype plates for use in conventional printing presses.[7]

Stereotyping had particular attractions for table makers. Stereotypes represented an immutable form of information capture that offered immunity from the inherent vulnerability of moveable type to derangement during printing or storage. Particular editions of tables acquired their cachet for reliability over decades and reputable volumes of early tables remained in demand sometimes for centuries. Type, the costliest part of a printer's equipment, was rarely retained in its set form and it was common practice to redistribute type for reuse after the edition was out, and reset type from scratch for later editions of the same work. However, resetting type broke the line of integrity with the original edition. Augustus de Morgan, whose annotated bibliographies of tables are a rich source of contemporary attitudes and practice, observes that

> a second edition derives no authority from the goodness of the first, because the printer, who is, as already observed, as important a person as the author in the matter of tables, has again stepped between the latter and the public.[8]

Storing the stereotypes provided an economical way of preserving the investment in typesetting and proof reading, and extended confidence in highly reputed tables to subsequent editions.

The benefits of stereotyping to printers were less clear cut and the introduction of the technique suffered the prejudices of compositors and type founders protective of their trades. Resetting type for a new edition from scratch was welcome new work, and tying up expensive type, once set, in storage for future editions increased trade for type founders. For the printers, committing expensive type to storage represented unused capital and few books were kept set up in type as a result. Stereotyping released type stock for reuse in new work. A more subtle benefit to the printer of the reuse of type was the distribution of wear. Continuous redeployment of type spread wear more uniformly across the stock, prolonged its useful life, and deferred the costs of replacement raising yields on capital.[9] For a combination of economic and information management reasons stereotyping became the preferred technique for reducing the risk of corruption in the press and keeping standard books in print. For table makers stereotyping became a hallmark of accuracy and in bibliographies of tables, if an edition was the product of stereotyping, this was noted as an implied recommendation. So strongly did de Morgan feel about the benefits of stereotyping tables that he recommended enforcement of its use.

Calculating and transcribing

Humans are fallible and the risk of error was a central and continuing concern for table makers. The appeal of machinery was the promise of certainty. But the essential informational processes—calculation, transcription, verification, typesetting, and proofreading—continued to rely on the mental and physical labour of people. Errors were an unavoidable fact of life and vigilance alone was insufficient to prevent them. Charles Babbage commented that errors occurred 'in works where neither care nor expense were spared' but he exculpates the table makers. He wrote that it was 'but just to the eminent men who presided over the preparation of these works... to observe that the real fault lay not in them but *in the nature of things*'.[10]

Before turning to one of the most ambitious attempts to mechanize tables production it may be useful to review the separate manual processes involved and their respective vulnerabilities to error. In the production of mathematical tables, calculation was typically divided between at least two distinct groups, mathematicians and computers. Mathematicians chose the formulae, fixed the range of the table (the start and end values of the argument) and decided the decimal precision. Instead of substituting incrementally increasing values of the argument into the mathematical formula and redoing the calculation to produce each new tabular value of the function, interpolation techniques were used to find intermediate values between more widely spaced pivotal values.

A common technique for such subtabulation used the method of finite differences in which each next tabular value is generated by a series of additions from a precalculated set of starting differences. Most major non-polynomial functions can be approximated in a sufficiently small domain by a polynomial expansion and this allows them to be tabulated using finite differences.[11] The advantage of the method is that it reduces the arithmetical process to simple repeated addition and avoids the need for multiplication and division that would ordinarily be needed to evaluate the function. One benefit of the technique is that it deskills the calculation process and allows the repetitive work to be carried out by computers with only rudimentary arithmetical skills.[12] In cases where subtabulation using finite differences was used the mathematicians also calculated the pivotal values and the set of starting differences for the computers. Pivotal values were chosen sufficiently far apart to reduce their overall number (and so transfer the greater burden onto the computers), and close enough to ensure that the interpolated values stayed true to the requisite number of end digits. Subtabulation runs of one hundred to two hundred values between pivotal values were not uncommon.

Repetitive calculation was a hideous drudgery and the prospect of using mechanical aids to relieve the worst of its burdens features as a motivational drive in some of the earliest attempts to use machines. In 1685 Leibniz wrote that 'it is unworthy of excellent men to lose hours like slaves in the labour of calculation which could be done by any peasant with the aid of a machine'.[13] The reference to 'excellent men' and 'worthiness' implies a hierarchy of values in which abstract, analytical and philosophical activity is ranked higher than repetitive task-specific activity. The contrast between 'excellent men' and 'peasants' or 'slaves' also reflects a class difference which pre-echoes Babbage's description of de Prony's three groups amongst whom the labour was divided, as 'classes' with the computers referred to as the 'third class' performing the 'lowest processes of arithmetic' and ordinality providing a scale of superiority.[14]

The numbing grind of routine calculation is described with vividness and conviction in contemporary writing: 'the intolerable labour and fatiguing monotony of a continued repetition of similar arithmetical calculations'; 'that wearisomeness and disgust, which always attend to monotonous repetition of arithmetical operations'; 'the dull and tedious repetition of many thousand consecutive additions and subtractions'; the 'mental drudgery' of constructing tables.[15] Luigi Menabrea, at the time an engineer, who attended Babbage's seminar in Turin on the Analytical Engine in 1840, wrote of the stultifying effect of such drudgery on higher thought. 'And what discouragement', he asks, 'does the perspective of a long and arid computation cast in the mind of a man of genius'.[16] In the absence of machines, mathematicians and scientists were more than happy to offload routine calculation onto the computers.

The burden of calculation was further increased by the need to work to a larger number of decimal places than required for eventual printing. One reason for this is to ensure that rounding errors did not accumulate during the successive additions to the point at which they affect the least significant digit. If fifteen significant digits were needed in the answer the calculation would often be carried out to twenty or twenty five decimals.[17] The generic precision of tables for some applications was in any event high. De Prony's cadastral table of natural sines was tabulated to twenty five decimal places, and one set of logarithms to nineteen places. The inclusion of additional digits to ensure correctness to large numbers of places compounded the volume of routine numerical calculation as well as increased the burden on the mathematicians who were responsible for specifying the number of such 'guard digits' sufficient to ensure the integrity of the end digits in the printed version.

Checking

With the calculations complete manuscript results were checked for errors. A standard technique was for the same set of calculations to be performed by different computers without collaboration, and verification took the form of consistency checks performed on the separate outcomes. If each had done their work perfectly the separate sets of results would be identical. The effectiveness of the technique relies on the independence of the computers from each other. In 1770 two computers were instantly dismissed when, hired to assist in the preparation of the *British Nautical Almanac*, they were found copying from each other.[18]

The technique of double computation was not foolproof and it was not unknown for computers who, despite insulation from each other, produced the same incorrect result. Such occurrences would elude detection using coincidence checks. Dionysius Lardner, a prolific popularizer of science, moralizes with characteristic puff that 'falsehood in this case assumes that character of consistency, which is regarded as the exclusive attribute of truth'.[19] He reports that on occasions de Prony found 'three and even a greater number of computers, working separately and independently, to return him the same numerical result, and *that result wrong*'.[20] To further reduce the risk of error the separate computers were sometimes given computationally different but mathematically identical formulae to find the same result.[21] The intention here is to disrupt shared habitual mental patterns or common psychoperceptual tendencies originating this class of error.

The technique of subtabulation using differences has self-verification features which assisted error detection and lightened the burden of verification. Since each value is used in the calculation of the next tabular value it follows that the last value depends on all prior values. The task of verification can therefore be reduced to a single final-value check. This applies to manual methods as well as subtabulation by automatic machines.[22] The pivotal values provide a further check. If subtabulation is continued one value beyond the end of the run then the new tabular value should be the same as the next pivotal value. Since the pivotal values are calculated independently the coincidence of the two values provides confirmation of correctness. Finally, there is verification by differencing i.e. by reversing the process of subtabulation by repetitive addition (numerical integration by differences) and subtracting successive tabular values already calculated. Lower order differences can sometimes be checked by visual inspection and mental arithmetic and any deviations signal that something is amiss. The technique is very sensitive. Small errors produce wild fluctuations in the higher order differences.[23]

The two independently computed sets of de Prony's grand *Tables du Cadastre* survive in manuscript and a visual comparison gives an indication of the frequency and nature of corrections made to reconcile the two independently calculated sets or results. The first observation to make is that the number of corrections is very large and it is clear from this that the contribution of errors from the combined processes of calculation and transcription was substantial. There are many corrections to single entries by overwriting and these are episodic i.e. their distribution does not follow any evident pattern. By far the largest number of corrections are to systematic errors. Typical of this category is the overwriting of one digit in the same position for all entries on the same page. A common form of this would be say to increment all the third digits in a column of entries by '1' down the whole page. In an extreme case a run of forty seven consecutive double-page openings has either two or three columns of each page overwritten in one digit position incrementing the original digit by '1'.[24] Without the computers' worksheets it is impossible to tell whether these are systematic calculation errors from the method of differences correctly transcribed, or systematic transcription errors from correctly calculated values. Overall, the profusion of corrections confirms the wisdom of duplicating the calculations and of verifying by consistency checking as insisted on by the table makers.

Typesetting

The corrected and transcribed manuscript results were then given to the printer where the first process was to set the results by hand using moveable type. Here the compositor reads from the authored source and sets each digit of each result in loose type to form the groups, lines, and blocks of numbers to make up the page. Compositors retrieved individual digits from boxes filled with samples of the same character and the correct selection of type is guided by the habitual action based on the location of the box in the matrix of boxes. Selection depends on the position of the box in the matrix and in general an individual piece of type is not inspected as it is used. Errors can arise by retrieval of characters from the wrong box (mental lapse) or where the wrong species of type is distributed in the boxes before retrieval ('foul case').

It would seem that typesetting numbers rather than language text would pose special difficulties in that while groups of letters make up recognizable words, numbers have no immediate meaning, and the compositor has no intuitive sense of whether one digit has any sensible relationship to the one before or after. Surprisingly, typesetting does not appear to have been a source of great anxiety to table makers. Augustus de Morgan comments on how, in the case of literary texts, a reputable compositor will improve the integrity of a manuscript source and professes surprise at the accuracy with which 'first-rate London printers can turn out their proofs, even where the manuscript is criminally bad'.[25] It is clear that the high standards of accuracy for literary texts extended to numerical tables. Indeed the standards of top printers appear to have been disconcertingly high, to the extent that the absence of error led to a sense of insecurity about the quality of error checking:

We have frequently looked at page after page of table-matter more times that we should otherwise have thought necessary, merely because the total absence of detected error left it an unsettled point whether it was the excellence of the proof, or a temporary suspension of our own quickness of perception, which caused the absence in question.[26]

Proof reading

The first stage of printing was the production of proof sheets for checking against the manuscript. Proof reading is an exacting and tedious process and this was reflected in the high pay rates for such work.[27] It is curious that while meticulous care is urged for each stage of the process, the most rigorous controls are reserved for checking the stereotyped proofs rather than for the proofs taken from results set in moveable type. De Morgan urges that 'the strictest investigation should take place in the proof which is taken from the stereotype, ordinary pains being taken with the previous proofs'. The reason for this uncharacteristic relaxation of the highest standards of care throughout is that however scrupulous the checking of proofs from type, the integrity of the printed result depends on the moveable type remaining entirely undisturbed during the whole printing process and there were clearly anxieties about the vulnerability of type to derangement between the printing of the first proofs and producing the stereotype plates. One hazard was the 'drawing of type' which occurs when the ink is too thick or the type not secure and the adhesion between the type and the inking roller or leather inking balls draws the type from the frame. It was not unknown for type to simply fall out and be replaced incorrectly. A further reason why checking stereotype proofs attracted superordinate priority was that any errors found could still be corrected in moveable type, whereas the stereotype, once cast, was amenable to only minor correction.

For any substantial correction the mould of the whole page would need to be redone after correction of the type.[28] Checking proofs from the stereotypes was the safeguard of last resort, and this is reflected in the variety, subtlety and ingenuity of the practices and ploys devised to counter fatigue and ensure effectiveness in error detection.

De Morgan recommends a checking process involving three people, one to read aloud from the manuscript, the other two to check, without collaboration, identical printed proof copies from the stereotypes. He urges a *modus operandi* using multiple listeners where possible, but offers special advice for solo checking in cases where the luxury of multiple listeners is unaffordable. He counsels that the manuscript and the proof copy are brought into close proximity so that the two numbers being compared are contained in the same unbroken field of view, and he disdains physical wear on the manuscript recommending that it is folded as frequently as every two or three lines 'so as always to have both manuscript and proof under the eye in one position'. The wisdom of this practice is articulated by Glaisher who writes that 'it is well known that the number of errors ... is proportional to the distance the eye has to carry the numbers'.[29] De Morgan warns against the tendency to mistake double figures (744 for 774, for example) and against transposition (012 for 102). If either proof or manuscript is harder to read, he suggests reading the easier first as 'for the mind is apt to allow knowledge derived from the more easy to give help in interpreting the more difficult'. Alternating the datum for each reading is advised i.e. reading a result from the manuscript, making the comparison with the proof, then taking the next reading from the proof and so on. Different checkers appear to have had different preferences, some favouring visual checking in silence, others by vocalising the numbers, combining ear and eye.[30] Whatever the preference, the checker is advised to break patterns to relieve tedium. Shifting bodily position, moving hands and feet, and varying the pitch or tone of voice in the spoken repetitions, are recommended best practices.

Another technique used, especially for inexperienced checkers, was the use of 'author traps'. This involved the printer being requested to make, at his discretion, a fixed number of deliberate mistakes in every page. The location of these is concealed from the author-checker but carefully registered by the printer. Failure by the checker to find the prescribed number of errors per page serves as a warning of faltering attention or fatigue, and serves as a quality control alarm. Clearly, the hazards of fatigue and tedium were many and real, and however ingenious the measures, insecurities remained. De Morgan observed:

> It is hardly credible, to those who have not tried, how much the perceptions are dulled by the monotonous comparison of one column of figures with another, how many and how gross errors both eye and ear, when tired, will suffer to pass.[31]

Rewards were offered for the detection of errors after publication, both as an advertisement of confidence in accuracy and as a way of improving subsequent editions through 'user-testing' in the field. Glaisher reports that Vega offered a reward of a ducat for every error found in his table of 1794, including errors of a unit in the last digit. It turns out that Vega was fortunate that no one took up the challenge. An analysis of the relative frequency and magnitude of last-digit errors published in 1851 by Gauss would have cost Vega dear: there are large numbers of hereditary errors from Vlacq on whose tables Vega's are based.[32] It is also possible that a few deliberate errors were left as a protection against unauthorized copying.

As many errors as possible were corrected prior to printing and the process of progressive improvement continued after publication through the distribution of errata sheets and the incorporation of discovered errors into later editions and revisions. Knowing which tables were reliable was an issue of connoisseurship and publishers sought testimonials from eminent mathematicians to preface new editions. The genealogy of specific editions of tables was an essential determinant in their suitability as a source and the annotated bibliographies of tables, which served as comprehensive consumer guides, were near valueless without a record of 'pedigree' or 'stable', and many entries are accompanied by a critique of accuracy. Poor tables were not spared damning reviews. The response of the table makers and users to error-ridden tables was to uncork the bottle of red ink and correct them. Errata sheets would be published, reviews written, and bibliographies annotated.

For the most part the knowledge of deficiencies in tables was confined to the community of tablemakers and users. When the unreliability of tables overflowed into scientific and political arenas, the problems were seen as failures of superintendence, poor leadership, or negligent recruitment of appropriate staff—organizational shortcomings in the execution of established practice. The decline of the *Nautical almanac* under John Pond is a case in point. Pond succeeded Nevil Maskelyne as Astronomer Royal in 1811 and the appointment carried with it responsibility for the annually issued *Almanac*. So battered was the Almanac's credibility by 1818 that it was seen as an embarrassment to English science especially on the Continent, and the affair was debated in Parliament.[33] The scandal focussed on issues of organization and leadership, the deficiencies of which resulted in lowered standards and higher error rates.

Just how bad were the tables? A frequently quoted source is Lardner's lengthy semi-popular article on Babbage's engine published in 1834.[34] Lardner cites a survey of a collection of tables and uses published errata as the barometer of accuracy. He reports that in a random selection of forty volumes of arithmetical and trigonometric tables (nautical and astronomical tables were excluded) selected from a collection of 140 volumes over 3700 acknowledged errors were found as evidenced in the errata sheets.[35] He compounds this apparently shocking statistic with details of equally egregious errors in tables for astronomical navigation citing seven folio pages of errata to supplementary *Almanac* tables which themselves contained more than 1100 errors. The errata were corrected by further errata and his triumph is ecstatic at the finding that the *Nautical almanac* of 1836 would require an erratum of the erratum of the errata.

The implication of Lardner's survey is that tables were generically flawed and that the large numbers of errors in the host tables discredited the printed errata sheets as a corrective process being itself vulnerable in the selfsame ways. The evidence of errata sheets is undeniable but there were those, the Astronomer Royal, George Biddell Airy included, who asserted that tables were already sufficiently accurate. The problem was not the known errors, but the unknown ones. Babbage observes that 'the multitude of errors really occurring is comparatively little known'.[36] Arguing from the premise of known errors was less than conclusive and the inability to definitively quantify residual errors left the issue of reliability open to differences in expert opinion. Maritime safety was a central and continuing concern and shipwrecks featured routinely in newspaper reports which were often luridly vivid. John Herschel, writing to the Chancellor, Henry Goulburn in 1842, captures the anxiety of the unknown: 'an undetected error in a logarithmic table is like a sunken rock at sea yet undiscovered, upon which it is impossible to say what wrecks may have taken place'.[37] It is curious that though astronomical navigation features prominently in the advocacy for the need for accurate tables, none of the protagonists of mechanical solutions cites an instance of a navigational error that resulted in the loss of a ship being attributed to a tabular error. The unreliability of charts, doubts about observational procedures, and errors in calculation blurred issues of accountability. Most British ships lost at sea foundered on the coasts in storms.[38]

Mechanized table production

The most ambitious attempts to mechanize tables production in the nineteenth century were those of Charles Babbage whose efforts are increasingly well known (Fig. 6.2). The genesis occurred in 1821 when he and Herschel were checking manuscript calculations for astronomical tables commissioned by the newly founded Astronomical Society. Dismayed at the number of errors Babbage famously exclaimed 'I wish to God these calculations had been executed by steam'. 'Steam' was a metonymic reference to the 'untiring action and unerring certainty of mechanical agency'[39] and he devoted the major part of the rest of his working life to the realization of this mechanical epiphany. While errors in calculation for tables were the first stimulus for his work it is an oversimplification to see the engine designs as a 'solution' to a widely recognized and clearly defined 'problem'. Experts disagreed whether

Fig. 6.2 Charles Babbage (1791–1871), c. 1832. (Science Museum/Science & Society Picture Library)

the deficiencies in tables were serious and whether or not the engines would be of any use. The cost benefits of building engines for table making compared to manual methods were disputed at the highest level[40] and the ambiguous successes of Scheutz and Wiberg support these reservations.

The clearest coupling of tabular errors as 'problem' with machines as 'solution' is found in Lardner's article of 1834 on Babbage's Difference Engine No. 1 and this has had a defining influence in the historiography of Babbage's work. Here the parade of known tabular errors is paired with the unabashed promotion of Babbage's engine. It is known that Babbage collaborated with Lardner on the article and it is easy to assume that Lardner was Babbage's mouthpiece—that the views Lardner expressed were Babbage's, perhaps elaborated and dramatized for the purposes of public consumption. However, the timing and content of Lardner's article needs to be examined in the political context of the collapsing fortunes of the construction project which after a decade of engineering design and massive public expense had little in the way of credible progress to show. Lardner's own agenda is also relevant. He enjoyed a lucrative career as author and public lecturer. Babbage's calculating engine represented fresh material for his lecture tours and writing from which he derived substantial fees. Creating a 'scandal' of tabular errors may have been a way of simplifying the complex motivational uses of the machines into a problem–solution pairing—the act of a publicist with an astute sense of what was needed for public appeal, and a way of promoting Babbage's damaged interests at the same time.

In developing his engines Babbage was led from mechanized arithmetic and tabulation to fully fledged general-purpose programmable automatic computation, and the technical and intellectual challenges of this programme soon consumed his interests. Tabular errors feature alongside other arguments in his advocacy. Benefits to mathematics, the heuristic value of the machine in the discovery of new mathematical series, the stimulus to new forms of analysis, the use of computational method as an alternative to analytical solution to equations, and the saving of mental and physical labour, are prominent in the earliest expressions promoting the utility of his machines. Eliminating errors in tables provided the entry point to a new discourse of computation and as the work progressed tabular errors were subsumed under the larger ambitions of general-purpose computation. Table making remained a preoccupation though not always a central one.

Babbage's engines, for the most part unbuilt in his lifetime, had negligible influence on table-making practice. However, they represent the most elaborate and ambitious attempt to mechanize tables production, and the detail and physical scale of his undertaking provide clear insights into the size and difficulties of the task. In the study of the designs it is the calculating mechanisms that have commanded the lion's share of attention.[41] Yet integral to Babbage's conception of table-making machines is not only error-free calculation but the extension of the supposed infallibility of machinery to the task of printing results. An elegant illustration of the scope and scale of Babbage's intentions is his Difference Engine No. 2 which he designed between late 1846 and 1849. Plans for the new engine, all of which survive, were offered to the Government in 1852 through Lord Rosse who acted as an intermediary, but the offer was rebuffed and no further attempt was made in Babbage's lifetime to construct the machine.[42] The Engine has since been built to the original drawings at the Science Museum in London. The calculating section was completed for the bicentenary in 1991 of Babbage's birth, and the printing and stereotyping apparatus was completed in March 2002.[43] The construction of the machine gives insights into the engineering task that faced Babbage. Moreover, the availability of a working engine gives clues about how the machine might have been operated and managed, and the extent to which such a device might have been practically useful.

Security against error is provided by a number of devices and mechanisms. The Engine is a decimal digital machine: it recognizes only integral decimal whole numbers as individual digits in a number and the machine is designed to jam in the event that any figure wheel, print wheel or stereotyping wheel punch takes up a position that is indeterminate i.e. somewhere between integer values for any given digit position. Jamming is a form of error detection and Babbage boasts that the machine will calculate correctly or break, but never deceive.

Typesetting is automatic and the layout of results on the stereotyped page can be programmed by the operator. Formatting features that can be varied include line height, margin widths, page borders, number of lines on a page, leaving blank lines between groups of lines, and options to stereotype in one, two, three, or four columns. Results can be stereotyped down the page (line-to-line) and automatically column wrap to the top of the next column, or, across the page (column-to-column) and automatically line wrap to the start of the next line. Program options are selected before the start of a run. Programmability is provided by 'pattern wheels'. These are flat wheels with teeth cut into the circumference, and the separation of the teeth determines the amount the tray advances to receive each next result. One set of pattern wheels controls the line-to-line operation and the 'factory issue' version of the machine comes with four different pattern wheels with variations of tooth pitch for different line spacing.

The printer and stereotyping apparatus is also part of the control system of the whole tabulation process. From the point at which the handle is first turned the internal state of the machine has altered from the initial conditions and is unknown to the operator. The printing and stereotyping apparatus is at the opposite end of the machine and the furthest away from the operator. If the operator overruns beyond the end of a stereotyped page, the start state of the engine for the next page will be incorrect and the first result stereotyped in the fresh tray will be spoiled. There is an ingenious device that prevents this happening. The end-of-page condition is detected by the stereotype apparatus and automatically disengages the drive handle. The handle suddenly runs free and the Engine is halted automatically in exactly the state it needs to be to start a new page. With the Engine halted the operator has the opportunity to replenish the stereotype trays for the start of a new page.

There are several practical and operational issues that have become evident during early trials of the apparatus. The force with which the stereotype punches are lowered has been found insufficient to make any meaningful impression on lead or copper, two of the materials used by Babbage in contemporary trials of other machines. Soaked cardboard and plaster of Paris are being experimented with (Fig. 6.3).[44] The construction of the printwheels for the inked impression is such that leading zeros cannot be suppressed without dismantling a section of the apparatus, and this was clearly not intended. Since the printed record is a one-off and intended as a checking record only, this is of little significance. Suppressing leading zeros in the stereotyped record by replacing the wheel punches with blank spacers is possible but not practical, and it is likely that leading zeros would have been cut or machined off the printing plate after it was cast.[45]

Finally, there is the question of how tabulation of the argument would have been carried out. If the full thirty digits are used in the printed result then generating the argument would require a separate calculation run using only one difference to increment each next value. A feature of the machine is that the calculating section can be split horizontally to isolate upper and lower sections of the machine. This is done by inhibiting the

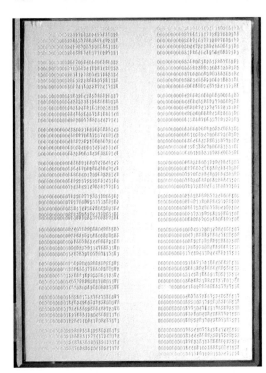

Fig. 6.3 Difference Engine No. 2. Experimental stereotype mould. Plaster of Paris. Figures are indented into the surface. (Science Museum/Science & Society Picture Library)

carriage of tens at any digit position in the thirty-one digits on each of the eight columns of figure wheels, and the mechanism to do this is simple and easy to operate. Separate calculations can then be carried out in each of the isolated sections with the thirty digits apportioned between the two sections at will. It follows that one section of the machine can perform the tabulation of a polynomial function of up to the seventh order, and a separate section can tabulate the argument using only one difference to provide the linear increment of the argument. The argument and result will be printed and stereotyped on the same line alongside each other. The use of the machine to generate the argument and the result in the same run by distributing the full digit capacity of the machine between the argument and result provides a speculative answer to the question of why Babbage designed his machines to operate with so many digits.

The machine in its entirety is a sumptuous piece of engineering sculpture. In operation it is an arresting sight. Its complexity and physical size testify to the difficulty of bringing the 'unerring certainty' of machinery to the task of producing error-free tables.

Babbage's Difference Engine No. 2: overview

Difference Engine No. 2 was designed between 1846 and 1849. The built machine weighs five tonnes and consists of 8000 parts equally divided between the printing and stereotyping apparatus, and the calculating section. The side view of the machine (Fig. 6.4) is one of the main drawings in a set of twenty, and one that is most evocative of its overall shape.[46] The drawing shows a machine eleven feet long and seven feet high with the depth varying between eighteen inches and four feet. The design has an elegant economy: it is about three times more efficient than Difference Engine No. 1 in having three times fewer parts than the earlier design for a similar computing capability (Fig. 6.5).

The Engine is operated by turning a crank handle by hand. The handle drives, via a set of bevel gears, a set of twenty-eight conjugate cams arranged in a vertical stack.[47] The cam stack orchestrates and harmonizes the lifting, turning, and sliding motions required to execute the repeated additions required for the method of finite differences. The crank handle also drives the printer and stereotype apparatus from a long shaft which runs the length of the underside of the machine.

The Engine automatically calculates and tabulates any seventh order polynomial to thirty-one decimal places using the method of finite differences, prints the result to thirty places on a paper roll, and impresses the same results on soft material to produce stereotype moulds from which printing plates can be made. The results are stereotyped in two font sizes, large and small, at the same time, and the layout of the results in the stereotype trays can be formatted by programming mechanical options. Calculation and typesetting are completely automatic eliminating the need for manual transcription.

Numbers are stored and operated on using figure wheels engraved with the decimal numbers '0' through '9'. A thirty-one-digit number is stored in a column of figure wheels thirty-one high with units at the bottom, tens next above, hundreds next above, and so on. The machine has eight columns each with thirty one figure wheels, one column for each of seven differences, and a column for the tabular result. Negative values are represented by complements. For a seventh order polynomial the machine adds the constant difference on the seventh difference column to the sixth difference column alongside, the sixth difference to the fifth difference, and so on. Initial values are set from a precalculated table by rotating the figure wheels after disabling the security mechanisms. During each cycle the machine executes seven thirty-digit decimal additions to produce each next value in the table which appears on the column containing the tabular value. For lower order polynomials the higher order difference columns are set to zero and play no part. The Engine, once run in, produces ten results per minute.

Babbage's Difference Engine No. 2: overview *cont.*

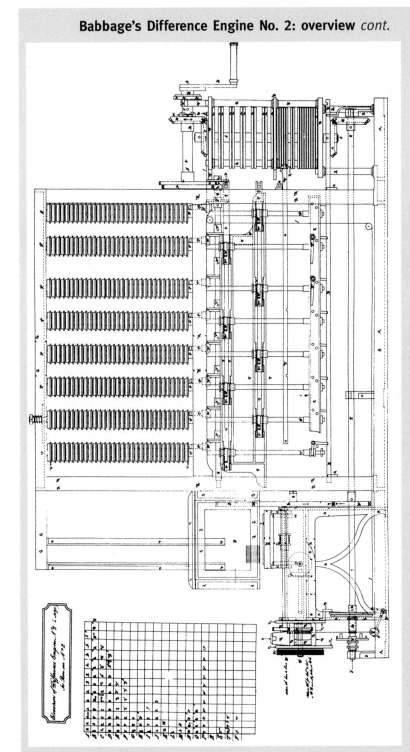

Fig. 6.4 Design Drawing, Difference Engine No. 2, Side Elevation. (Science Museum/Science & Society Picture Library)

Fig. 6.5. Charles Babbage's Difference Engine No. 2. Completed 2002. The machine weights 5 tonnes and consists of 8,000 parts. Foreground: Printing and stereotyping apparatus. Rear: calculating section. (Science Museum/Science & Society Picture Library)

Difference Engine No. 2: printing and stereotyping

A central feature of Babbage's calculating engine designs is the automatic printing and stereotyping of results. Difference Engine No. 2 incorporates an apparatus which prints in ink a one-off copy of results on a print roll as well as producing a stereotype mould for the production of printing plates for use in a conventional printing press. The printing and stereotyping apparatus is

Difference Engine No. 2: printing and stereotyping *cont.*

directly coupled to the calculating section and each new tabular value is transferred automatically from the results column, via a system of racks, pinions, and spindles, to the print wheels, and at the same time to two sets of punches for stereotyping (Fig. 6.6). Typesetting is automatic and each result is printed and stereotyped during the calculating cycle in which it is generated: there is no buffering or storage of the result and there is no time overhead to print, i.e. printing and stereotyping is accommodated in the calculating cycle and takes no additional time. Each cycle leaves an inked impression of a thirty-digit result on the paper print roll which advances automatically to provide-fresh paper for each line (Fig. 6.7). This hardcopy printout is intended as a

Fig. 6.6(a) Difference Engine No. 2, printing and stereotyping apparatus (computer simulation), showing transfer of results from figure wheel column (right) to print wheels and two sets of stereotyping wheel punches. Trays below receive impressions.

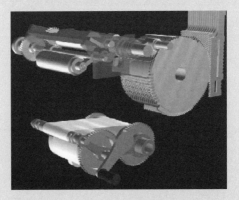

Fig. 6.6(b) Difference Engine No. 2, printing apparatus (computer simulation). Paper drums (bottom). Inking mechanim and print wheels (upper). (Science Museum/Science & Society Picture Library)

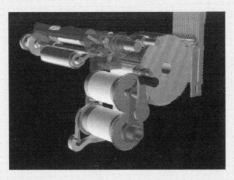

Fig. 6.6(c). Difference Engine No. 2, printing apparatus (computer simulation). Paper drum in raised position taking an inked impression from the pintwheels. (Science Museum/Science & Society Picture Library)

record and checking copy only and the line spacing and format is not alterable by the operator (Fig. 6.8). The production of multiple copies for distribution is achieved through stereotyping.

There are two sets of stereotype punches (Fig. 6.9), one with a larger and one with a smaller font. Each set consists of thirty wheels, one for each digit of the result, with hardened steel number punches fixed at the circumference. The wheels making up the punches have two degrees of freedom only: rotational (to register each digit of the result), and vertical (lowering to impress the result, and raising to return to receive the next result). The wheel punches are assembled on the same shaft and are lowered and raised together as a set. Below each set of punches is a bronze tray to take the soft material to receive the impression (Fig. 6.10).

During each calculating cycle the punches are driven downwards into the tray below and the material is impressed with all thirty digits at the same time in one action. After each impression the tray advances automatically and repositions itself to receive the next result on a new line or column and the distance between the impressions (the line spacing) is determined by the incremental advance of the tray. The smaller tray advances by a proportionately smaller amount to give a reduced line spacing in keeping with the smaller font size. The platform or carriage supporting the stereotype trays has two horizontal degrees of freedom which allow the trays to advance down the page (line-to-line) or across the page (column-to-column). The carriage is driven by falling weights which are rewound automatically at the end of a column or page. The layout of results can be programmed by the operator. Programmable formatting features include line spacing, number of columns, margin widths, number of lines on a page, and blank lines between groups of lines.

Difference Engine No. 2: printing and stereotyping *cont.*

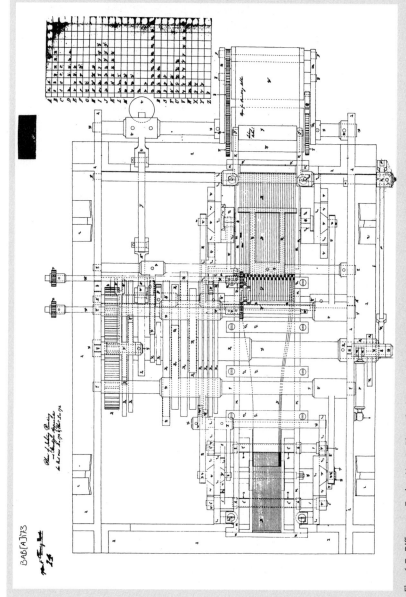

Fig. 6.7. Difference Engine No. 2, Plan of Inking, Printing and Stereotyping Apparatus. (Science Museum/Science & Society Picture Library)

UNERRING CERTAINTY OF MECHANICAL AGENCY

```
675000003394498660735540644241
676000003441718270860028917/7
677000003484414330048692002005
678000003521567587684282505589
6790000035712416940844428/593
680000003616361600494056707081
6810000036860009759198127629807
6820000037061302736148720070785
68300000375074665838908579329
6840000037968626894657860030303
685000003843481704284127170981
686000003890607701671593260117
687000003935624429213846520020
```

Fig. 6.8 Difference Engine No. 2, experimental printout of checking copy for tabulation of seventh-order polynomial. Argument is left-hand three figures. (Science Museum/Science & Society Picture Library)

Fig. 6.9 Difference Engine No. 2, stereotyping wheel punches for large font. (Science Museum/Science & Society Picture Library)

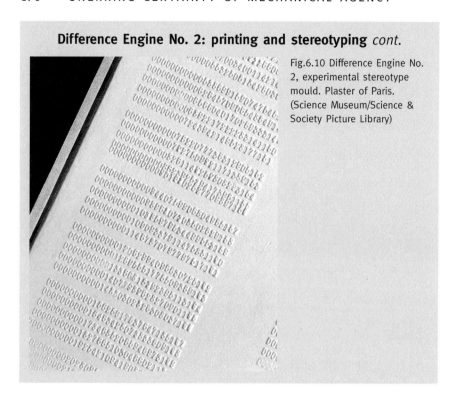

Difference Engine No. 2: printing and stereotyping *cont.*

Fig.6.10 Difference Engine No. 2, experimental stereotype mould. Plaster of Paris. (Science Museum/Science & Society Picture Library)

Further reading

There is no single reference that comprehensively treats the influence of machinery on table making, and material tends to be drawn from the histories of the separate relevant technologies, particularly the history of mechanical calculation, of book production, printing, and papermaking. An excellent work on computational devices and systems is Michael Williams's *A history of computing technology*, Englewood Cliffs, N.J.: Prentice-Hall, Inc, 1985. A concise and authoritative account of scientific computing which seamlessly blends calculating devices with organizational and operational practice is Mary Croarken's *Early scientific computing in britain*, Oxford: Clarendon, Press, 1990.

Philip Gaskell's *A new introduction to bibliography*, Oxford: Oxford University Press, 1972, is an impressive source on the history of the processes involved in book production. For a detailed economic anlaysis see Alexis Weedon's *Victorian publishing*: Continuum International Publishing Group, 2002.

Much of the material on contemporary table making is 'in the cracks'. Augustus de Morgan's articles in *The Penny Cyclopaedia of the Society for the Diffusion of Useful Knowledge*, 496–501, London: Charles Knight, 1842, and the extended version in *The English Cyclopaedia: A Dictionary of Universal Information*, 976–1016, London: Bradbury, Agnew, 1861, are rich sources which serve both as annotated bibliographies of tables as well as revealing repositories of information on contemporary practice and attitudes. The prefaces and editors' introductions to published tables tend to be a neglected source which sometimes offer rich pickings, often inadvertent.

The standard primary-source reference for Charles Babbage's published work is *The Works of Charles Babbage*, 11 vols, London: William Pickering, 1989, edited by Martin Campbell-Kelly. Uniquely authoritative accounts of the workings of Babbage's engines are by Allan Bromley in various issues of *IEEE Annals of the History of Computing*. A recent accessible account of Babbage's life and the evolution of his thinking on calculating engines is *The cogwheel brain: Charles Babbage and the quest to build the first computer*, London: Little, Brown, 2000 by Doron Swade. The last third of the book gives an account of the late twentieth-century construction of Babbage's Difference Engine No. 2. An uncompromisingly internalist description of this engine is 'Charles Babbage's Difference Engine No. 2: Technical Description, Science Museum Papers in the History of Technology No. 5.', Science Museum, 1996, by the same author.

Notes

1. See Mike Williams, Chapter 5, and George Wilkins, Chapter 11, in this volume.
2. See, S. Johnston, 'Making the Arithmometer Count.' *Bulletin of the Scientific Instrument Society* 52, no. March (1997), 12–21.
3. D. Baxandall, *Calculating machines and instruments*. Revised and updated by Jane Pugh (ed.), Science Museum, London, 1975. See p. 11 for detailed account of operation.
4. P. Gaskell, *A new introduction to bibliography*, Oxford University Press, Oxford, 1972, p. 274.
5. For detailed analysis see, A. Weedon, Chapter 3, 'Trends in book production costs' in *Victorian publishing*: Continuum International Publishing Group, 2002.
6. Gaskell, *op cit.*, p. 228.
7. Ibid. pp. 201–4.

8. A. de Morgan, 'Table', in *The Penny Cyclopaedia of the Society for the Diffusion of Useful Knowledge*, 496–501, Charles Knight, London, 1842, p. 500. For an extended version of the article see 'Table' in *The English Cyclopaedia: A Dictionary of Universal Information*, 976–1016. Bradbury, Agnew, London, 1861.
9. Gaskell, *op cit.*, p. 201.
10. C. Babbage, *Passages from the life of a philosopher*. Longman, Green, Longman, Roberts & Green, London, 1864, p. 138.
11. See Michael Williams, Chapter 5, in this volume.
12. See Ivor Grattan-Guinness, Chapter 4, in this volume. Babbage observes that the 'third class' of computers who were capable of no more than addition and subtraction made fewer mistakes than the mathematicians higher up in the hierarchy of skills. See C. Babbage, *On the economy of machinery and manufactures*, 1st edn, London: Charles Knight, 1832. Reprinted, (ed. M. Campbell-Kelly), *The works of Charles Babbage*, 11 vols, William Pickering, London, 1989, vol. 8, p. 138, henceforth 'Works'.
13. E. Martin, *The calculating machines (Die Rechensmashinen): their history and develoment*, edited by Peggy Aldrich Kidwell and Michael R Williams, *Charles Babbage Institute Reprint Series for the History of Computing*, Cambridge, MIT Press, Mass., 1992, p. 38.
14. For Babbage's reference to 'classes' and 'lowest processes' see, *Works*, Vol. 8. pp. 138, 141 respectively.
15. Sources in order: Babbage, 1822, *Works*, vol. 2 p. 6; *Ibid*. p. 15; Francis Baily, 1823, *Works*, vol. 2, p. 45; Juris, Judex, 1861, *Works*, vol. 1, p. 3.
16. Luigi Menabrea (1842), *Works*, vol. 3, p. 113.
17. An extreme example of this is given by Babbage's son. Calculations for logarithm tables were taken to twenty decimal places for correctness to seven figures. See, B. H. Babbage, 'Babbage's Calculating Machine or Difference Engine':Science and Art Department, 1872, p. 8. Reprinted: *Works*, vol. 2, p. 232.
18. E. Forbes, 'The Foundation and Early Development of the Nautical Almanac.' *Journal of the Institute of Navigation* 18, no. 4 (1965): 391–401, p. 394.
19. D. Lardner, 'Babbage's Calculating Engine' *Edinburgh Review* 59 (1834): 263–327. Reprinted: *Works*, vol. 3, p. 134.
20. *Ibid*. Italics original.
21. See Ivor Grattan-Guinness, Chapter 4 in this volume, p. 114.
22. For a clear nineteenth-century statement of this feature, see, C. Babbage. *Passages from the life of a philosopher*, Longman, Green, Longman, Roberts & Green, London, 1864, p. 50.
23. For a worked example see, M.V. Wilkes. 'Babbage's Expectations for the Difference Engine.' *Annals of the History of Computing* 9, no. 2, 203–5, p. 203. Checking by differencing was used in the late eighteenth century in the preparation of the *Nautical almanac* and differencing summed groups of entries was used, for example, in the production of eight-figure logarithm based on

de Prony's manuscripts and published in 1891 for the Geographical Service of the Army. See, Preface to *Tables des logarithmes a huit décimales des nombres entiers de 1 a 120 000 et des sinus et tangentes de dix secondes en dix secondes d'arc dans le système de la division centésimale du quadrant publiées par ordre du Ministere de la Guerre*, Impremerie Nationale, Paris, 1891, pp. III and IV. George Biddell Airy suggested in 1856 that the major use of automatic calculating engines might well be for error-checking existing tables by differencing.

24. See, for example, tabular entries for log 14 050 to log 18 790 in the set at the *Biblioteque de l'Institut*.
25. A. de Morgan, 'Table' in *The English Cyclopaedia: A Dictionary of Universal Information*, 976–1016. London: A. Bradbury, 1861, p. 1015. This is a variant of the same claim carried forward from the version of the article published in 1842.
26. A. de Morgan (1842), *op* cit. p. 501.
27. Babbage reports in 1822 that the rate for checking proofs was three guineas a sheet. See, C. Babbage, *A letter to Sir Humphrey Davy, Bart., President of the Royal Society, on the application of machinery to the purpose of calculating and printing mathematical tables*. Cradock and Joy, London, 1822. Reprinted: *Works*, vol. 2, p. 13, footnote.
28. Minor changes to stereotypes were possible depending on the nature of the correction. See C. Babbage, *On the economy of machinery and manufactures*, 4th edn, Charles Knight, London, 1835. Reprinted: *Works*, Vol. 8, see p. 53, Section 94.
29. J. W. L. Glaisher, 'Report of the Committee of Mathematical Tables.', in *Report of the Forty-Third Meeting of the British Association for the Advancement of Science, held at Bradford in September 1873*, (ed. J. W. L. Glaisher), John Murray, London, 1874, p. 135.
30. Later practice questions the effectiveness of reader–listener pairs. For work at the NAO see, L. J. Comrie, 'Computing the Nautical Almanac.' *Nautical Magazine* July (1933), p. 16.
31. De Morgan (1842), *op cit.*, p. 500; (1861), *op cit.*, p. 1015.
32. Glaisher (1874), pp. 138–9.
33. M. B. Hall, *All scientists now: The Royal Society in the nineteenth century*, Cambridge University Press, Cambridge, 1984, p. 11.
34. D. Lardner, 'Babbage's Calculating Engine' *Edinburgh Review* 59 (1834), 263–327. Reprinted: *Works*, vol. 2, pp. 118–86.
35. *Ibid., Works*, p. 129. The private individual is almost certainly Babbage. See M. Campbell-Kelly, 'Charles Babbage's Table of Logarithms' *Annals of the History of Computing* 10, no. 3 (1988), 159–69, 162.
36. *Passages*, p. 138.
37. Herschel to Goulburn, September 1842. Royal Society Herschel Archive, Box 27, Item 51.

38. R. Larn, *Shipwrecks of Great Britain and Ireland*, David & Charles, London, 1981.
39. The phrase is Lardner's. See *Works*, Vol. 2, 169.
40. See, D. Swade, *The cogwheel brain: Charles Babbage and the quest to build the first computer*, Little, Brown, London, 2000, pp. 37–8, 141–7.
41. The major work on Babbage's technical work is by Allan Bromley. See, A. G. Bromley, 'Charles Babbage's Analytical Engine, 1838' *Annals of the History of Computing* 4, no. 3 (1982): 196–217. 'Difference and Analytical Engines' in *Computing before computers*, (ed. W. Asprey), Ames: Iowa State University Press, 1990. 'The Evolution of Babbage's Calculating Engines', *Annals of the History of Computing* 9, no. 2 (1987), 113–36. 'Babbage's Analytical Engine Plans 28 and 28a—The Programmer's Interface', *IEEE Annals of the History of Computing* 22, no. 4 (2000), 5–19.
42. Swade (2000), pp. 174–7.
43. *Ibid*. See Chapters 12–18 for an account of the construction of the calculating section of the engine.
44. Babbage records that he used specially prepared thick paper and plaster of Paris in his stereotyping experiments. See *Passages* p. 46.
45. This technique was used in the Scheutz prototype completed in 1843. See, M. Lindgren. *Glory and failure: the difference engines of Johann Muller, Charles Babbage, and Georg and Edvard Scheutz*. Translated by C. G. McKay, 2nd edn, MIT Press, Cambridge, Mass., 1990, p. 150.
46. For an index of Babbage's technical work see, A. G. Bromley. *The Babbage papers in the Science Museum: a cross-referenced list*, Science Museum, London, 1991.
47. For a detailed description of the engine see, D. Swade, *Charles Babbage's Difference Engine No. 2: Technical Description, Science Museum Papers in the History of Technology No. 5*: Science Museum, 1996.

VENUS, 1855.

For Washington Mean Noon and Meridian Transit.

Mean Solar Time of Meridian Transit.			Sidereal Date	Apparent Right Ascension.		Apparent Declination.		Log. of Motion in a Sidereal Minute.		Log. of Factor for Second Diff's.	
				At Mean Noon.	At Transit.	At Mean Noon.	At Transit.	In R.A.	In Dec.	In R.A.	In Dec.
d.	h.	m.		h. m. s.	m. s.	° ′ ″	′ ″				
Sept. 2	1	57.7	245	12 43 24.03	43 27.62	−11 5 32.1	6 47.9	+8.48371	−9.8077	−4.23	+5.06
3	1	54.4	246	12 44 4.34	44 7.27	11 20 32.6	21 42.2	8.40654	9.7838	4.25	5.08
4	1	51.1	247	12 44 37.41	44 39.68	11 34 42.8	35 46.4	8.30969	9.7576	4.26	5.10
5	1	47.6	248	12 45 3.04	45 4.67	11 48 0.1	48 57.7	8.18089	9.7280	4.27	5.12
6	1	43.9	249	12 45 21.04	45 22.05	12 0 21.9	1 13.5	7.99178	9.6948	4.27	5.14
7	1	40.2	250	12 45 31.24	45 31.67	12 11 45.4	12 31.0	+7.63752	9.6567	4.28	5.16
8	1	36.3	251	12 45 33.49	45 33.37	12 22 7.9	22 47.5	−7.08961	9.6132	4.28	5.18
9	1	32.2	252	12 45 27.67	45 27.03	12 31 26.6	32 0.3	7.83728	9.5624	4.29	5.20
10	1	28.1	253	12 45 13.66	45 12.54	12 39 38.7	40 6.6	8.10052	9.5023	4.29	5.22
11	1	23.8	254	12 44 51.36	44 49.80	12 46 41.4	47 3.9	8.26423	9.4293	4.29	5.24
12	1	19.3	255	12 44 20.74	44 18.79	12 52 32.0	52 49.3	8.38309	9.3380	4.30	5.26
13	1	14.7	256	12 43 41.77	43 39.51	12 57 7.9	57 20.2	8.47632	9.2177	4.30	5.27
14	1	10.1	257	12 42 54.50	42 51.99	13 0 26.5	0 34.2	8.55252	9.0430	4.30	5.29
15	1	5.3	258	12 41 59.00	41 56.30	13 2 25.1	2 28.7	8.61674	−8.7315	4.29	5.30
16	1	0.2	259	12 40 55.39	40 52.55	13 3 1.2	3 1.1	8.67159	+7.0457	4.28	5.31
17	0	55.1	260	12 39 43.86	39 40.96	13 2 12.6	2 9.2	8.71913	8.8035	4.27	5.32
18	0	49.9	261	12 38 24.63	38 21.75	12 59 57.5	59 51.3	8.76056	9.0947	4.25	5.33
19	0	44.6	262	12 36 58.00	36 55.21	12 56 14.3	56 6.0	8.79694	9.2691	4.23	5.33
20	0	39.0	263	12 35 24.31	35 21.67	12 51 2.2	50 52.5	8.82855	9.3940	4.20	5.34
21	0	33.4	264	12 33 44.07	33 41.66	12 44 20.6	44 10.2	8.85587	9.4912	4.16	5.33
22	0	27.8	265	12 31 57.80	31 55.69	12 36 9.7	35 59.5	8.87934	9.5701	4.12	5.33
23	0	22.1	266	12 30 6.09	30 4.34	12 26 30.7	26 21.2	8.89927	9.6358	4.06	5.32
24	0	16.2	267	12 28 9.60	28 8.26	12 15 25.1	15 17.2	8.91576	9.6920	3.99	5.31
25	0	10.2	268	12 26 9.06	26 8.19	12 2 55.5	2 49.9	8.92903	9.7396	3.90	5.29
26	0	4.3	269	12 24 5.21	24 4.84	11 49 4.9	49 2.3	8.93931	9.7810	3.79	5.27
26	23	58.3	270		21 58.99		33 58.3				
27	23	52.3	271	12 21 58.84	19 51.48	11 33 57.2	17 42.5	8.94656	9.8170	3.61	5.24
28	23	46.2	272	12 19 50.79	17 43.18	11 17 37.0	0 19.8	8.95074	9.8480	−2.59	5.21
29	23	40.2	273	12 17 41.94	15 34.96	10 60 9.4	41 55.7	8.95193	9.8748	+2.58	5.17
30	23	34.1	274	12 15 33.19	13 27.75	10 41 40.0	22 36.7	8.95004	9.8978	3.59	5.13
Oct. 1	23	28.1	275	12 13 25.46	11 22.41	10 22 15.3	2 29.7	8.94497	9.9172	3.67	5.07
2	23	22.2	276	12 11 19.65	9 19.82	9 62 2.3	41 42.3	8.93677	9.9332	3.83	4.99
3	23	16.2	277	12 9 16.62	7 20.70	9 41 8.8	20 22.3	8.92562	9.9460	3.93	4.89
4	23	10.4	278	12 7 17.12	5 25.78	8 79 42.7	58 37.2	8.91143	9.9555	4.01	4.78
5	23	4.7	279	12 5 21.89	3 35.79	8 57 51.7	36 34.8	8.89395	9.9622	4.08	4.61
6	22	59.1	280	12 3 31.66	1 51.46	8 35 43.5	14 22.5	8.87281	9.9665	4.14	4.36
7	22	53.5	281	12 1 47.15	0 13.41	7 73 25.7	52 7.8	8.84751	9.9685	4.19	+3.68
8	22	48.1	282	11 60 9.05	58 42.21	7 51 5.9	29 58.1	8.81794	9.9681	4.23	−4.12
9	22	42.8	283	11 58 37.84	57 18.30	7 28 51.6	8 0.6	8.78390	9.9652	4.25	4.49
10	22	37.6	284	11 57 14.01	56 2.07	6 66 50.1	46 22.2	8.74462	9.9598	4.27	4.68
11	22	32.5	285	11 55 57.93	54 53.87	6 45 8.3	25 9.0	8.69933	9.9521	4.29	4.80
12	22	27.6	286	11 54 49.97	53 54.01	6 23 52.3	4 26.8	8.64639	9.9424	4.30	4.88
13	22	22.9	287	11 53 50.45	53 2.74	5 63 8.0	44 20.7	8.58389	9.9303	4.31	4.95
14	22	18.3	288	11 52 59.58	52 20.22	5 43 0.5	24 55.5	8.50872	9.9163	4.32	5.00
15	22	13.8	289	11 52 17.54	51 46.58	5 23 36.6	6 15.7	8.41660	9.8999	4.33	5.05
16	22	9.5	290	11 51 44.44	51 21.86	4 64 54.7	48 25.0	8.29816	9.8812	4.34	5.08
17	22	5.3	291	11 51 20.33	51 6.10	4 47 4.6	31 26.5	8.13412	9.8603	4.34	5.11
18	22	1.2	292	11 51 5.23	50 59.28	4 30 7.3	15 23.1	7.86736	9.8370	4.34	5.13
19	21	57.4	293	11 50 59.12	51 1.38	4 14 5.8	0 17.4	−7.05383	9.8113	4.33	5.15
20	21	53.6	294	11 51 1.98	51 12.35	3 59 2.6	46 11.7	+7.70615	9.7828	4.33	5.16
21	21	50.0	295	11 51 13.74	51 32.06	3 44 59.5	33 6.0	8.05008	9.7514	4.33	5.17
22	21	46.6	296	11 51 34.28	52 0.35	3 31 57.9	21 3.1	8.23714	9.7169	4.32	5.18
23	21	43.2	297	11 52 3.43	52 37.07	3 19 58.9	10 3.3	8.36526	9.6791	4.31	5.18
24	21	40.0	298	11 52 41.02	53 22.04	3 9 3.6	0 6.9	8.46209	9.6371	4.30	5.18
25	21	37.0	299	11 53 26.86	54 15.07	2 59 12.1	51 14.3	8.53983	9.5897	4.29	5.19
26	21	34.1	300	11 54 20.79	55 15.95	2 50 24.6	43 25.4	8.60417	9.5365	4.28	5.19
27	21	31.3	301	11 55 22.58	56 24.44	2 42 41.3	36 39.7	8.65854	9.4766	4.27	5.19
28	21	29.6	302	11 56 31.97	57 40.32	2 36 1.5	30 56.7	8.70544	9.4079	4.26	5.18
29	21	26.0	303	11 57 48.74	59 3.36	2 30 24.4	26 15.7	8.74646	9.3274	4.25	5.18
30	21	23.7	304	11 59 12.68	60 33.34	2 25 49.4	22 35.8	8.78305	9.2284	4.23	5.17
31	21	21.4	305	12 0 43.56	2 10.03	− 2 22 15.6	19 56.1	+8.81583	+9.1034	+4.21	−5.16

7

Table making in astronomy

ARTHUR L. NORBERG

For millennia moving around on the surface of the Earth, especially the oceans, presented difficulties for mankind. Astronomers, seamen, naval personnel, explorers, and boatmen overcame some of these difficulties through the use of one or more of the following aids: position tables of celestial objects as a function of time, topographic maps, charts of shorelines and currents, and the various instruments for making accurate measurements of the position of objects in the heavens and on Earth. These aids appeared after a great amount of research on technique and laborious observation and measurement. This was especially true of the position tables of celestial

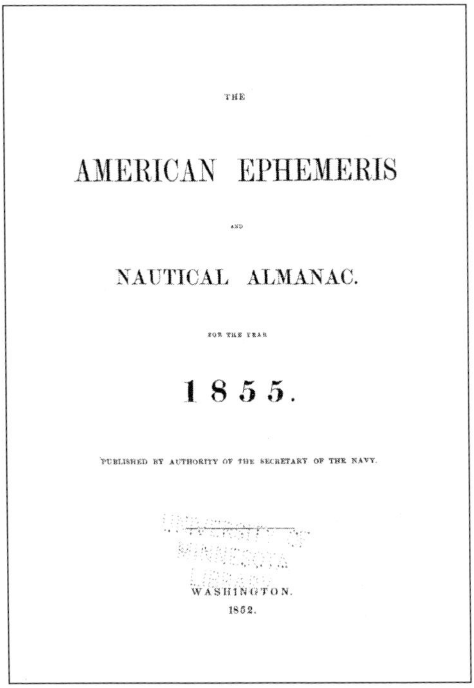
THE
AMERICAN EPHEMERIS
AND
NAUTICAL ALMANAC.
FOR THE YEAR
1855.
PUBLISHED BY AUTHORITY OF THE SECRETARY OF THE NAVY.
WASHINGTON.
1852.

Fig. 7.1 The *American ephemeris and nautical almanac for the year 1855* was the first issue from the newly established (American) Nautical Almanac Office. The table shown predicts the positions of Venus for that year. (Courtesy of the University of Minnesota Libraries.)

objects as a function of time used to calculate the longitude and latitude of a particular position on the surface of the Earth. A significant amount of calculation is needed to predict celestial positions, which is complicated by the fact that relative positions of all objects change because of the moving platform, the Earth or satellite, from which the positions are measured. Without this motion, the problem would have been straightforward once the principal of gravitation was developed. All we would have to do is set up and solve an equation of motion using the principle of gravitation. Alas, the problem is much more complicated.

Throughout history, astronomers have studied the motions of the bodies of the solar system. In the second half of the seventeenth century, these studies took a monumental step forward with Isaac Newton's (1642–1727) formulation of the laws of motion and the principle of universal gravitation. The precise application of the Newtonian formalism to the solar system involved calculation of the interactions of no less than eighteen bodies in the solar system: given the positions and velocities of all these bodies at an instant of time, and assuming that the only force among them was gravitational, one should be able to develop equations of motion and to predict the future positions and velocities of any one of the bodies.

While we now accept that the only force is gravitational, this was not obvious at the beginning of the eighteenth century. At that time, astronomers had few conceptual problems in formulating the equations of motion of planetary objects themselves. Each of these equations contained a term for the basic two-body gravitational interaction (between central body and satellite), and a term representing the perturbative effects produced on this motion by one or more other bodies. The development of analytical mechanics in the eighteenth century opened the way for a complete treatment of all the gravitational effects produced within the solar system, as required to predict the motions of these bodies. Gravitational astronomers, who formulated equations of motion for the interactions of the planets, discovered that these equations could not be integrated in closed form, and developed approximation techniques for the successful integration of the equations.

The history of predicting planetary positions using theories developed in celestial mechanics was a search for a complete unified set of precepts for use in computing future planetary positions. However, this search was occasionally beset by related difficulties due to inadequacies in astronomers' planetary observations and understanding of planetary interactions.

Therefore, it took almost 200 years after Newton to realize the goal. The drama has a cast of epic proportions and its denouement occurred at the beginning of the twentieth century in Paris with a major assembly of astronomers at an international conference. In this essay, we examine some highlights in this history, namely, the search, the problem of seemingly unstablizing interactions, and the difficulties with observation, focusing on only a few of the principal actors and institutions. Primary emphasis in this chapter is on the first stage in the process of predicting future planetary positions, namely, preparation of precepts, rather than the following stage of computing predicted positions.

Solving gravitational equations using perturbation theory

Newton and his immediate successors investigated the simple Keplerian two-body problem using a geometric approach, and made some attempts to include a third body. If only the Sun and one planet interacted, that planet would follow an elliptical orbit with constant orbital elements, with predictable velocity values over the entire orbit. However, in the three-body problem, where a second planet is added, the mutual interaction between the two planets alters the velocities of the planets. This effect on velocity can be positive or negative, depending on the relationship of the directions of motion and the instantaneous positions of the two planets relative to each other. Furthermore, if the orbits of the two planets are not in the same plane, there will also be adjustments in the angle between the two orbits, hence a change in the inclination of each orbit with respect to the ecliptic, the plane of the Sun's orbit. In a major advance in the eighteenth century, Leonhard Euler (1707–1783) and his contemporaries approached the three-body problem analytically. To account for the perturbing effects of a third body, Euler altered the equations of motion to include a disturbing function, described using a series of trigonometric functions to express angle and position changes as a function of time.

Euler and his successors solved the equations of motion expressed in these variables for the heliocentric position coordinates of the primary planet, expressing the position of the body in terms of the position in a Keplerian orbit and the variation of this position due to the perturbations produced on the planet. Next the computer subtracted the resulting equations from

> ### III. PRECEPTS FOR THE USE OF THE TABLES.
>
> To use these tables, the date for which the tabular quantities are required must be expressed in terms of the Julian calendar for any epoch before 1500; in terms of the Gregorian calendar for any epoch after 1600; and in terms of either calendar for the century 1500–1600. In Tables I and II a tabular year is used, in which the count of days differs from that of the calendar year only through the first two months of leap years—that is to say, the zero day is taken to begin with December 31, Greenwich mean noon of the year preceding in common years, and January 1 of the year itself in leap years.
>
> If the century is not the twentieth, enter Table I with the century, taking out all the numbers there given if a complete and rigorous computation is to be made. If the year is a zero one of the century, we may enter either with the year itself or with the century next preceding it. For example, a computation for any epoch during the year 1800 may be made by entering Table I with 1700 or with 1800.
>
> Enter Table II with the year of the twentieth century, or with the corresponding year of the given century, and write down the values of the arguments under those from Table I. In the case of the zero year of a century, Table II must be entered with the year 1900 when the century itself was used in Table I, but with the year 2000 if the number of the preceding century was used. But in the latter case the varying arguments must be diminished by their motion for one day, or Table III must be entered with a date one day earlier than the actual one, if the year is not a bissextile year. The two methods should give the same results.
>
> Enter Table III with the calendar month and day, and write the day of the year under the variable Arguments K, A, and S, and the values of l and θ under those from the preceding tables. If the epoch is not a Greenwich mean noon, take from Table IV the motion of l for the hours, minutes, and seconds of Greenwich mean time, and the fraction of a day to be added to the value of the variable Arguments K, A, and S.

Fig. 7.2 Simon Newcomb published precepts for the planet Venus, and the precepts for other planets, in *Astronomical papers for the use of the American ephemeris and nautical almanac*. This selection and the tables referred to for calculation run for 110 pages in volume 6 (1894).

the equations for the Keplerian orbit and obtained a new set of equations in the second derivative. After some substitutions of terms to arrange the equations of motion only in terms of the variations, the equations of motion could be integrated to obtain the equations for the variations in terms of time. When direct integration of these more complicated equations failed, Euler and his successors resorted to approximation techniques. From the equations for the variations, one can calculate positions in orbits. These equations contain many repetitive collections of terms. Mathematical astronomers constructed tables, either single or double entry, containing calculations of subsets of these collections. A set of instructions, called precepts, described how to use these tables. Using the precepts, the computer needed only to refer to the tables and sum the selected entries for a given case. A complete set of precepts and the tables constituted a 'theory' for the planet under investigation. Use of the precepts makes the final calculation of future planetary positions easier.

But the tables of planetary motion involved more than simply an analytical theory. The expressions for the heliocentric position coordinates generated from the solution of the equations of motion, contained coefficients that were functions of the orbital elements. The orbital elements had to be derived from precise observations of the planet to which they referred. These observations, corrected for such interferences as aberration, parallax, etc. (effects that were themselves in need of theoretical refinement), served also as initial conditions in the equations for the position coordinates.[1] The initial values to specify the initial elliptical orbit were found through calculations using the same theory. Uncertainties existed in the observations, and it is not surprising that the tables could produce only approximations to the true positions of the body to which they applied. Examining the effectiveness of the predicted positions as time passed provided a test for the precepts, or theory.

Gravitational astronomers forced the terms to represent these observations faithfully, by making any needed adjustments in the values of the orbital elements employed in the calculations of the coefficients of each periodic term. Much of this was done from hindsight, by knowing which terms from the theoretical calculation could be associated with which observed inequalities in the motion of the body under consideration. For example, the amount of rotation of the line of nodes of the lunar orbit[2] could be calculated from observations of the Moon's latitude above the reference plane, and this value could be used in the theoretical terms that exhibited such a variation. This whole process—the development of the theory of motion of the planet, the calculation and correction of the orbital elements to determine their initial values, and the solution of the equations of motion—involved large amounts of calculation. Any possibility to reduce this calculation—either by truncating series, or developing new methods—without negative effects on the accuracy of prediction was quickly employed in the generation of tables of predicted positions.

For short periods, perhaps a decade, the predicted positions were usually effective (though decreasingly so), and, therefore, the results were considered a positive test of theory. However, these sets of precepts of the eighteenth century were not effective producers of ephemerides for long periods. As the eighteenth century wore on, gravitational astronomers recognized that to improve the precision of the tables many more terms in the series employed in the equations of motion would have to be retained. Thereafter, with each succeeding generation of astronomers, the number of terms retained increased and the quantity of calculations rose at a rapid rate.

In spite of the retention of more terms, a certain disunity of the approach to the solution of the problem of perturbative motion contributed to the lack of long-term effectiveness of the position tables. The theories, i.e. the

Orbital elements

Astronomers use several reference spheres and planes for locating celestial objects, depending on the interaction to be examined. One reference sphere is a simple projection of the Earth's latitude (0° to +90°; 0 to −90°) and longitude (0° to 360°) lines against the celestial background. In this sphere, an object's position is located where right ascension, equivalent to longitude, and declination, that is, latitude, are the two reference angles. The plane of the Sun's orbit, called the ecliptic, defines a second reference sphere. (The ecliptic is analogous to the equatorial plane (plane of 0° latitude) in the Earth's reference sphere.) The place of a planet's orbit is related to the ecliptic, which is related to the Earth's equatorial plane for predicting future positions. The indicators, called elements, of a planetary orbit give information about the size (the semi-major axis, a) and shape (the eccentricity, e) of the orbit, the orbit's position in space relative to the ecliptic (the inclination, i), and the planet's position in the orbit (specified by angles from certain reference points). Referring to Fig. 7.3(a), the eccentricity, e, is defined as the ratio (2CS/AA′), where (1/2)AA′ is the semi-diameter a of the ellipse. Angle f is the true anomaly of the planet P. A is the perihelion, i.e. the closet planet P gets to the Sun and A′, the farthest position is the aphelion. The line AA′ is the line of apsides.

The methods described by Lagrange, Laplace, and others, used features of the shape of a planet's orbit, that is its size, a, its deviation from a circle, e, the orientation of the orbit in space, i, with respect to the Sun's orbit, and the position of the planet at some time, t. The positions in Fig. 7.3(b) S, the

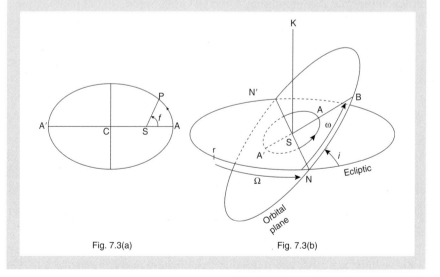

Fig. 7.3(a)   Fig. 7.3(b)

position of the Sun, and A and A′ are the positions of the planet's perihelion and aphelion, identical to the positions in Fig. 7.3(a). In locating the orbit, astronomers use as a reference the points, called nodes, labeled N and N′, where the orbit of the planet intersects the orbit of the Sun, the ecliptic. The node at which the planet moves above the Sun's orbit, the ascending node, N, is the main point of reference for citing the planet's position, known as small omega, ω. Over time this line between nodes oscillates, and the motion of the planet, that is, the time between crossings of a chosen line of longitude, varies each day. Astronomers calculate the mean value of the motion and use it in many calculations of position. Angle i represents the inclination of the planet's plane with respect to the Sun's orbital plane, the ecliptic. Angle Ω identifies the position of the ascending node of the planet's orbit as the planet moves counterclockwise in this figure, measured from the first point of Aires. The line NN′ is the line of nodes. AA′ is the line of apsides, identical to Fig. 7.3(a). Projecting the position of the planet at time t onto the celestial sphere at B, we obtain the angle ω, the longitude of perihelion from the ascending node. Ω, i, ω, along with a, e, and f from Fig. 7.3(b), are the orbital elements. Expansions of periodic terms in the equations of motion contain all of these elements, and methods differ depending on which element is chosen as the variable in solving the many equations for a single planet.

precepts and tables, were constructed in a semi-empirical manner, wherein techniques were used to force the theory to fit one body's motion, with little concern for the motions of the other bodies involved. Such a difficulty entered the process in the following way. The final series expressions for the heliocentric position coordinates, resulting from integration of the equations of motion in which the perturbing function R was expressed in terms of the orbital elements, contained terms that were functions of the orbital elements of the disturbed and undisturbed planets. The values of the coefficients in these terms were determined using orbital elements calculated from observations of each of the planets involved. Before the theory was used to generate an ephemeris, however, the values of the coefficients in the terms of the series expressions in the tables were checked by comparing the equations with past observations of the disturbed planet. The orbital elements in these coefficients were then adjusted to fit the early observations, and in most cases the elements of the disturbing planet's orbit were the quantities altered without respect for the values these elements must have to satisfy the primary motion of the disturbing planet. As a result, one planet would have several sets of orbital elements, depending on which planetary theory was being analysed. The values of certain orbital elements, for example, the mass of a planet, used as constants in the equations to calculate Uranus' effect on Saturn were often different than those used to calculate its effect on Neptune.

One further complication in the way of obtaining precise predictions lay with possible errors in the early observations, which errors would then be transferred to the coefficients in the series expressions for the heliocentric coordinates. The effect of these three difficulties—the arbitrary elimination of terms in the series for R, the alteration of parameters in the final expressions for the position coordinates by comparing the expressions with observation, and any errors in observation—was discrepancies between predicted and observed positions, which appeared after only a short time, making repeated corrections to the tables necessary.

Euler, in the 1740s, had the idea that some of this computation could be reduced. Instead of performing the above operations for all the equations of motion, he treated the equations for the radius of the orbit r and longitude in orbit λ in the same way but isolated the equation for the latitude and replaced it by two equations that expressed the perturbing forces in terms of the variation of the line of nodes and the inclination of the planet's orbit.[3] This was the germ of the method, developed to a fine degree toward the end of the century by Joseph Louis Lagrange (1736–1813), called the method of the variation of arbitrary constants.[4] Published by Lagrange in preliminary form in 1766 and refined in the 1780s, it eventually superseded all the other approaches. This method involved expressing the heliocentric coordinates in terms of the elements of the orbits in which the planets moved, and the perturbing forces as functions of the elements, instead of the coordinates, as described above. Euler, and later Lagrange, found that from knowledge of the position and velocity of a perturbed planet, hence the elements of an instantaneous orbit, the rates of change of the elements of its orbit could be found, and that these variations would then be useful in calculating the actual elements at some time t. Once the expressions for the elements were in hand, they could be used to determine the heliocentric position coordinates of the planet. The amount of computation using this method was no less, but the results seemed easier to obtain than by earlier methods.

Fig. 7.4 Joseph Louis Lagrange's scientific studies in mathematics and astronomy occurred primarily during his adult years in his native Turin and later at the Berlin Academy. He interacted with all the major mathematical astronomers of the period, especially Laplace and Euler, and focused on all the significant problems in the field, winning several prizes. He moved to Paris just before the Revolution and was instrumental in rebuilding science in this period.

A threat to the system

The pursuit of a unified set of precepts for computing predicted positions was interrupted in the last third of the eighteenth century by a new problem, which seemed to threaten the long-term stability of the planetary system. In the solution of Lagrange's planetary equations by integration of the equations of motion in terms of the orbital elements, the final terms displayed two types of time effects resulting from perturbations produced on a planet. One class of terms—the periodic inequalities—affected the perturbed body in a periodic manner, with a net effect of zero over the complete cycle. The periods varied in length, but could usually be associated with times of the order of the revolution of the perturbed and/or the perturbing body around the Sun. Some of these inequalities, however, possessed much longer periods, up to 900 years. The second class came to be known as secular inequalities, because they exhibited time effects that were cumulative, and it was this group that received considerable attention in the eighteenth century.

Short-period variations were easy to identify, and were, for the most part, well known empirically by 1760. However, most of the long-period inequalities were discovered in the eighteenth century, and with these discoveries came a great deal of confusion about the distinction between the secular variations and long-period effects. Moreover, there was concern that if the secular variations did exist the solar system was unstable. As a result of this latter concern, certain astronomers, notably Pierre Simon Laplace (1749–1827) and Lagrange, launched a theoretical effort to analyse the secular variations. These investigations were guided by the belief that the secular inequalities were really long-period effects, and that profound investigations of the secular variations using Newtonian gravitation would reveal how they affected the solar system in a periodic way.

Laplace, stimulated by several early studies by Lagrange, began his analyses of the secular variations by returning to the beginning of the solution of the equations of motion and extending the series expansions beyond the terms kept earlier by Euler and Lagrange. He applied this analysis first, in a memoir written in 1773, to the interactions of Jupiter and Saturn, trying to isolate the inequalities responsible for the differences between the predicted and observed longitudes of each planet.[5] Since the longitude terms depended primarily on the mean motions of the planets, the irregularity in the longitude was thought to result from uncertain knowledge about the mean motion effects. The differences in the longitudes of Jupiter and Saturn

Fig. 7.5 Pierre Simon Laplace is seen as one of the most influential scientists of all time. Laplace contributed to many areas of astronomy, physics, chemistry, mathematics, cosmology, theory of games of chance, and causality. His career spanned the turbulent years of the French Revolution, yet he was influential in institution building, such as the founding of the famous École Polytechnique, the definition of the scientific disciplines, and the establishment of the metric system.

appeared cumulative, but from an analysis of the variation of the elements by the method of arbitrary constants the variation of the mean motion turned out to be zero. This led to a search for some periodic terms that could account for the irregularities in these longitudes.

In the same year, both Laplace and Lagrange began investigating the secular variations to see whether or not a long-period inequality might account for them. Laplace analysed the limits over which the numerical values of the eccentricities and aphelia, that is, the points at which planets are furthest from the Sun, varied. In a report delivered to the Academy in 1773 and published in 1776, Laplace found that the variation within these limits would have little effect on the basic configuration of the solar system, and that this result was valid as long as the major bodies moved in the same direction, which, of course, they do.[6] Thus, the system was stable; and this lawful regularity confirmed the belief of Laplace and others that one could predict the motions of the planets over long periods of time using Newtonian gravitation.

Lagrange punctuated this work of Laplace with memoirs of his own. Indeed, the work of each man stimulated that of the other. He, too, found that the inclinations of the planetary orbits oscillated about mean values. In 1774, Lagrange announced that the secular variations would cause the inclinations of the orbits to oscillate about mean values from which they would deviate by only small amounts.[7] Later, in 1776, Lagrange applied the method of arbitrary constants to the irregularities in the mean motions of Jupiter and Saturn.[8] He demonstrated that the mean distances of the planetary orbits from the Sun were not subject to secular variations. In this manipulation of the variation of the mean motion for the changes in the other orbital elements, Lagrange came tantalizingly close to solving the problem of the irregularities in the mean motions of Jupiter and Saturn.

A few years later, Lagrange published a memoir, which included the final form of his development of the method of arbitrary constants.[9] He applied the technique to three cases: the interaction of Jupiter and Saturn; the mutual interactions of the system of Mercury, Venus, Earth, and Mars; and the perturbative action of Jupiter and Saturn on each of the four inner planets. Here, again, within the limit of his series expansions, he demonstrated that variations in the eccentricities and inclinations would be limited to periodic inequalities, thus narrowing the problem to secular variations in the motions of the line of apsides,[10] the line of nodes, and the longitude of the planet. And so it continued over the next decade.

The multitude of papers on perturbations was not in a form useful for the practical astronomer intent on constructing tables of the motions of these bodies, i.e. a unified set of precepts. Practical astronomers were more concerned with short-period terms in order to improve the usefulness of tables. The rapid developments in perturbation theory and their application in the eighteenth century required a synthesis before further definition of their quantitative usefulness to solar system astronomy could be achieved. Laplace, in his *Mécanique céleste* provided such a synthesis, by incorporating the work on perturbations in a general discussion of perturbation theory starting from first principles of Newton's mechanics as developed by Euler.

Theoretical development continued in the years during which Laplace prepared the *Mécanique céleste* (published in 5 volumes between 1798 and 1825).[11] Lagrange revived his investigations of the mean distances of the planets from the Sun. Simeon Poisson (1781–1840) initiated similar studies in about 1805, and Johann K. Burckhardt (1773–1825) and Jacques P. M. Binet (1786–1856) calculated the terms in the disturbing function to the sixth power of the eccentricity and the inclination. In addition, Burckhardt and Johann T. Burg (1766–1834) were both developing precepts for tables of the moon, and each, in the process, tried to isolate more inequalities so that their respective tables would be more accurate. Laplace included all this work in several of the appendices to the *Mécanique*. Laplace's treatise came into immediate use, and astronomers made rapid advances in planetary theory in the years following.

Tables

While Lagrange and Laplace were busily preparing the groundwork for a new programme for the development of astronomical tables, the tables and precepts that attracted the most attention were those of the Moon, because of their practical value in determining longitude positions on the Earth. The best tables for the Moon's motion were those of Tobias Mayer, developed in the 1750s.[12] With these tables lunar positions could be predicted within 1.25 minutes of arc of observation. Using predicted positions of the moon based on these tables, a skillful calculation would yield a position at sea within 27 miles of the true position. Hence, these tables constituted a splendid aid to navigation. A measure of the difficulty involved in the analysis of the Moon's motion, or the motion of any of the other bodies of the solar system for that matter, can be seen by noting that instruments of this period could measure positions with a precision of *two seconds* of arc. Since observations were so much better than predictions from theory, it is no surprise that the appearance of Laplace's *Mécanique*, with a new programme for practical astronomy, stimulated a drive in the beginning of the nineteenth century to produce new tables for the motions of all the major planets of the solar system and a new table for the motion of the Moon.

The success of Laplace's main thesis in the early volumes can be judged by the number of new tables that were published in the few years after 1803. Even before the last volume of the *Mécanique* appeared (1825), Alexis Bouvard (1767–1843) and Jean B. J. Delambre (1749–1822) were in the process of constructing tables, at the behest of Laplace. Franz X. von Zach (1754–1832) in 1804 and Delambre in 1806 published tables of the Sun, Bernhard A. von Lindenau (1780–1854) presented tables of Mercury (1813), Venus (1810), and Mars (1811), and Bouvard prepared tables of Jupiter and Saturn in 1808 and of Uranus in 1821. Burckhardt published tables of the Moon in 1812. Some of these tables, with corrections, were still in use in the middle of the century in the British and the then new American almanacs.

The French Bureau of Longitudes, among whose members in 1805 were Lagrange, Laplace, Delambre, Bouvard, and Burckhardt, sponsored the tables, and published them as they became available. The Bureau published Part One, of a projected series, in 1806; this contained Delambre's 'Tables of the Sun' and Burg's 'Tables of the Moon.'[13] Sprinkled into the section 'on the Construction and use of the Tables' for both sets were references to Laplace's *Mécanique*, but the authors attributed no general credit to him. Both men based their tables on observations of James Bradley (1693–1762) and Nevil Maskelyne (1732–1811) of Greenwich.

Each set of tables that appeared in the hundred years after 1750 showed improved agreement with existing observations, and great hopes were expressed each time for their continued usefulness. The success of the tables in this period was based partly on the improvements in theory, which increased the computational accuracy, and partly on an increase in the quality of observations.

It was not an easy task to compare the resulting ephemeris or predicted position tables with observation. Ephemerides of this period did not furnish geocentric positions for any body on a particular date, as is done today. At that time, in order to know the position, it was necessary to extract information from several tables and add, subtract, etc. until one arrived at a final position.

An example of the type of computation required in astronomy using these tables

To illustrate the content and use of tables, at the beginning of the nineteenth century, consider first the calculation for the true longitude of the Sun for a given date, say $16^h\ 7'19''$ November 13th, 1805, using Delambre's tables. The starting point for such a calculation was the mean longitude on January 1st of the year of interest. Now these values depended on past observations, and when Delambre drew up his tables of the Sun, in 1806, he possessed 50 years of good equinox and apogee positions, which yielded yearly changes in the mean longitude. With these changes, he generated a table of the mean longitude of the Sun for January 1st of each year. Next the change in longitude from January 1st to $0^h\ 0'\ 0''$ November 13th was extracted from the proper tables and added. But to this value one did not add the remaining advance of the body to the true time, i.e. up to $16^h\ 7'\ 19''$. First, the true time had to be altered by the equation of time (the slowness or fastness of the Sun), the secular variation (rate of change of precession per century), and for perturbations on the Earth due to the planets (though the latter perturbations made a correction of only 2.8 seconds out of 57 100 seconds in Delambre's tables). The result of these corrections was the mean time, and using the mean time, one could look up the correction to the mean longitude for the true time on the given day.

But then further corrections due to changes in the perigee position had to be made to this value of the longitude. From the value for the advancement of the perigee to November 13th, 1805, another table yielded the equation of the centre, which expresses the amount of eccentricity of the elliptical orbit, a correction to the mean longitude. To this one added the indirect effect of the planets and the satellites. Finally, the mean longitude was corrected for the perturbations of the planets on the Sun. Added to all this was the necessity to correct the obliquity, i.e. the angle of inclination of the ecliptic and the celestial equator, and calculate the perturbations of the radius vector and the latitude, the remaining two heliocentric position coordinates of the Sun. In sum, to obtain the mean heliocentric longitude, sixteen calculations were necessary, which made use of fifty entries from the tables. Similar calculations yielded the other two coordinates—heliocentric latitude, and the logarithm of the radius vector (reduced to the ecliptic). A simple conversion, in the case of the solar tables, then yielded the geocentric position of the Sun (or, in other words, the heliocentric position of the Earth). The data needed for these calculations appeared in 35 tables. Using these tables and repeating the above calculations, a computer could generate an ephemeris containing heliocentric and geocentric position coordinates of the planet for as small an interval of time between positions as desired.

Problems with observed positions of planets and stars

Turning now to the increase of the quality of observations, we note that some of the best observations available at the beginning of the nineteenth century for use in the equations of condition came from the work in the eighteenth century of James Bradley and Nevil Maskelyne, at Greenwich. Using these observations, along with selected others, the tables developed with Laplace's programme exhibited greater accuracy in the predicted positions. However, like their predecessors, the positions calculated from these new tables showed substantial discrepancies after only a few years. These discrepancies could have arisen in two ways: either the observations were inexact or the theoretical developments were incomplete. As the theoretical developments, outlined above, increased the astronomer's ability to calculate positions more accurately, a reinvestigation of the treatment of observations became necessary.

Friedrich W. Bessel (1784–1846) attacked this problem in a number of ways from 1806 until his death. At the suggestion of Wilhelm Olbers (1758–1840), Bessel reduced 3222 previously unreduced stellar observations of Bradley, made between 1750 and 1762 at Greenwich, when Bradley was Astronomer Royal. Bessel recognized that he must first analyse the theories of applying corrections to stellar observations for precession, nutation, and aberration effects produced by the motion of the Earth.[14] For his analysis, he determined the latitude of Greenwich for 1755 from modern observation and used Bradley's observations and those of Giuseppe Piazzi (1746–1826) to calculate proper motions.[15] As a result, Bessel presented a new theory of precession in 1815, a significant asset to the unified theory search, and the analysis of Bradley's observations appeared shortly after in 1818. These star reductions provided a new reference system for making observations of any celestial object.

Further, Bessel pointed out the necessity of continual observation of stars until enough information existed to predict their motion, in place of occasional measurements that resulted in no sure comparisons. He also called attention to the treatment of instrumental errors, and the necessity for an observer to correct observational positions for all these problems immediately after they were taken. In all, Bessel set a new standard in the treatment of observations in order to improve the accuracy of their use in equations of motion.[16]

Several specific aspects of Bessel's investigations were of great significance to nineteenth-century astronomy. Besides his improvements of the theories of precession, refraction,[17] and aberration, Bessel recognized another brand of error: the personal equation. He noted that each observer possessed a different rate of perception when looking through a telescope observing the motion of a celestial object, and that this rate should be accounted for when reducing observations. Bessel's investigations into perturbations were also significant, because they resulted in new values for the mass of Saturn and of its ring, based on his researches of Saturn's satellites. Bessel improved the methods of reduction and published them in 1830 in his *Tabulae Regiomontanae*.[18] These methods advanced accuracy and increased the ease of application; as a result, the methods were adopted everywhere.[19]

Another important observer in the Besselian vein was George Biddell Airy (1802–1892), Astronomer Royal from 1835 to 1881. Airy was exceedingly meticulous in his observations and reductions. Indeed, his penchant for orderliness was the bane of his computers. The demands he placed on himself and the staff at Greenwich resulted in large amounts of data. He has been seen as a man of little imagination, but his contributions to astronomy were large because of his demand for accurate data and his long tenure.

Airy's effect on positional astronomy was similar to Bessel's, in that he and his computers reduced large amounts of data that were useful in detecting difficulties in planetary theory. As early as 1827 he compared Delambre's solar tables with 86 solar observations and found that several elements required alteration.[20] In the following year, he obtained corrections to the elements of the Sun's apparent orbit in a discussion of 1200 observations made at Greenwich between the years 1816 and 1826.[21] His discussion of the corrections needed in the elements of the orbits of the planets Mercury, Venus, and Mars led him to conclude that the mass of Mars was too high by a factor of 30 per cent. (Burckhardt had made corrections to Delambre's tables in 1812. Airy's corrections agreed with Burckhardt's for Venus, but not for Mars.) Airy also discussed the coefficients of the lunar equation and found that there were several anomalies in the mean longitude of the Moon's position. These he ascribed to an inequality of long period that was produced by Venus. Airy suggested that this inequality might be large enough to account for the errors in Burckhardt's tables of the Moon, when coupled to the effect of the Earth. Airy's corrections were subsequently incorporated into calculation schemes for the British *Nautical almanac*.

Airy became Astronomer Royal and Director of the Greenwich Observatory in 1835. The Greenwich Observatory was the principal world producer of positional data from its founding, but gradually over the eighteenth century scientific supervision of the data decreased.[22] This was due to the Admiralty's interest in the data for navigational purposes only. However, a refocusing took place in the nineteenth century, partly due to the heightened interest in tables for astronomical use, and partly due to Airy's background. Educated as a mathematician, Airy spent several years as a professor at Cambridge, during part of which time he served as head of the Cambridge observatory, where he introduced Bessel's methods for the reduction of observation. With this training and his exceptional abilities as an administrator, he transformed Greenwich into a highly efficient institution. Even before Airy took command of Greenwich, he proposed the reduction of the mass of observations lying dormant since 1750. This task, begun in 1833 and completed in 1846, constituted a permanent set of positions for the correction of planetary theory.[23] The Observatory added to this set yearly, after Airy's succession, by rapid publication of reduced observations.

Besides attending to observations of position, Airy, like Bessel, also made a number of studies into the values of certain orbital elements of planetary theory. For example, he recalculated the mass of Jupiter and the elements of the Moon's orbit and he analysed data to further specify the position of the ecliptic. In addition, he published a number of papers on the improvement of instrumentation at Greenwich.[24] While he did not noticeably advance perturbation theory, he did provide the basic data for others. Urbain J. J. Leverrier (1811–1877) and John Coach Adams (1819–1892) both used Greenwich observations in their investigations of Uranus. Later, at least half of the observations Simon Newcomb (1835–1909) used in his analyses of planetary motion were made at Greenwich. Airy also made an important indirect contribution to lunar astronomy before 1860: through his intercession, the Admiralty awarded money to Peter Andreas Hansen (1795–1874), which allowed the German astronomer to complete and publish his 'Tables of the Moon' (1858).[25]

With the improvements in reduction, hence better observations, came a consequent increase in the discrepancies between observation and theory. Tables of the planets proved less useful as time passed and a general discomfiture about them infected gravitational astronomers. In the late 1830s, theory again became uppermost in their minds in an effort to preserve the notion of stability of the solar system, and this led to a series of memoirs by Hansen and by Leverrier.

Fig. 7.6 Urbain Jean Joseph Leverrier. A graduate of the École Polytechnique, Leverrier published several articles on celestial mechanics in the 1830s, which marked him as an important astronomer. Not long after he turned to study of the planets, taking on the task of recalculating all the tables for use in the French ephemeris. He is also considered one of the founders of modern meteorology, having reorganized the meteorological service in France, which other nations used as a model for their activities. An authoritarian head of the Paris Observatory, he often was at odds with the staff. Nevertheless, he achieved a great reputation among the world's scientists for his work. (Courtesy of the Academie des Sciences, Paris.)

The contributions of Hansen and Leverrier to planetary theory

The early works of Hansen and Leverrier grew out of a renewed concern for the older problems of the irregularities in the motions of Jupiter and Saturn. Hansen took up the problem of Jupiter and Saturn, and, during this investigation, he made one of the nineteenth century's most significant contributions in the articulation of perturbation theory. In the late 1820s, he began a study of the coefficients of the perturbing function expressing the perturbations of Jupiter and Saturn on each other, and, to treat the problem, he invented a new method for the analysis of perturbations, especially of orders higher than the first. The problem Hansen faced here was similar to that of Euler and Lagrange before him. Instead of employing Lagrange's method of arbitrary constants, Hansen developed a new method for ease in handling the higher-order terms that emerged from the series expansion of the perturbing function R.[26] This method became an important element in the work of Newcomb later in the century.

In the late 1830s, Leverrier, in his first astronomical memoir, took up the same question of the stability of the solar system.[27] He reanalysed the earlier investigations of Lagrange on this problem and found that, in calculations beyond a hundred years, Lagrange's formulae drifted. While his analysis was similar to its predecessors (though he ignored the new developments of Hansen) in that his expansions were carried out only to the first power of the eccentricity and inclination, it served as a prelude to a second memoir in which Leverrier expanded the coefficients of the equations to the third power. Previously, the contribution of these higher order terms had been considered negligible, but Leverrier found that after a short time they indeed became noticeable. As higher powers were included in the expansions, the treatment of perturbations became more difficult. Through his investigations on the secular inequalities, Leverrier was led to consider the difficulties of the motion of Mercury. This work led to a new set of precepts for Mercury's motion, precepts that included collections of terms to the eighth power.[28]

In a sense, Leverrier was repeatedly sidetracked onto other astronomical projects. He attended to short-period comets; he was involved in the discovery of Neptune, and the many controversies surrounding it and the plotting of its orbit; in 1853, he became director of the Paris Observatory. These other activities prevented him from completing a unified theory by the end of his life. Nevertheless, in computing new precepts for planetary orbits, especially the more difficult ones of Mercury, Uranus, and Neptune, he employed the latest observations, which due to their accuracy, increased the amount of calculation required by Leverrier and his staff in the production of new precepts and positions. Though he did not provide a unified synthesis, his was path-breaking work,[29] and it led the way for Newcomb later. In fact, the mass of analyses left by Hansen and Leverrier set many of the problems Newcomb would later deal with at the American Nautical Almanac Office in his successful attempt to develop a set of precepts with common constants for all the solar system bodies.

Ephemerides in the nineteenth century

Before completing the history of the development of a unified set of precepts, it is instructive to examine the similarities and differences among the computed tables of planetary positions in the nineteenth-century ephemerides. Basically, the three principal ephemerides (i.e. British, French, and United States) resembled one another.[30] The French almanac was meant to serve the nautical community, while the British and United States ephemerides attempted to fulfill the needs of both the nautical and astronomical communities. Each contained the apparent right ascension (RA) and declination (δ), the equation of time, and tables for the semi-diameter and the heliocentric rectangular coordinates of the Sun. These values were given for Paris mean noon (the beginning of the Julian day) and Greenwich mean and apparent noon (in separate tables), in the respective almanacs. The Americans, however, divided their ephemeris, first published in 1855, into two parts: one in Greenwich mean time for the use of mariners and one section in Washington mean time for the benefit of astronomers in the United States.[31]

Information on the Moon's motions included the RA and δ for Greenwich mean time, the semi-diameter, the horizontal parallax, the lunar distances (distance of the moon from the Sun, a planet, or a fixed star, which is used in calculating longitude at sea), moon culminations (the crossing of the moon through certain meridians), and the time of meridian passage. The British and Americans gave more information for the fixed stars, those used as the reference positions, than the French. While the French gave the apparent positions for a list of 100 principal stars, the others tabulated the constants for reduction (precession, nutation, etc.), Bessel's formulae of reduction using the notation of the catalogue of stars of the British Association for the Advancement of Science (1845), information on two stars of Ursae Minoris, and tables for nutation effects on these stars.

The ephemerides for the planets were essentially identical, though once again the French offered less information. For their respective meridians, the almanacs gave the time of transit of a planet, the RA and δ of the planet, plus correction factors. The inner planets' heliocentric positions were recorded at 5-day intervals, but as the radii increased and the arc traveled per day decreased, the intervals rose to 20 days. Only the Americans offered an ephemeris of Neptune, and only the British prepared tables of the minor planets.

All the almanacs contained information on the phenomena for that year: eclipses of the Sun, Moon, and Jupiter's satellites; the orientation of Saturn's rings and the apparent discs of Venus and Mars. They printed a list of the latitudes and longitudes of the principal observatories and tables to convert from mean solar time to sidereal time and vice versa. The almanacs also included the necessary data for observing occultations (or the disappearance from view) of stars as the moon passes in front of them, though again the French information was less complete.

Judged from another perspective, that of the information on which the ephemerides were based, the French also came in a distant second. The French used almost exclusively the products of French astronomers, ignoring the publications of foreign astronomers. They employed the tables published by the Bureau of Longitudes, prepared by members of the Board. I hasten to add that this does not characterize these French works as defective, merely not as good as they could have been. Therefore, the following remarks are restricted to the British and American almanacs.

The mean positions of the fixed stars for both almanacs came from Airy's Greenwich 12-year catalogue,[32] and the catalogue of the British Association for the Advancement of Science. Both used Johann F. Encke's (1791–1865) analysis of the transits of Venus of 1761 and 1769 to obtain the equatorial horizontal parallax (or angular diameter) of the Sun at the Earth's mean distance.[33] In 1855, the British still employed for the ephemeris of the Sun tables developed in Milan in 1833, while the Americans adopted Francesco Carlini's (1783–1862) later tables, which contained Bessel's improvements.

The bases for the lunar tables in these two ephemerides provide an interesting anomaly. The British still used Burckhardt's 1812 tables of the Moon. For the horizontal parallax and semi-diameter of the moon the office staff corrected these tables slightly. The Americans, on the other hand, generated manuscript tables incorporating the latest observations and reductions of constants. They followed Airy's theory, which was based on Giovanni A. A. Plana's (1781–1864) lunar theory modified with Hansen's inequalities and Airy's corrections of the lunar elements. To this, Benjamin Peirce (1809–1880) and Joseph Winlock (1826–1875) added Hansen's analyses of the secular changes of the Moon's mean motion and of the motion of the position where the Moon is furthest from Earth (lunar perigee), and Miles Longstreth's corrections of the equations for the Moon's longitude. The positions generated from this theory compared more favorably with observation than did the tables of Burckhardt.

Certainly the newness of the office in the United States encouraged the staff to choose the latest modifications in theories for their calculations. But in noting that they adopted the latest tables, we should point out also that the improvement in a number of cases was marginal. For the inner planets, the British employed von Lindenau's tables dating from the 1810s, and for the outer planets they used Bouvard's of the same years. The Americans had recourse to many of the same tables, but they made efforts to correct them with the latest data. In every case, they developed manuscript tables out of the earlier published works. For Mercury they used Leverrier's theory of 1848. The ephemerides of Venus and Mars were updates of von Lindenau's tables. Similarly, the tables of Jupiter and Saturn came from Bouvard, with corrections from Bessel. Only the elliptical solution for the orbit of Uranus came from Bouvard, and to this the staff added Leverrier's corrections for the perturbations of Jupiter and Saturn, and Peirce's corrections for the perturbations of Neptune.

The astronomical community praised the changes made by the American Almanac Office because of the more accurate predictions that resulted. In the late 1850s, the British and French incorporated a few of these modifications into their almanacs, particularly Peirce and Sears Walker's (1805–1853) theory and tables of Neptune. As the observatories at Greenwich, Paris, Poulkova, and Washington added more planetary observations, especially of Uranus and Neptune, to the bulk already reduced, comparisons revealed steadily increasing defects in the tables in use. Leverrier, Hansen, Delaunay, Airy, and Peirce planned to remedy the situation by providing new tables. In the next few years only Hansen succeeded in providing new tables for the Moon. Meanwhile, Leverrier began a new set of tables for the motions of each of the planets. While Leverrier followed closely the Laplacian programme, Hansen deviated from the Laplace methods in many ways. None of these men lived long enough to complete their planned work, though Leverrier came close.

Simon Newcomb and the completion of the planetary programme

If an astronomer desired to obtain one set of elements useful in all the orbits, so as to improve significantly on the earlier works, he had to substitute and cross-substitute the calculated elements until one set satisfied as many past observations as possible. This greatly increased the number of computations required before the generation of each ephemeris began. Such a programme required much technical and organizational skill to make it a success. Few astronomers were willing to invest such monumental efforts into a task whose end result might prove to be little better than those of earlier tables. Newcomb had not only the ability to manage such a task, even more than Leverrier, he had the ambition to contribute to astronomy in as significant a manner as possible. Newcomb reached the height of his powers in astronomy in the decade when both Hansen and Leverrier died, leaving an incomplete legacy for a unified set of precepts.

When Newcomb began his studies of Uranus' motion in 1858, he had to decide which method was best for computing the general perturbations. He had before him the works of Laplace (1798–1825), of Hansen on Jupiter and Saturn (1831), of Peirce on Neptune (1852), of Leverrier on the perturbative function (1855), and of Hansen on the minor planets (1857). Newcomb examined the three principal methods of developing the perturbative function. As we have seen, there were:

(1) the analytic method of Lagrange, presented by Laplace and used by Leverrier;
(2) the older mechanical method, also in Laplace's *Mécanique*; and
(3) the Hansen method, wherein method (2) was modified such that the motion of one of the planets is described in terms of the sweep of the radius vector using the mean angular velocity of the planet in an ellipse, after which the final positions were referred to a fixed plane.

Fig. 7.7 Simon Newcomb was perhaps the most internationally known scientist at the end of the nineteenth century. He was a prolific author, who published works in astronomy, mathematics, economics, psychology, education, religion, and science fiction. Leaders of various foreign governments repeatedly engaged his services in the development of scientific activities in their countries. When he died in Washington, DC, in 1909, he received the equivalent of a state funeral. (Courtesy of the Library of Congress Manuscripts Division.)

There was also a threefold choice of method to integrate the equations of motion.

(1) There was the method of the variation of arbitrary constants we described above. The integration resulted in an expression for the variation of the elliptic orbital elements due to the perturbations, and when numerical values calculated from these expressions were added to the elliptic values of the elements a new planetary position could be predicted.

(2) The integration of the equations of motion, including perturbative forces expressed in terms of the coordinates, yielded expressions for the variation in the coordinates directly, from which new positions of the planet could be obtained.

(3) There was Hansen's method, where the perturbations were expressed in terms of a disturbed mean radius motion in the instantaneous ellipse, and the results integrated to obtain variations in the coordinates in terms of this mean motion.

In his early attempts on Uranus, Newcomb followed Leverrier's technique, both items (1) of the above choices, almost exactly. He expanded the perturbative function in terms of the distances of the planets along their orbits (using as reference the node of intersection of the orbits of the two planets), the logarithms of the radius vectors of the planets, and the mutual inclination of the two orbits. Then, he expressed the equations of motion in terms of the longitude, radius vector, and latitude of the planet under consideration. The integration of the equations with these constants, which Newcomb performed before 1870, was a straight application of the Lagrange method of variation of arbitrary constants.

Newcomb ran into trouble when he tried to use the major theories available for Neptune, which were based on insufficient observational data—Neptune having been observed for only eight years at the time the tables were cast. So, in 1864, he decided to rederive the Neptune theory using the ten further years available. Once again, his method was almost identical to that of Leverrier.

At the time Newcomb published his Neptune tables, they represented the observed positions excellently. The American *Nautical almanac* and the British *Nautical almanac* adopted this computational method, which remained in use for the next 30 years. Alas, the tables themselves did not fare so well. They were replaced by a new set of Leverrier's tables for Neptune published in 1876, which met newer observations better.

None of this satisfied Newcomb's ambition to complete the Laplacian programme. In a new programme begun in 1878, Newcomb first proposed a revision of the *American ephemeris*, to improve its usefulness, and to free some time of the computers for other work. He submitted to the astronomers of the United States fifteen suggested changes. The new structure of the *Ephemeris* required less calculation, so the office personnel, along with some new help, could spend some of their time on the new planetary programme. George W. Hill (1838–1914) was assigned the work of Jupiter and Saturn. He coordinated this work with Newcomb's intentions for the entire programme, but, essentially, he was free to generate the tables in his own way. This was a massive task and it took Hill ten years to complete. Newcomb divided the rest of the work into two parts: the inner planets and Uranus and Neptune. For the last two planets, he planned only on updating his earlier work.[34]

Fig. 7.8 George William Hill was a very accomplished mathematician. He worked for the Nautical Almanac Office, but did his calculations on the farm where he was raised in West Nyack, New York. Besides his work on Jupiter and Saturn mentioned in the text, Hill made significant contributions to lunar theory, secular perturbation studies, subjective geometry, comet orbits, and the calculus of finite differences. His collected papers fill six substantial volumes. (This photograph originally appeared in F. Schlesinger, *Publications of the Astronomical Society of the Pacific* 49 (1937), p.5. Copyright 1937, Astronomical Society of the Pacific; reproduced with permission of the Editors.)

Newcomb's programme for the analysis of the motions of the inner planets can be summarized as follows. First, he discussed the computation of the general perturbations of the planets. For the four inner planets, fourteen pairs of planets had to be considered, because of the effects of Jupiter and Saturn on the inner planets, as well as their own interactions. One important element in sorting out the perturbations of the planets on each other was the mass of Jupiter. To determine this value, Newcomb used the interaction between Jupiter and the asteroid Polyhymnia, discovered in 1854. This asteroid made a near approach to Jupiter in 1885, and reached perihelion in 1888, so Newcomb was able to use observations of it from 1854 to 1889.

Second, he undertook the re-reduction of the older observations, and a discussion of the later ones to reduce them to a uniform system. He re-reduced Maskelyne's Greenwich observations from 1765 to 1811, using modern data; and he did the same for Piazzi's observations at Palermo for 1791 to 1813. To save time, he carried over Arthur Auwer's (1838–1915) reductions of Bradley's observations of 1750 to 1762, and Airy's reduction of John Pond's (1767–1836) observations of 1812 to 1830, as well as Leverrier's reduction of the Paris observations from 1800 to 1875, and Bessel's reduction of the observations at Königsberg from 1814 to 1845. To these Newcomb added his own reductions of later observations made at over a dozen observatories. The total number of observations ultimately used for the four inner planets was 62 030!

Third, since Leverrier's tables were in use in the ephemerides of Europe from 1864 on, Newcomb decided to calculate places for the planets prior to 1864 from these tables also. He began with Leverrier's formulae, partially reconstructed them, and recomputed the positions of the four inner planets and the Sun. From these manuscript tables, Newcomb calculated a set of elements, which he employed as the provisional elements in the development of his own theory. Several supplementary tables were also generated, such as those for changing the Sun's positions in longitude and latitude to right ascension and declination, and changing the latter from Greenwich mean noon to Greenwich apparent noon and then to apparent noon at the observatories whose observations were to be used. His final results embraced a complete revision of the planetary orbits and the masses of the planets, as well as the principal constants of reduction, such as solar parallax, precession, aberration, and nutation, in short, a unified set of precepts and constants for the planetary system—the Laplacian dream.[35]

Newcomb proposed that his new fundamental star catalogue and the orbital constants and masses of the planets developed in the American office be adopted internationally. While there was some bickering in the astronomical community at the end of the 1890s,[36] the tables of the Sun, the constants, and the masses developed in the American office were adopted universally by 1903, having been endorsed at an international conference in Paris in 1896. These astronomical constants and tables with occasional modifications remained in general use until the advent of electronic computers and Earth satellites in the 1950s. Evidence from satellite observations and computer calculations revealed necessary adjustments in the constants, and these were made and adopted at a conference in 1961.

But this did not lead to common use of either Leverrier's or Newcomb's tables in the generation of ephemerides. Leverrier's tables served as the basis for the *Connaissance des temps* until 1984; tables used in the construction of the British *Nautical almanac* were based on Leverrier's. In spite of a somewhat greater accuracy obtainable with Newcomb's tables, only the American *Nautical almanac* employed them, again until 1984. Beginning in 1960, satellites contributed to improved observational data, and electronic computers allowed for computational methods that are more accurate as more terms can be used in calculating positions. The changes resulting from the use of these new tools swept away the tables developed in the nineteenth century, though not the spirit of the Laplacian programme, now embedded in the computer programmes. Today's almanacs are more accurate and easier to use, thus making the endeavours of sailors and mariners safer and more effective.[37]

Further reading

Treatments of the history of astronomy in the past 150 years have reflected the emphases in astronomy at the time. Robert Grant in his *History of physical astronomy*, (London, 1852; reprint edition, Johnson Reprint Corp., New York, 1966) presents a highly detailed account of the concerns and accomplishments of the theoretical astronomy shortly after the publication of the work of Laplace and Lagrange. Writing 100 years later, Anton Pannekoek, *A history of astronomy*, (Allen & Unwin, London, 1961) focused more on the developments in astrophysics that happened after Grant's time. A more up-to-date work is that of John North, *The Norton history of astronomy and cosmology*, (Norton, New York, 1995), which offers a better context for the developments in the old and new astronomy. For a good brief history consult Arthur T. Berry, *A short history of astronomy*, (reprint edition, Dover, New York, 1961). For the history of celestial mechanics, the elegant essays in M. Hoskin (general editor), *A general history of astronomy*, vol. 2, (Cambridge University Press, Cambridge, 1995) are a necessity, especially those by C. Wilson on Lagrange, B. Morando on Laplace, and Morando on developments in celestial mechanics in the nineteenth century.

Those interested in the details of astronomical techniques can consult, besides the *Mécanique* of Laplace, Dirk Brouwer and Gerald M. Clemence, *Methods of celestial mechanics* (Academic Press, New York, 1961), from which the discussion of Hansen's work in this essay was taken. Technical details of navigational astronomy can be found with ample diagrams in Charles H. Cotter's *A history of nautical astronomy*, (Hollis & Carter, London, 1968).

For a charming and telling presentation of events and personalities in astronomy of the nineteenth century see Simon Newcomb, *Reminiscences of an astronomer*, (Riverside, New York, 1902).

Notes

1. The finite velocity of light gives rise to an apparent motion of a star or planetary object as the Earth turns in its orbit, called the aberration. The direction of a celestial object seen from the Earth is not the same as the direction seen by an hypothetical observer on the Sun, called the stellar parallax; another form of parallax, geocentric parallax, results from the finite size of the Earth. The centre of coordinates is the centre of the Earth and the parallax must be accounted for by an observer on the surface.
2. The line of nodes is that line joining the two points at which the orbit intersects a given plane, especially the plane of the ecliptic or of the celestial sphere.
3. L. Euler, *Récherches sur la question des inégalités du mouvement de Saturne et de Jupiter*, Paris, 1749.
4. J. L. Lagrange, 'Sur la méthode des variations,' *Misc. Soc. Turin*, vol. 4, 1766–69. *Oeuvres de Lagrange*, (Paris, 1867–92), vol. 2, pp. 37–63. See also Note 8.
5. P. S. de Laplace, 'Sur le principe de la gravitation universelle, et sur les inégalités séculaires des planets qui en dependent,' *Mémoires de mathématique et de physique, présentés . . . par divers sçavans* (often cited as *Savants étrangers*), vol. 7, 1776.
6. P. S. de Laplace, 'Sur les solutions particulières des équations différent celles et sur les inéqualities séculaires des planètes,' *Mém. (Paris)*, 1772, pp. 343–77. (This essay was begun in 1772, read in two parts (1773 and 1774), published in 1775 in the volume for year 1772.)

7. J. L. Lagrange, 'Sur les equations séculaires du mouvement des noeuds et des inclinations des orbites des planets,' *Mémoires Academie des Sciences (Paris)*, 1774; *Oeuvres de Lagrange*, vol. 6, pp. 635–709.
8. J. L. Lagrange, 'Sur l'altération des moyens mouvements des planètes,' *Mém. Academie des Sciences de Berlin*, 1776; *Oeuvres de Lagrange*, vol. 4, pp. 255–71.
9. J. L. Lagrange, 'Théorie des variations séculaires des élémens des planètes,' *Nouveaux Mémoires de l'Académie Royale des Sciences et Belles-Lettres (Berlin)*, 1781, pp. 199–276, 1782, pp. 169–292; *Oeuvres de Lagrange*, vol. 5, pp. 125–344.
10. The line of apsides is the major axis of an elliptical orbit.
11. P. S. de Laplace, *Traité de mécanique céleste*, 5 vols., Chez J. B. M. Duprat, Paris, 1798–1825.
12. T. Mayer, 'Novae tabulae motuum solis et lunae,' *Commentarii Societatis Royale Scientiarum Göttingensis*, 2 (1753), 383–430. At this time, Mayer sent manuscript tables to N. Maskelyne for use in calculating lunar positions. After Mayer's death, Maskelyne published an English corrected set of the tables to honour Mayer. See *New and correct tables of the moon and Sun, by Tobias Mayer...*, London, 1770. For a full treatment of Mayer's work, see E. Forbes and C. Wilson, 'The solar tables of Lacaille and the lunar tables of Mayer,' in *The general history of astronomy, Volume 2, Planetary astronomy from the Renaissance to the rise of astrophysics: Part B: The eighteenth and nineteenth centuries* (ed. R. Taton and C. Wilson), Cambridge University Press, Cambridge, 1995.
13. Bureau des Longitudes, *Tables astronomiques*, Première Partie, Courcier, Paris, 1806.
14. Precession, akin to a wobbling of the Earth's polar axis, results from the gravitational attractions of the Sun and Moon on the asymmetrical Earth—it is pear shaped. The effects of the Sun and Moon on the Earth produce another smaller wobble of the polar axis called nutation—an oscillation of the pole about the position it would have if only precession were present.
15. E. Lebon, *Histoire abrégée de l'astronomie*, Gauthier-Villars, Paris, 1899.
16. R. Grant, *History of physical astronomy*, Henry G. Bohn, London, 1852, p. 344.
17. Refraction of light from a star or planet is a change of direction in the light's path as it moves from an area of one density of matter to another, such as it does when it enters the Earth's atmosphere.
18. F. W. Bessel, *Tabulae regiomontanae reductionum observationum astronomicarum ab anno 1750 usque ad annum 1850 computatae*, Regiomonti Prussorum, 1830.
19. A. M. Clerke, *A popular history of astronomy during the nineteenth century*, 4th edn, London, 1902. See also, W. Fricke, 'Friedrich Wilhelm Bessel,' *Dictionary of scientific biography*, vol. 2, 1972, pp. 97–102.
20. G. B. Airy, 'Remarks on a correction of the solar tables, required by Mr. South's observations,' *Philosophical Transactions of the Royal Society of London*, 117 (1827), 65–70.

21. G. B. Airy, 'On the corrections in the elements of Delambre's solar tables required by the observations made at the Royal Observatory, Greenwich,' *Philosophical Transactions of the Royal Society of London* 118 (1828), 23–34.
22. O. J. Eggin, 'George Biddell Airy,' *Dictionary of scientific biography*, vol. 1, 1970, 84–7.
23. Clerke, *Popular history* (note 19), p. 79.
24. For Airy's bibliography, see W. Airy, *Autobiography of Sir G. Airy*, Cambridge University Press, Cambridge, 1896.
25. P. A. Hansen, 'Tables de la Lune construits d'après le principe newtonien de la gravitation universelle,' *Journal des mathématique pures et appliqués* 3 (1858), 209–19.
26. For details of the method, see P. A. Hansen, *Untersuchungen über die gegenseitigen Störungen des Jupiters und Saturns*, Akademie der Wissenschaften zu Berlin, Berlin, 1831.
27. U. J. J. Leverrier, 'Sur les variations séculaires des elements des orbites, pour les sept planètes principales, Mercure, Vénus, la Terre, Mars, Jupiter, Saturne et Uranus,' *Additions à la connaissance des temps pour l'an 1843*, Paris, 1840, pp. 1–66.
28. U. J. J. Leverrier, *Theorie du mouvement de Mercure*, Paris, 1845.
29. Leverrier's many results can be found in the volumes of the *Annales de l'Observatoire de Paris*, beginning in 1855.
30. *Astronomical Ephemeris and nautical almanac for* [year], (beginning in 1767) London; *Connaissance des temps ou des mouvements céleste, á l'usage des astronomes et des navigateurs* [year] (beginning in 1679), Paris; *American ephemeris and nautical almanac for the year* [date] (beginning in 1855).
31. The American office was established in 1849 and published its first ephemeris for the year 1855.
32. G. B. Airy, *Catalogue of 2156 stars, formed from the observations made during twelve years, from 1836 to 1847, at the Royal Observatory, Greenwich*, London, 1849.
33. J. F. Encke, *Die Entfernung der Sonne von der Erde, aus dem Venusdurchgange von 1761 hergeleitet*, Gotha, 1822; *Der Venusdurchgang von 1769, als Fortsetzung davon*, Gotha, 1824.
34. Newcomb and his colleagues published the results of this work in a new publication of the (American) Nautical Almanac Office called the *Astronomical papers prepared for the use of the American ephemeris and nautical almanac*, beginning with volume 1 in 1882, US Government Printing Office, Washington, DC. Newcomb retired as head of the Almanac Office in 1894, but continued on to complete his work on the unified planetary theory with additional appropriations from the US Congress.
35. A brief summary of the work and some results was published separately. S. Newcomb, *The elements of the four inner planets and the fundamental constants of astronomy* (Supplement to the *American ephemeris and nautical almanac for 1897*), US Government Printing Office, Washington, DC, 1895.

36. Details of the controversy are discussed in A. L. Norberg, 'Simon Newcomb's role in the astronomical revolution of the early nineteen hundreds,' in *Sky with ocean joined* (ed. S. J. Dick and L. E. Doggett), US Naval Observatory, Washington, DC, 1983.
37. I am grateful to Dr P. L. Frana for help with preparation of the figures.

Ab—Ad

ALPHABETICAL LIST OF OCCUPATIONAL TERMS SHOWING THE CODE NUMBER OF THE HEADING TO WHICH EACH IS TO BE REFERRED.

The figures denote the Code Number of the Heading under which the Occupation will be found.

If an Occupation is expressed by more than one word, and is not found under the first, it should be sought under each successive word.

Sometimes an Occupation which does not appear in the Alphabetical List may be found under a description which is only a slight variant from it.

A

Term	Code Number
A.B.	735
able-bodied seaman	735
able seaman	735
abrasive wheel fireman	092
abrasive wheel kiln man	092
abrasive wheel maker	098
abrasive wheel maker's helper	099
abrasive wheel mixer	098
abrasive wheel turner	098
absorber man	143
accessories assembler	309
accessories mounter	309
accessories preparer	319
accessories worker (sewing machines)	279
accident inspector	714
accompanist	886
accordion maker	648
accordion pleater	428
accountant, assistant	939
accountant (bank)	791
accountant (book-keeper)	939
accountant (chief)	931
accountant, chief	931
accountant, cost	933
accountant, costing	933
accountant, turf	890
accountant-clerk	939
accountant's clerk	939
account book binder	532
account book ruler	533
account man, owner's	056
accounts clerk	939
accounts clerk (railway)	939
accoutrement clicker	342
accoutrement maker, army	348
accumulator assembler	302
accumulator attendant	954
accumulator box and stand maker	474
accumulator case bender	619
accumulator case cementer	619
accumulator case maker	618
accumulator charger	954
accumulator erector	302
accumulator installer	302
accumulator maker	302
accumulator paster	302
accumulator plate maker	188
accumulator repairer	302
accumulator sawyer	319
acetate of lime maker	142
acetate of sodium maker (artificial silk)	143
acetate acid distiller	142
acetic acid maker	142
acetone maker	142
acetone purifier	142
acetone recoverer	142
acetone recovery man	142
acetone rectifier	142
acetone rectifying man	142
acetone weigher	149
acetylene burner	249
acetylene burner and welder	249
acetylene gas apparatus fitter	210
acetylene gas maker	953
acetylene welder	249
acetylene welder and cutter	249
acetyliser	143
achromatic hand	138
achromatic lens worker	138
achromatic man	138
achromatic worker	138
acid bath tenter	361
acid bath worker	126
acid blower	149
acid boiler man	951
acid boiler man, sulphuric	951
acid burner	143
acid burner man	143
acid chamber man	143
acid column man	143
acid concentrator	143
acid de-arsenicator	143
acid dipper	250
acid furnaceman	143
acidifier (candles)	158
acidifier (oil)	158
acid kiln man	143
acid labourer (explosives)	149
acid maker	143
acid maker, acetic	142
acid maker, nitric	143
acid maker, oxalic	143
acid maker, stearic (candle)	159
acid maker, wood	142
acid man	449
acid man, carbolic	143
acid man, concentrating (acids)	143
acid man, concentrating (explosives)	142
acid man, de-nitrating	142
acid man (dyes)	149
acid man (galvanised sheet)	149
acid man, hydrochloric	143
acid man, muriatic	143
acid man, nitric	143
acid man (oil)	158
acid man, rectified sulphuric	143
acid mixer	143
acid oil attendant (margarine)	143
acid pan man	143
acid polisher	126
acid pot man (sugar refining)	449
acid purifier	143
acid refiner (acids)	143
acid refiner (oil)	158
acid shop worker (celluloid)	148
acid tower man	143
acid worker (chemicals)	143
acid worker, picric	143
acid worker (pottery etching)	108
acid worker (pottery finishing)	118
acid worker (xylonite)	148
acid works labourer	149
acoustic instrument maker	683
acrobat	899
acting guard	702
acting manager	880
action assembler (gun)	239
action assembler (piano)	642
action busher	649
action filer (gun)	233
action filer (piano)	649
action finisher (concertinas)	642
action finisher (pianos)	642
action fitter (gun)	239
action fitter (pipe organs)	642
action fitter-up (gun)	239
action forger, gun	190
action forger (gun)	190
action freer	239
action jointer	239
action machiner	200
action maker, bottom (player pianos)	642
action maker (concertinas)	642
action maker, French (pianos)	642
action maker (gun)	239
action maker (organ, piano)	642
action maker, top (player piano)	642
action part maker (pianos)	642
action rail finisher	498
action rail maker	486
action regulator (pianos)	642
action screwer-down (pianos)	649
action viewer (small arms)	278
actor	885
actor, cinema	885
actor manager	885
actor vocalist	885
actress	885
actress, cinema	885
actuarial clerk	939
adapter (artificial teeth)	684
adapter, gas and candle fittings	305
adding machine inspector	267
adding machine mechanic	267
adding machine operator	939
addresser, circular	789
addresser, envelope	789
addressograph operator	939
adhesive seal maker	559
adjuster	323
adjuster	210
adjuster, average	799
adjuster, balance	260
adjuster, binocular	322
adjuster, claims	793
adjuster, compass	322
adjuster, field glass	322
adjuster (gramophones)	648
adjuster, instrument	309
adjuster, keyless	324
adjuster, lens	322
adjuster, load	706
adjuster, scale	260
adjuster, ship's compass	322
adjuster, sound board (pianos)	498
adjuster, spring balance	260
adjuster, travelling	260
adjuster, unit	210
adjuster, watch and clock	323
adjuster, weighing machine	260
adjuster, weight	260
adjuster (Weights and Measures Department)	260
adjustment operator (sewing machine)	210
advance agent	880
advertisement agent	779
advertisement canvasser	774
advertisement contractor	779
advertisement drop curtain painter	594
advertisement fixer	989
advertisement inspector	779
advertisement manager	779
advertisement manager (newspaper)	779

379

8

The General Register Office and the tabulation of data, 1837–1939

EDWARD HIGGS

In the 1861 census of England and Wales householders were asked to give eight pieces of information on each member of their household—name, relationship to head of family, marital condition, age, sex, profession, birthplace,

Fig. 8.1 Page from an occupational dictionary used by the General Register Office in 1921. (Ministry of Labour, A dictionary of occupational terms, HMSO, London.)

and certain information on medical disabilities. To help them, householders had printed instructions on their household schedules, and an example of how to fill in the form. This made it quite clear that 'relationship to head of family' was to be understood in terms of kinship, or non-kin terms implying a pecuniary status: 'servant', 'apprentice', 'lodger', and 'boarder'. However, Jasper Messenger, an agricultural labourer living in a shed at the bottom of a farmer's garden in the village of South Cerney in Gloucestershire, described his relationship to the household head as 'friendly'.[1] Jasper may have been foolish, or having a little joke, but the census authorities, and modern historians, would have almost certainly said that his entry had been filled in 'incorrectly'. Certainly, if one looks in the nineteenth-century *Census reports*, 'friendly' does not appear on any of the axes of tables relating to relationship to head. But why is the term 'wrong'? After all a 'friendly relationship' is a perfectly sensible English usage. But in this case it was not what the census authorities wanted.

Data truncation in a world of scarcity

Jasper Messenger's 'incorrect' census schedule entry points to the central paradox of all census-taking, along with many other social processes—that they are as much about the truncation of information flows as their collection. The construction of forms and tables involves decisions about what is important and what is not, and what should be collected and presented, and what can be ignored. It also involves a decision about what information means, and what it does not mean. A census schedule today, as well as in the nineteenth century, demands only certain pieces of information about individuals; the rest is regarded as irrelevant to the task at hand. To the bureaucratic mind we can be male or female but nothing in between, and who we love is irrelevant. Citizens are not asked what they find important, or what to them reveals the 'state of the nation'. Similarly, the two-way tables given in nineteenth-century *Census reports* select merely a few variables to hold up for inspection: occupation cross-tabulated by sex and age but not by the social class of one's mother, or by ethnicity. Someone makes decisions about what is 'important', and that person is someone in a position of authority. But these decisions can also change over time, reflecting the changing concerns of such people.

However, there is nothing necessarily sinister about such processes of truncation within survey methodologies. They may merely reflect the need to do certain tasks in as efficient a manner as possible given the scarcity of means: information collection, analysis and presentation needs to be curtailed to fit the resources available. Such scarcity of resources applies as much to the public that has to read tables as to the institutions constructing tables, since the human ability to digest information is limited. In the case of the organization considered here, the General Register Office (GRO), the resources available for the processes of tabulation were indeed limited. The GRO was established in 1837 to supervise the secular system for registering births, marriages, and deaths set up locally in England and Wales under the provisions of the 1836 Registration and Marriage Acts. The GRO maintained a central 'database' of copies of the certificates of registration of

Fig. 8.2 Search Room of the General Register Office, 1861. (*Illustrated Times*, April 20, 1861, p. 259.)

Fig. 8.3 Dr William Farr, compiler of statistics and superintendent of statistics at the General Register Office, 1838–79. (The Mansell Collection Ltd.)

vital events issued by local registrars, from which it produced weekly, quarterly, annual, and decennial reports on medical and demographic trends. A key element in this process was the analysis of cause of death data supplied on death certificates, which helped underpin advances in both medical science and public health. The GRO's first superintendent of statistics, William Farr, was at the forefront of the development of medical and sanitary statistics from his appointment in 1839 to his retirement in 1879.[2]

From 1840 the GRO was also responsible for the administration of the decennial censuses, gathering information on the size and distribution of the population, on housing, household structure, and on the characteristics of individuals already noted above. This material, aggregated at the district, regional, and national levels, and cross-tabulated in various ways, was disseminated in published parliamentary papers. The census produced important statistical information in its own right but it was also the basis of much of the GRO's medical and demographic work—knowledge of the size of the population allowed the calculation of national, local, and cause-specific death rates; age and occupational figures allowed the calculation of mortality statistics and life tables for age cohorts and particular trades; and so on.

The processes of collecting and analysing census and civil registration data involved the handling of vast amounts of information. During the English census of 1901, for example, 10 pieces of information were gathered on 32 527 842 individuals, whilst registered vital events in the same year totalled just over 2 000 000. In order to undertake this work the GRO had precious few hands. For most of the nineteenth-century the GRO had fewer than 90 staff, and of these never more than 25 percent were employed in the Statistical Department which undertook the tabulation of data.[3] At the time of the census the GRO was able to employ temporary clerks to undertake the work of data processing. However, since in the nineteenth century the necessary Census Acts were usually passed only some nine months prior to Census Night, there was little time to organize and train such staff. Moreover, temporary clerks were the dregs of the clerical profession over whose selection the GRO had little control. The quality of the clerks can be judged from an official complaint to the Treasury in 1889 that in the previous census:

> There were altogether 98 clerks supplied to the Census Office, by Treasury nomination, independently of some few writers transferred from the GRO. Of these 98, four were in such bad health that no work could be got from them, and two of those died. Of the remaining 94, no less than 49, or more than a half, have marked against their names in the private register kept as to their conduct and qualifications either 'indifferent', or 'bad' or 'very bad'.[4]

The Office had made similar protests to the Treasury in 1861 and 1871.[5]

These temporary clerks, and the established staff of the Office, only had simple, manual technologies to undertake this vast task of data processing and presentation. Until 1911 the GRO's clerks had to add up and present results via the use of tabling sheets and the 'ticking' method. In the case of occupational abstraction by age, for example, the tabling sheets were large pieces of paper with occupational headings down one side and age ranges across the top. These headings were ruled across the sheet, creating a matrix of boxes into which the census clerks were to place a tick for an occurrence in the enumerators' returns of a person of the relevant age and occupation. The ticks in the columns were then added up, and the results placed in another series of columns on another sheet, giving the raw numbers of people under particular occupational headings within particular age ranges. Sheets were created in this manner for each registration sub-district.[6]

In order to create tables for registration districts, the sheets for sub-districts had to be folded at the column to be totalled and then lined up so that they overlapped, and the figures were then read off on to district sheets. Figures were transferred from district to county sheets in a similar manner. This was extremely fiddly work, and the constant leaning over tables was fatiguing for the tabler, necessitating frequent stops for rest.[7] In 1856 the GRO reported to the Treasury that an experienced clerk had just completed 'tabling the ages and diseases of the females of Lancashire 1854, comprising 310 abstract sheets containing an aggregate of 29,063 "ticks"', and that this had taken him four days.[8] This cumbersome process helps to explain why the GRO sought to keep the census and civil registration processes as simple as possible in the nineteenth century, and why its statistical output was restricted to tables with simple cross-tabulations.

At its inception the GRO was also under pressure to produce quick and easily comprehensible results because the Treasury was initially doubtful about the usefulness of the Office having a statistical function at all. Thomas Lister, the first registrar general, approached the Treasury in 1838 for permission to employ someone for the 'difficult and important duty' of drawing up an abstract of the causes of death. However, the Treasury expressed itself very doubtful about making an appointment at all, on the not insubstantial grounds, 'that unless the information proposed to be given can be afforded with the certainty of entire accuracy, it would be better that it should be omitted, as being otherwise calculated to mislead in a matter of very great importance'. It was only because Lister complained that he had already prevailed upon the Presidents of the Royal Colleges of Physicians and Surgeons, and the Master of the Society of Apothecaries to circulate medical practitioners with an exhortation to provide accurate information on the death certificates for statistical purposes, that the Treasury grudgingly gave way.[9] Civil servants, the medical profession and the general public were woefully innumerate then as now, and there was plainly a danger that data presented in a complex manner would baffle and confuse them. A newly formed institution such as the GRO, with a distinct scientific and political agenda, could not afford to fail to get its message across.

Forms of data truncation in the GRO

Given these constraints the GRO required simple and easily understood means of presenting data, and in this the two-way table was its prime statistical device. But the adoption of this essentially two-dimensional form of representation entailed a necessary simplification of the complex natural and social phenomena which the Office sought to study. These forms of complexity reduction can be summarized in the following manner:

- the restriction of data collection to a narrowly defined set of questions;
- the reduction of multi-faceted variables to unitary entities;
- the subsumption of complex responses under general headings in classification schemes informed by ideological assumptions; and
- the assumption of homogeneous populations.

Taking these in turn, the information collected by the GRO was intentionally circumscribed, partly via the development of the forms and schedules used in the census and civil registration process, but also by limiting the numbers and scope of the questions asked. On the death certificate, for example, all that was asked for was cause of death, and no attempt was ever made by the Office to obtain information about morbidity (i.e. sickness rather than death—'mortality'). In the case of the census, the questions asked were almost identical from 1851 through to 1881, and thereafter the GRO fought to prevent other questions being inserted into the schedule. When a Treasury Committee on the Census in 1890 supported calls from social scientists for the insertion of a question on employment status, the GRO claimed that it had no legal authority to collect such information. It had to be ordered by the Local Government Board, its parent department, to do so, and the lead in the subsequent insertion of new questions was taken by other government departments.[10] Even in the twentieth century the number of questions asked by the British census was always significantly lower than that in the US census.[11]

The use of simple two-way tables also encouraged the Victorian GRO to present variables as undifferentiated, unitary entities, that helped to form cells whose meanings appeared unambiguous. People could then be neatly placed once, and once only, in one of the boxes. As a result, the numbers of cases in various census tables would be the same, thus avoiding difficult questions over comparability. Census respondents, for example, could only have one occupation in the published tables, although they were encouraged to give multiple occupations on their census schedules. Many did so, in forms such as 'Farmer, Maltster and Brewer', although the census clerks were instructed to only abstract the 'most important', usually the first, during the processes of tabulation. The wording on the census schedule also spoke of people having a 'Rank, Profession or Occupation', which may have precluded the recording of much casual, and seasonal work, especially that of women and children, from the decennial enumeration.[12]

Similarly, the mortality tables in the GRO's *Annual reports* assumed that everybody died of a 'primary' cause of death, rather than from multiple causes, enabling the dead to appear only once in each annual table. However, on their death certificates doctors defined 'primary' in several ways, either chronologically, or in terms of the most important with regard to the termination of life.[13] The later introduction on the certificate of the concept of a 'secondary' or 'contributory' cause of death caused even more confusion, since it then became difficult to tell whether any 'secondary' cause was regarded as a consequence of the primary, or as of independent origin, but contributing appreciably to the death. These difficulties led the GRO to begin to doubt the very objectivity of the concept of a primary cause of death. As it noted in the *Statistical review* for 1927

it was very difficult to say, in any instance, how far the quest for the 'primary cause' should be pushed. If a cancer, for instance, could be referred to some definite irritation, as by tar or oil, was the cancer or the irritant the primary cause?[14]

Just as the census clerks had rules for choosing one occupation amongst many, so the registration clerks were given rules for choosing which causes of deaths were to be given precedence. According to the 1911 instructions to the GRO's clerks, violence was to be preferred to general diseases as a cause of death, general diseases to local diseases, and local diseases to ill-defined causes. Terms for general diseases were then placed into four classes in order of preference for selection: smallpox, for example, came in Group I, scarlet fever in II, influenza in III, and diabetes in IV. Where multiple causes of death came from the same group, then those of the longest duration were to be given precedence. Chronic diseases were to take precedence over non-chronic, and if all else failed the first given was to be taken as the cause of death for the purposes of abstraction.[15] As a result of these processes of truncation, citizens were presented as living simple lives, and dying uncomplicated deaths.

Allied to these issues were the conceptual difficulties arising from the elaboration of various classification systems in terms of medical nosologies (classifications of diseases), occupational classification systems, and socio-economic groupings. These all worked by placing reported causes of death and occupations under broader headings that were then the terms placed on the axes of tables. In the occupational classification system, for example, all the various jobs and occupational titles involved in the production of cotton goods would be subsumed under the broader headings such as

Causes of Death, 1871.

LONDON.—CAUSES of DEATH at different Periods of Life in the Year 1871—MALES—*continued.*

Class.	CAUSES OF DEATH.	ALL AGES.	MALES.															
			Total under 1 Year.	1	2	3	4	Total under 5 Years.	5–	10–	15–	20–	25–	35–	45–	55–	65–	75–	85–	95 & upwds.
I.	**ORDER 1.**																			
	1 Small-pox	4126	558	223	185	221	226	1413	636	205	296	432	641	314	133	41	12	3	–	–
	2 Measles	719	155	270	130	75	36	666	46	2	–	–	4	1	–	–	–	–	–	–
	3 Scarlet Fever (Scarlatina)	978	64	137	184	161	113	659	233	30	19	11	18	3	4	1	–	–	–	–
	4 Diphtheria	173	15	23	25	23	17	103	46	3	4	1	3	4	4	3	2	–	–	–
	5 Quinsy	10	–	–	–	–	–	–	1	–	1	1	1	3	1	1	1	–	–	–
	6 Croup	279	55	70	52	50	30	257	22	–	–	–	–	–	–	–	–	–	–	–
	7 Whooping-cough	1045	397	351	152	69	45	1014	30	1	–	–	–	–	–	–	–	–	–	–
	〔 Typhus Fever	172	1	1	4	4	5	15	15	13	15	15	30	28	16	16	8	1	–	–
	8〈 Enteric or Typhoid Fever	410	16	16	19	17	15	83	45	39	47	38	48	34	31	23	18	4	–	–
	〔 Simple Continued Fever	195	12	14	18	19	11	74	39	12	10	10	12	10	13	7	5	3	–	–
	9 Erysipelas	286	86	3	2	3	–	94	2	1	5	10	14	31	37	51	31	9	1	–
	10 Puerperal Fever (Metria)	–	–	–	–	–	–	–	–	–	–	–	–	–	–	–	–	–	–	–
	11 Carbuncle	23	–	–	–	–	–	–	–	–	–	1	–	5	4	2	10	1	–	–
	12 Influenza	8	4	–	–	–	–	4	1	–	–	–	–	–	1	–	1	1	–	–
	13 Dysentery	56	5	6	1	–	2	14	–	1	1	4	7	6	7	5	7	4	–	–
	14 Diarrhœa	2029	1499	330	36	13	5	1883	11	1	–	2	8	4	16	30	42	27	4	1
	15 Cholera	125	62	22	5	4	1	94	5	–	–	–	3	5	7	6	3	2	–	–
	16 Ague	13	–	–	–	1	–	1	1	1	2	–	1	–	1	4	1	1	–	–
	17 Remittent Fever	4	–	–	–	–	–	–	1	1	–	–	1	–	–	1	–	–	–	–
	18 Rheumatism	206	–	–	–	–	4	4	14	24	14	15	38	37	20	21	17	2	–	–
	19 *Other Zymotic Diseases*	13	5	2	1	1	2	11	1	1	–	–	–	–	–	–	–	–	–	–
	ORDER 2.																			
	1 Syphilis	194	151	15	3	1	–	170	–	–	1	2	6	3	7	5	–	–	–	–
	2 Stricture of Urethra	63	–	–	–	–	–	–	–	–	–	1	6	8	18	15	13	2	–	–
	3 Hydrophobia	–	–	–	–	–	–	–	–	–	–	–	–	–	–	–	–	–	–	–
	4 Glanders	1	–	–	–	–	–	–	–	–	–	–	1	–	–	–	–	–	–	–
	ORDER 3.																			
	1 Privation	22	–	–	–	–	–	–	–	–	–	–	2	7	1	6	4	2	–	–
	2 Want of Breast-milk	180	179	1	–	–	–	180	–	–	–	–	–	–	–	–	–	–	–	–
	3 Purpura and Scurvy	73	10	17	5	–	–	32	5	3	2	6	8	8	6	2	1	–	–	–
	4 Alcoholism 〔 *a* Del. Trem.	72	–	–	–	–	–	–	–	–	–	1	;14	28	18	9	2	–	–	–
	〔 *b* Intemp.	20	–	–	–	–	–	–	–	–	–	–	1	3	8	5	3	–	–	ᵋ
	ORDER 4.																			
	1 Thrush	91	88	3	–	–	–	91	–	–	–	–	–	–	–	–	–	–	–	–
	2 Worms, &c.	9	–	8	2	–	–	2	–	–	–	3	2	–	1	1	–	–	–	–
II.	**ORDER 1.**																			
	1 Gout	101	–	–	–	–	–	–	–	–	–	–	1	·4	23	42	21	9	1	–
	2 Dropsy	206	18	5	2	6	1	32	13	3	3	4	12	19	25	36	37	18	4	–
	3 Cancer	459	–	–	–	–	–	5	1	2	5	4	19	65	106	142	84	22	4	–
	4 Cancrum Oris (Noma)	6	1	2	1	1	1	6	–	–	–	–	–	–	–	–	–	–	–	–
	5 Mortification	83	5	–	–	–	–	5	–	–	–	–	.1	3	7	15	21	23	7	1
	ORDER 2.																			
	1 Scrofula	232	77	41	13	9	4	144	26	8	13	5	17	7	4	6	1	1	–	–
	2 Tabes Mesenterica	659	367	159	48	20	12	606	31	9	6	3	2	2	–	–	–	–	–	–
	3 Phthisis	4765	95	75	32	24	23	249	76	55	254	486	1199	1131	816	382	104	12	1	–
	4 Hydrocephalus	774	320	197	89	37	48	691	61	11	1	2	3	5	–	–	–	–	–	–
III.	**ORDER 1.**																			
	1 Cephalitis	451	107	65	40	32	31	275	57	9	15	10	22	25	15	13	8	2	–	–
	2 Apoplexy	952	42	14	12	4	6	78	18	7	7	12	42	89	168	211	95	14	–	–
	3 Paralysis	702	1	1	–	2	1	5	1	5	2	4	22	72	94	131	215	133	18	–
	4 Insanity	31	–	–	–	–	–	–	–	–	2	1	1	7	6	4	4	4	2	–
	5 Chorea	4	–	–	–	–	–	–	1	–	–	–	1	–	–	1	–	–	–	–
	6 Epilepsy	170	3	2	–	4	2	11	6	6	13	12	31	25	28	11	16	8	3	–
	7 Convulsions	1609	1260	217	70	25	14	1586	17	1	–	–	1	1	1	2	–	–	–	–
	8 *Brain Disease, &c.*	430	32	17	9	6	7	71	12	10	6	12	26	41	55	81	84	32	–	–
	ORDER 2.																			
	1 Pericarditis	65	–	3	–	–	1	4	8	8	6	2	7	9	11	7	3	1	–	–
	2 Aneurism	134	–	–	–	–	–	–	–	–	–	2	3	42	37	24	16	8	–	–
	3 Heart Disease, &c.	1665	2	3	–	1	3	9	37	38	39	46	165	250	297	348	320	107	8	1
	ORDER 3.																			
	1 Laryngitis	243	82	54	7	17	11	171	14	1	2	3	13	12	13	9	4	1	–	–
	2 Bronchitis	4270	1003	473	167	56	37	1736	38	9	12	17	83	218	413	644	709	326	63	2
	3 Pleurisy	97	2	1	1	–	–	4	1	2	7	4	19	21	11	21	6	1	–	–
	4 Pneumonia	2069	606	395	160	54	35	1250	63	17	29	40	125	129	164	135	82	31	4	–
	5 Asthma	260	–	–	–	–	–	–	1	–	–	2	7	14	14	12	8	3	–	–
	6 *Lung Disease, &c.*	400	121	33	11	6	3	174	7	3	3	12	21	33	40	49	32	24	2	–
	ORDER 4.																			
	1 Gastritis	44	4	–	1	2	–	7	–	1	1	–	6	1	8	8	10	2	–	–
	2 Enteritis	140	49	7	–	4	1	65	9	8	6	3	8	8	5	10	9	6	3	–
	3 Peritonitis	117	12	4	–	1	2	19	7	10	12	6	13	11	19	11	6	2	1	–
	4 Ascites	44	–	–	3	–	1	5	1	–	–	1	6	7	7	7	3	7	–	–
	5 Ulceration of Intestines	82	5	–	–	–	1	6	1	1	4	5	14	14	14	12	8	3	–	–
	6 Hernia	76	–	–	–	–	–	–	2	–	4	4	5	7	16	15	15	4	1	–
	7 Ileus	72	12	4	–	–	–	17	1	–	–	3	4	6	8	11	10	6	2	–
	8 Intussusception	27	8	1	–	1	1	11	–	1	1	–	3	2	2	2	3	1	–	–
	9 Stricture of Intestines	33	–	–	–	–	–	–	–	–	–	–	3	3	6	8	10	2	–	–
	10 Fistula	14	–	–	–	–	–	–	–	–	–	1	1	1	2	1	4	4	–	–
	11 *Stomach Disease, &c.*	116	14	4	–	–	3	21	–	–	–	4	1	8	19	20	20	15	7	1
	12 *Pancreas Disease, &c.*	–	–	–	–	–	–	–	–	–	–	–	–	–	–	–	–	–	–	–
	13 Hepatitis	81	1	–	–	–	–	1	–	–	–	1	11	17	15	21	11	2	1	–
	14 Jaundice	95	50	1	1	–	–	52	1	–	3	4	7	4	5	7	7	9	5	–
	15 *Liver Disease, &c.*	392	14	–	1	1	–	16	3	3	7	9	28	57	116	92	43	18	–	–
	16 *Spleen Disease, &c.*	9	–	2	–	–	1	3	–	–	–	1	–	–	1	2	1	1	–	–

(continued at page 158.)

Fig. 8.4 Part of a GRO table showing deaths by age in 1871. (*Annual report of the Registrar-General (Abstracts of 1871)*, HMSO, London, 1927, p. 156.)

An Alphabetical List of Diseases, and Causes of Death, nearly all of which have been met with in the Registers— with References (by figures) to the Statistical Nosology

The insertion of names in this list must not be considered as by any means sanctioning their use. For the names which it is recommended should be used, see the first column of the previous *nosology*, to which the figures always refer. The queries in the second column will remind the informants and registrars of points which should be borne in mind in assigning the causes of death.

Abdominal effusion	Ascites, 30f.
Abdominal inflammation (vague.)	Query, whether peritonitis? or enteritis? 80, 81.
Abortus	Abortion, 104b.
Abrasion of the mucous membrane of stomach and intestines	(A bad term.) Poisoning? Ulceration? 137 85.
Broken heart	Mental distress. Sometimes improperly used for rupture of the heart.
Bronchi (abscess of) (bad)	Phthisis? Pneumonia? 71–73.
Bronchi (inflammation of)	Bronchitis. 69.
Bronchial fever	Influenza? Bronchitis? 13; 17; 69.
Bronchial ulceration extending to the lungs	69, 73.
Bronchitis	69.
Bronchitis and broken rib	How was the rib broken? 69, 144.
Bronchocele	33.
Bruised corn	144. Phlegmon?
Bursa (inflammation of)	105.
Bursting of a blood-vessel	See 28.
Bursting of a fallopian tube	Under what circumstances? 104.
Cachexia	Scrofulous? Cancerous? 33, 35.
Ciccutn (stricture of)	84.
Caecum (biliary)	89b.
Calculus (urinary)	Stone, 97.
Cancer	35. Of what kind? of what part?
Cancer (chimney-sweepers')	35.
Cancrum oris	39 d.
Canker	Of what part? Cancrum oris? 39d.
Carbuncle	132.
Carbunculus	Carbuncle, 132.

Carcinoma	Cancer, 35.
Carditis	58.
Caries	Of what bone? 111.
Catacausis	40.
Catalepsy	54.
Catarrh	69.
Catarrah of the bladder	98b.
Catarrahal fever	Influenza? 13.
Catarrahus epidemicus	Influenza, 13.
Catarrahus vesicae	Catarrah of the bladder, 98b.
Cauliflower polypus of the womb	37c; 103.
Cellular dropsy	30.

General Register Office, 4th annual report of the registrar general of births, marriages and deaths for 1840–41, HMSO, London, 1842, pp. 166–86.

'Cotton Goods Manufacture'. This allowed the size of the tables to be radically reduced, thus aiding the processes of tabulation and public understanding. But the use of such systems inevitably introduced into the tables certain ideological principles that have bedevilled subsequent analysis.

The GRO's system of socio-economic groupings in the early part of the twentieth century highlights these issues. T. H. C. Stevenson, the GRO's senior medical statistician from 1909 to 1931, constructed this classification in order to study the class-specific fertility of married women, as revealed by questions in the 1911 census on completed family size. He assumed that the social status of a family could be determined by assigning it to one of a series of socio-economic groups on the basis of the occupational designation of the male household head. The classification was based on the concept of a social gradation of skill and intelligence which was assumed to be reflected in occupational biographies, and to be constant both spatially and over time. The taxonomy used was basically tripartite at it formal inception, with 'the upper- and middle-class' in Class I; 'those occupations of which it can be assumed that the majority of the men classified to them at the census are skilled workmen' in Class III; and Class V containing 'occupations including mainly unskilled men'. Two groups of intermediate occupations were then placed between these classes, making five in total. Thus, the term 'physician' would be placed in Class I, 'vet' in Class II, 'cotton weaver' in Class III, and 'general labourer' in Class V. Stevenson used this classification to 'show' that marital fertility declined the 'higher' one progressed up the social scale. He explained this in terms of the gradual diffusion of knowledge on mechanical contraception from the middle classes 'downwards' in society. The GRO's system was subsequently taken up widely within the social science research community as a means of stratifying households into social classes.[16]

This system of socio-economic groupings based on job titles, however, masks all sorts of local and occupational anomalies in family size, and obscures the dynamics of fertility decline. These cannot be simply read off from occupational designations but need to be seen in terms of the negotiations which went on between men and women within families over sexual and reproductive matters. It was not levels of intelligence which determined fertility but concrete matters such as the availability of work for married women, the relative cost of having children, and the gendered balance of power within marriage. The assumption that the occupation of the male 'breadwinner', rather than total family income, determined status, and the lumping together of various occupational groups with very different social and economic experiences, have all been criticised by historians.[17]

Similar issues could be raised with respect to the occupational classification systems used by the GRO in the Victorian period. These were not based on self-evident and concrete economic structures but arose from particular intellectual models which appeared relevant at the time. The occupational classification systems were based on the grouping of occupational titles according to materials being worked up. These, in turn, were taken as affecting the morbidity and mortality of those who undertook these occupations. In addition, William Farr also appears to have followed classical precedents in believing that working with particular materials affected the character of workmen. Those who failed to indicate the material upon which they worked in the census schedule, or those, such as clerks, who did not shape or tend particular substances or animals, were fitted into the material-based classification in residual categories. Thus 'labourers' who neglected to indicate their branch of employment were abstracted under the heading 'General Labourers'. This occupational category included over half a million men in 1871. Similarly, clerks, warehousemen, and many dealers, were placed in the material categories relating to the establishments within which they worked. Thus a clerk in an iron mill would be abstracted under the heading 'Iron Manufacture Service', rather than under 'Commercial Clerk.' The retirement of Farr, and his replacement as superintendent of statistics by William Ogle, led to an important shift in the principles of occupational classification used in the 1881 census, with clerks, and other tertiary workers, now being abstracted under their own distinct headings (see Fig. 8.1(a) at the beginning of this chapter). This shift may lie behind the apparent tertiary economic 'revolution' in the late nineteenth century.[18]

Similarly, what the right to life meant in practice was constrained by how the GRO and other public health bodies perceived death. As Christopher Hamlin has recently argued, the early public health movement under Victorian reformers such as Edwin Chadwick can be seen as an attempt to constrict what health implied. Rather than conceiving of the individual manifestations of disease in traditional terms as the impact of a myriad of external influences on unique human constitutions, Chadwick followed Neil Arnott and Thomas Southwood Smith in seeing disease as the result of the invasion of the body by specific chemical pathogens. This allowed Chadwick to narrow the concept of a right to health to the right to the removal of noxious human effluent from the cities, and to clean water supplies. Chadwick could thus outflank medical and Chartist claims that health had to be seen in broader constitutional terms such as access to food, rest, and tolerable working conditions.[19]

Farr agreed with Arnott and Southwood Smith that infectious diseases were caused by chemical blood poisoning. The whole cast of the medical nosologies drawn up by the early GRO, which placed each death in a single category according to the 'primary' cause of death, also helped to narrow how the concept of mortality could be understood. It was no longer possible to use terms such as 'Cold', or 'Damp clothes (putting on, or sleeping in)', as a cause of death—terms such as pneumonia and bronchitis needed to be substituted. In order to enforce this new concept of specific causes of death, which was ideally suited to the creation of the distinct cells of the two-way table, the GRO issued doctors with its nosologies, and sent death certificates back to them if they did not conform to the new form of classification. Plainly, this revolution in medical classification has brought vast benefits in terms of diagnosis and therapeutics but it has also detracted from the ability to see patients as whole human beings placed in a social setting. To exaggerate wildly, people do not have diseases because of the conditions within which they find themselves, they have become the medium through which causes of death manifest themselves.[20]

Equally problematic was the assumption made in the Victorian period that the populations in each of the cells of two-way tables were homogeneous. A central aim of the GRO was to create tables showing the number of deaths in distinct administrative districts. When combined with data from the census on the population of such units, this would enable the calculation of death rates, usually expressed as so many deaths per 1 000 live population.

These figures were then used to identify places with 'excessive' mortality, i.e. that in excess of a 'Healthy Districts Death Rate' based on a group of agricultural districts with low mortality. This provided ammunition for local sanitary and public health reformers. Under the 1848 Public Health Act, the central General Board of Health could also force a community to set up a local board of health if its annual death rate was above 23 per thousand. Crude death rates of this sort were easy to calculate from tables of mortality, and easy to understand. They were, however, deeply flawed, since they took no notice of the unique age and sex distributions of individual communities. This was especially serious in the case of resorts with high proportions of elderly residents, and relatively high mortality, whose local councils were highly critical of the GRO's tables.[21]

Transcending the limitations imposed by data truncation

Much of the subsequent technical history of the GRO can be seen in terms of attempts to overcome the limitations imposed by these forms of informational truncation through technological and methodological innovation. The former helped to speed up analysis and computation, and so allowed more, and more complex, tables to be created. Methodological innovations were introduced, however, in order to make the tables work better, or to transcend them altogether.

The general reluctance of the Victorian GRO to expand its data-processing operations stands in marked contrast to its willingness to experiment with new forms of computational technology. The GRO did not try to find new ways to reduce the mass of individual survey returns to quantitative results but it did attempt to find ways of manipulating the latter more easily when the old manual systems of data processing had created them. As Williams has described in Chapter 5, Georg Scheutz, a Swede with a background in printing and publishing, was attracted to the idea of producing error-free text via perfecting the printing unit of Charles Babbage's difference engine. The machine developed by Scheutz and his son was brought to England in the 1850s, won the approbation of Babbage and a committee of the Royal Society, and went on to win a gold medal at the Paris Exhibition of 1855. This apparatus was obtained by the GRO in 1858, at the considerable cost of £1 200, for the purpose of generating the actuarial tables for the Office's

ENGLISH LIFE TABLE, No. 3.

YEARLY TABLE:—MALES. Interest 3 per Cent.

VALUES OF ANNUITIES. SINGLE and ANNUAL PREMIUMS, which insure £1 on the LIFE of a MALE.

Age	Present value of an Annuity of £1 (first payment at the age x).	Annuity (first payment at age x) which £1 will purchase.	Sum which One Premium of £1 will insure over One Year.	Sum which One Premium of £1 will insure over the Whole Term of Life.	Sum which an Annual Premium of £1 will insure over the Whole Term of Life.	Single Premium which will insure £1 over One Year.	Single Premium which will insure £1 over the Whole Term of Life.	Annual Premium which will insure £1 over the Whole Term of Life.
x	$\frac{N_x}{D_x} = A_x$	$\frac{D_x}{N_x} = a_x$	$\frac{D_x}{C_x}$	$\frac{D_x}{M_x} = \Sigma_x$	$\frac{N_x}{M_x} = s_x$	$\frac{C_x}{D_x}$	$\frac{M_x}{D_x} = \Pi_x$	$\frac{M_x}{N_x} = \pi_x$
	£	£	£	£	£	£	£	£
0	19·1506	·052218	6·30	2·26135	43·3062	·158830	·44221	·023091
1	22·3518	·044739	16·02	2·86552	64·0494	·062425	·34898	·015613
2	23·5036	·042547	29·02	3·17027	74·5127	·034459	·31543	·013421
3	24·0316	·041612	43·19	3·33278	80·0920	·023155	·30005	·012486
4	24·3022	·041149	57·80	3·42267	83·1782	·017300	·29217	·012022
5	24·4367	·040922	75·79	3·46918	84·7752	·013194	·28825	·011796
6	24·4724	·040862	95·19	3·48172	85·2060	·010505	·28721	·011736
7	24·4410	·040915	112·45	3·47068	84·8270	·008893	·28813	·011789
8	24·3674	·041038	134·90	3·44506	83·9472	·007413	·29027	·011912
9	24·2536	·041231	159·33	3·40617	82·6120	·006276	·29358	·012105
10	24·1071	·041482	183·37	3·35736	80·9360	·005453	·29785	·012355
11	23·9347	·041780	203·59	3·30172	79·0256	·004912	·30287	·012654
12	23·7429	·042118	215·94	3·24191	76·9722	·004631	·30846	·012992
13	23·5374	·042486	218·71	3·18022	74·8541	·004572	·31444	·013359
14	23·3234	·042875	212·24	3·11839	72·7314	·004712	·32068	·013749
15	23·1052	·043280	199·11	3·05779	70·6510	·005022	·32703	·014154
16	22·8867	·043693	182·98	2·99944	68·6475	·005465	·33340	·014567
17	22·6710	·044109	166·10	2·94394	66·7421	·006021	·33968	·014983
18	22·4604	·044523	150·27	2·89172	64·9493	·006655	·34581	·015397
19	22·2568	·044930	136·26	2·84296	63·2752	·007339	·35175	·015804
20	22·0612	·045328	124·32	2·79767	61·7200	·008044	·35744	·016202
21	21·8743	·045716	115·36	2·75570	60·2789	·008220	·36288	·016590
22	21·6841	·046117	119·14	2·71426	58·8563	·008393	·36842	·016991
23	21·4904	·046532	116·79	2·67333	57·4509	·008562	·37407	·017406
24	21·2929	·046964	114·60	2·63284	56·0608	·008726	·37982	·017838
25	21·0913	·047413	112·45	2·59274	54·6842	·008893	·38569	·018287
26	20·8853	·047881	110·36	2·55303	53·3209	·009061	·39169	·018754
27	20·6748	·048368	108·34	2·51369	51·9701	·009230	·39782	·019242
28	20·4596	·048877	106·31	2·47469	50·6311	·009406	·40409	·019751
29	20·2394	·049408	104·28	2·43604	49·3041	·009590	·41050	·020282
30	20·0143	·049964	102·24	2·39774	47·9891	·009781	·41706	·020838
31	19·7840	·050546	100·16	2·35979	46·6862	·009984	·42377	·021420
32	19·5486	·051155	98·06	2·32221	45·3959	·010198	·43062	·022028
33	19·3078	·051792	95·90	2·28501	44·1185	·010428	·43764	·022664
34	19·0618	·052461	93·69	2·24819	42·8547	·010674	·44480	·023335
35	18·8105	·053162	91·43	2·21180	41·6049	·010937	·45212	·024036
36	18·5538	·053897	89·14	2·17582	40·3697	·011218	·45960	·024771
37	18·2918	·054669	86·74	2·14027	39·1494	·011528	·46723	·025543
38	18·0245	·055480	84·35	2·10521	37·9454	·011855	·47501	·026354
39	17·7521	·056331	81·91	2·07061	36·7576	·012209	·48295	·027205
40	17·4744	·057227	79·41	2·03650	35·5866	·012593	·49104	·028100
41	17·1916	·058168	76·92	2·00291	34·4331	·013001	·49927	·029042
42	16·9037	·059159	74·40	1·96983	33·2973	·013441	·50766	·030032
43	16·6108	·060202	71·84	1·93726	32·1795	·013920	·51619	·031076
44	16·3129	·061301	69·33	1·90525	31·0803	·014424	·52486	·032175
45	16·0102	·062460	66·80	1·87377	29·9994	·014970	·53368	·033334
46	15·7026	·063684	64·31	1·84284	28·9374	·015551	·54264	·034557
47	15·3902	·064976	61·82	1·81244	27·8939	·016175	·55174	·035850
48	15·0730	·066344	59·39	1·78260	26·8692	·016839	·56098	·037217
49	14·7511	·067792	57·00	1·75329	25·8629	·017545	·57036	·038665

Fig. 8.5 A page from the GRO's *English Life Tables* of 1864. (General Register Office, *English life tables*, HMSO, London, 1864, p. 52.)

English life tables. The machine's capacity to replace both human 'computers' and compositors, and thus produce reliable, error-free aids for the insurance industry, was seen as its most important advantage. By replacing clerks with machinery it was also hoped to halve the cost of producing tables. George Graham, the head of the GRO from 1842 to 1879, claimed in a letter to the Treasury in August 1860 that the machine had produced 153 pages of tables since February of that year at a cost of £77, whilst it would have cost £153 if the work had been done by two clerks.[22]

William Farr expressed his admiration for Scheutz's invention in his introduction to the *English life tables* but it is plain that in action the machine was very temperamental. Michael Lindgren, drawing on Scheutz's own correspondence, claims that the use of the machine was, on the whole, a failure, since only a quarter of the *English life tables* were calculated and printed with the machine. But even this rate of success was in fact beyond the capacity of Scheutz's invention since the Stationery Office informed the Treasury that only 28 pages of the completed work were entirely composed by the machine; 216 pages were partly so composed; and the remaining 520 pages were entirely set up by hand. This, and its limited application, almost certainly explains why the use of the difference engine was not taken up widely within government or commerce. Although the GRO continued to utilize the Scheutz machine for at least the next twenty years, the amount of tabulation it could perform was plainly limited by its specialist function and its unreliability.[23]

The real innovation in the GRO's computational arrangements came with the introduction of commercial mechanical calculators. The first commercially manufactured calculating machine, the 'arithmometer', had been designed by Thomas de Colmar in 1820. However, according to Andrew Warwick, although de Colmar's machine was exhibited and awarded a medal at the Great Exhibition of 1851, the large-scale mechanization of computation in British commerce appears to have only begun in the late 1860s. The first public debate in Britain on the utility of the arithmometer to the human 'computer' took place in the pages of the *Assurance magazine* in the early 1870s. The GRO's purchase of an arithmometer in 1870 for £20 was thus an extremely early example of the application of this technology in Britain. William Farr informed the Treasury that the machine had doubled the number of calculations a clerk could make in a set time, whilst enhancing their accuracy. By 1872 the GRO was obtaining Treasury permission to buy a second arithmometer, and further machines were purchased in 1873, 1877, 1881, 1893, and 1903. The GRO's use of such technology was far in advance of some other comparable government institutions.

According to Mary Croarken, for example, the Nautical Almanac Office at Greenwich was performing most of its computational work for astronomical tables by manual methods into the 1920s. It only began to introduce calculating machines in 1912.[24]

In the 1890s the GRO also began to employ various forms of slide rule, which the Office found both more useful and cheaper than arithmometers. Thus, in a letter of August 1895 the registrar general informed the Treasury that all calculations in the Statistical Department were made twice—once with an arithmometer and once with a slide rule. He continued that, 'it is only by the adoption of every possible mechanical appliance, for the saving of clerical labour, that the present staff of that Department can keep pace with the constantly increasing demand for calculated rates and percentages'.[25]

However, throughout the Victorian period the GRO stuck doggedly to the manual 'ticking' system of data processing. This technology was coming under increased strain in the early years of the twentieth century, as new questions were added to the census. The GRO was also anxious to ask the new questions about marital fertility already noted. In order to analyse the fertility data in the 1911 census, and that gathered by the other new census enquiries, the Office introduced Hollerith tabulators. These had been invented in 1890 for the US census of that year, and were being introduced into state statistical offices across Europe. This was a consequence of the increasing size and complexity of national census enumerations across the Western world as a whole in a period of widening state intervention into society.[26]

Machine tabulation broke data analysis down into two stages. First, information on individuals was punched on a card as a series of holes, and secondly, the information on the cards was read electronically. In essence, pads with spring-loaded pins were brought down on individual cards, and if the pins passed through a hole they completed a circuit through which electricity passed to move the dial of a counter on one position. The dials could be wired up via relay keys to undertake complex calculations. What this did was to separate data capture from data analysis, since the cards could be analysed in differing ways, and as many times as required. At a stroke the bottlenecks in the GRO's manual system of data processing were removed, opening up whole new possibilities for statistical manipulation. Although introduced for census purposes, the new machines also allowed the GRO to generate new medical tables, such as those showing the relationship between primary and secondary causes of death, and the places where people died.[27] The invention of the 'database' thus preceded that of the electronic computer by more than half a century.

Fig. 8.6 The use of Hollerith machines in the US census of 1890. (Cover, *Scientific American*, 30 August 1890.)

Perhaps less well known are the methodological innovations pioneered by the GRO in the late Victorian period and in the early decades of the twentieth century to overcome some of the limitations of the two-way table. These can be seen in terms of the reconceptualization of death; the standardization of populations; and the introduction of multivariate analysis. The concept of the 'primary' cause of death had allowed Farr and his successors to create two-way tables that claimed to reveal the face of death at any particular point in time. However, this was at the cost of portraying death as a state rather than as a process. In 1927 the GRO changed the instructions for the order of the statement on cause of death in the death certificate, by calling for the 'immediate' rather than the primary cause first, and then, in order, for any others of which it was the consequence. This then provided a more definite starting point for an analysis of the train of related causes. The GRO believed that this innovation led to significant changes in reporting. There was, for example, an increase of 20 per cent in the number of maternal deaths registered as from causes associated with childbirth other than puerperal ('childbed') fever in the period after 1927.[28] In essence the GRO was admitting that its published cause of death tables were merely one set out of a whole series of possible tables, rather than the best of all possible representations.

The numbers of deaths by place and cause were also increasingly adjusted by basing their calculation on a standard population with a defined age and sex structure. William Ogle, the GRO's superintendent of statistics from 1880 to 1893, began to use such standardization in the *Decennial supplement* covering 1871–80, published in 1885, and it was introduced into the *Annual report of the registrar general* for 1901. In his Letter in this volume, John Tatham, Ogle's successor as superintendent of statistics, noted that the more general introduction of death rates standardized by age had been made necessary because of changes in the structure of the population consequent upon the decline in the birth-rate in the late nineteenth century. This made comparisons of crude death-rates with those of earlier periods misleading.[29] Although William Farr has rightly been lauded as the father of modern medical statistics, his successors as superintendents of statistics in the GRO had to grapple with the deficiencies in his chosen tabular methods. Tables thus came to present worked rather than raw data.

This process of transcending the legacy of Farr was carried still further in the 1930s by the introduction of the modern forms of multivariate analysis associated with Francis Galton, Karl Pearson, and George Udny

Yule. This was consequent upon the appointment of Dr Percy Stocks as 'medical statistical officer' at the GRO in 1933. Stocks started out his career as a house physician at Manchester Royal Infirmary, and was assistant school medical officer in Bristol in the period 1918 to 1921, before going on to become Reader in Medical Statistics at the Galton Eugenics Laboratory under Pearson, from 1921 to 1933. In the years immediately preceding his appointment to the GRO he was a regular contributor to Pearson's *Annals of eugenics* but his work was mainly epidemiological rather than overtly eugenic. Stocks continued working at the GRO until 1950, although in the 1940s he was increasingly drawn into the work of the World Health Organisation (WHO).[30]

The advent of Stocks was associated with an increasing statistical sophistication in the GRO's work. In the *Registrar general's statistical review for 1934*, published in 1936, Stocks introduced the use of correlation coefficients and regression analysis to show the relationship between mortality, geographical latitude, housing density, and socio-economic class. The same *Review* also saw the introduction of local areal comparability factors (ACFs) to allow standardization between localities. The *Review* published in 1938 contained a discussion of the need to introduce measures of statistical significance in those cases where the number of cases of a particular cause of death was under 20. Stocks showed that the standard error decreased as a percentage of the number of deaths by cause as the number of cases increased—plus or minus 32 per cent with 10 cases, and plus or minus 3 per cent at 1000. The probability statistics of Galton, Pearson, and their successors, had finally arrived in the GRO, infusing the older statistical tradition of Farr with a new rigour and analytical power.[31]

Survey methodologies and information truncation

The story of table-making in the GRO reveals a general pattern—the reduction in informational complexity and the subsequent reintroduction of such complexity via improved technology and methodological advances.[32] Paradoxically, at least at first sight, the triumph of the GRO in the Victorian period lay not just in collecting and displaying 'data' but also in finding ways of truncating the collection and presentation of information via forms and techniques of tabular presentation. These allowed the Office to control the amount of information it had to process, so allowing it to create a coherent picture of the certain aspects of the external world. They also allowed the GRO to present information in a simple form that could be comprehended by a busy public.

These techniques allowed great advances in survey methodologies but also created considerable problems for subsequent analysis. The truncation of information flows involved in the GRO's survey methodologies led, inevitably, to the presentation of a partial view of the world, which others wanted supplemented or expanded. This led to further technical and methodological change within the Office's techniques of statistical production. But this is not a unique history, since all human development can be seen in terms of solutions to problems that create their own drawbacks, and thus necessitate further innovation. Thus, we advance as much by creating interesting problems as by finding solutions to them.

Further reading

A general history of the GRO can be found in Muriel Nissel, *People count*, HMSO, London, 1987. For the specific circumstances of its creation in the early nineteenth century see, Edward Higgs, 'A cuckoo in the nest? The origins of civil registration and state medical statistics in England and Wales', *Continuity and Change*, vol. 11, 1996, pp. 115–34. Edward Higgs, *A clearer sense of the census: the Victorian census and historical research*, HMSO, London, 1996, gives a general introduction to the nineteenth-century censuses, and the methods of data abstraction and tabulation used by the GRO in their analysis.

The standard text for the career and work of William Farr is John M. Eyler, *Victorian social medicine*, Johns Hopkins University Press, London, 1979. More details on Farr and several other of the figures mentioned here can be found in the *Dictionary of National Biography* and the standard obituaries of the period.

For the early nineteenth-century political, administrative, and intellectual context of the GRO innovations in disease classifications, see Christopher Hamlin, *Public health and social justice in the age of Chadwick: Britain, 1800–1854*, Cambridge University Press, Cambridge, 1998, and Simon Szreter, 'The GRO and the Public Health Movement in Britain, 1837–1914', *Social History of Medicine*, vol. 4, 1991, pp. 435–64. Simon Szreter, *Fertility, class and gender in Britain 1860–1940*, Cambridge University Press, Cambridge, 1996, provides similar context for the GRO's development of its socio-economic classification in the early twentieth century. The data processing implications of this innovation can be found in Edward Higgs, 'The statistical Big Bang of 1911: ideology, technological innovation and the production of medical statistics', *Social History of Medicine*, vol. 9, 1996, pp. 409–26.

The role that the truncation of communications plays in social systems is discussed in Niklas Luhmann, *Social systems*, Stanford University Press, Stanford, California, 1995. This is a very difficult text but one that merits the effort.

Notes

1. Public Record Office (hereafter PRO), 1861 Census Returns, RG9/1783, f. 35.
2. For the history of the GRO see, M. Nissel, *People count*, HMSO, London, 1987. For William Farr see, J. M. Eyler, *Victorian social medicine*, Johns Hopkins University Press, London, 1979.
3. E. Higgs, 'A cuckoo in the nest? The origins of civil registration and state medical statistics in England and Wales', *Continuity and Change* 11 (1996) 115–34; esp. 123.
4. PRO, GRO Letter Books, RG 29/3, pp. 78–81.
5. Martin Campbell-Kelly has argued that the GRO could control the selection of the census clerks via the system of Civil Service examinations but this is to confuse the recruitment for established posts within the Office with the temporary employment of casuals in the Census Office. The GRO instituted simple tests to weed out the worst of the Treasury nominees but the overall standard of the original population was inevitably low. M. Campbell-Kelly, 'Information technology and organizational change in the British census, 1801–1911', *Information Systems Research* 17 (1996), 22–36, esp. 27.
6. E. Higgs, *A clearer sense of the census: the Victorian census and historical research*, HMSO, London, 1996, pp. 155–6.
7. Higgs, *A clearer sense of the census*, p. 155.
8. PRO, Treasury Board Papers, T 1/6028B/12646, enclosure.
9. Higgs, 'A cuckoo in the nest?', p. 127.
10. PRO, Local Government Board and predecessors: Correspondence with Government Offices, MH 19/195, letters of 30 August, 8 September, and 15 September 1890; S. Szreter, *Fertility, class and gender in Britain 1860–1940*, Cambridge University Press, Cambridge, 1996, pp. 119–20.
11. C. Hakim, 'Social monitors: population censuses as social surveys', in M. Bulmer (ed.), *Essays on the history of British sociological research*, Cambridge University Press, Cambridge, 1985, pp. 39–51, esp. 41.

12. E. Higgs, *Making sense of the census. The manuscript returns for England and Wales, 1801–1901*, HMSO, London, 1989, pp. 80–9.
13. *First and second report of the select committee on death certification*, Parliamentary Papers (hereafter PP) 1893–4, XI, p. xvii.
14. General Register Office, *Registrar general's statistical review for 1927*, HMSO, London, 1929, p. 145.
15. PRO, Office of Population Censuses and Surveys and predecessors, Statistical Branch: Population and Medical Statistics: Correspondence and Papers, RG 26/88 Classification of Cause of Death. Rules of Selection from Jointly Stated Causes.
16. Szreter, *Fertility, class and gender in Britain 1860–1940*, pp. 67–282.
17. Ibid., pp. 443–602.
18. Higgs, *A clearer sense of the census*, pp. 161–6.
19. C. Hamlin, *Public health and social justice in the age of Chadwick: Britain, 1800–1854*, Cambridge University Press, Cambridge, 1998, pp. 52–66, 110–20.
20. E. Higgs, 'The linguistic construction of social and medical categories in the work of the English General Register Office', in S. Szreter, A. Dharmalingam and H. Sholkamy (eds), *The qualitative dimension of quantitative demography*, Oxford University Press, Oxford, forthcoming.
21. J. M. Eyler, 'Mortality statistics and Victorian health policy: program and criticism,' *Bulletin of the History of Medicine*, 50 (1976), 335–55.
22. General Register Office, *English life table, tables of lifetimes, annuities and premiums, with an intr. by W. Farr*, General Register Office, 1864, p. cxli; PRO, GRO Letter Books, RG 29/1, pp. 538–46, 600; PRO, GRO Letter Books, RG 29/5, p. 421. An example of the tables produced by the Scheutz machine can be found in PRO, Treasury Board Papers, T 1/6257B/12792.
23. General Register Office, *English life tables*, HMSO, London, 1864, p. cxlii; PRO, GRO Letter Books, RG 29/6, pp. 6–7; M. Lindgren, *Glory and failure: the difference engines of Johann Müller, Charles Babbage and Georg and Edvard Scheutz*, MIT Press, London, 1990, pp. 274–86. The Office obtained £12 from the Treasury in 1877 for its repair: PRO, GRO Letter Books, RG 29/2, p. 250; PRO, GRO Letter Books, RG 29/6, p. 179.
24. A. Warwick, 'The laboratory of theory or what's exact about the exact sciences?', in M. Norton-Wise, *The values of precision*, Princeton University Press, Princeton, 1995, pp. 328–9. Cortada also sees the emergence of a European market for calculating machines as being a feature of the 1870s: J. W. Cortada, *Before the computer. IBM, NCR, Burroughs, and Remington Rand and the industry they created, 1865–1956*, Princeton University Press, Princeton, 1993, p. 28. PRO, GRO Letter Books, RG 29/2, pp. 111–2, 149, 249–50; RG 29/3, pp. 129–30, 300; RG 29/6, p. 101, 233; RG 29/7, p. 80. PRO, Treasury Board Papers, T 1/7299b/11516, letter of 31 May 1872. M. G. Croarken, *Early scientific computing in Britain*, Clarendon Press, Oxford, 1990, p. 22.

25. PRO, GRO Letter Books, RG 29/3, pp. 149, 169; RG 29/7 pp. 93, 103.
26. E. Higgs, 'The statistical Big Bang of 1911: ideology, technological innovation and the production of medical statistics', *Social History of Medicine*, 9 (1996), 409–26.
27. M. Campbell-Kelly, *ICL. A business and technical history*, Clarendon Press, Oxford, 1989, pp. 8–13; General Register Office, *Seventy-third annual report of the registrar general for 1910*, HMSO, London, 1912, pp. vii–viii.
28. General Register Office, *Registrar general's statistical review for 1927*. HMSO, London, 1929, p. 145; General Register Office, *Registrar general's statistical review for 1936*. HMSO, London, 1938, p. 122.
29. General Register Office, *The registrar general's decennial supplement, England and Wales 1921. Part III. Estimates of population, statistics of marriages, births and deaths 1911–1920* HMSO, London, 1933, pp. xxxiii–xxxvii; General Register Office, *64th annual report of the registrar general for 1901*, HMSO, London, 1903, p. xvi. See also W. Ogle, 'Proposal for the establishment and international use of a standard population, with fixed sex and age distribution, in the calculation and comparison of marriage, birth and death rates', *Bulletin de l'Institute International de Statistique, Rome*, VI (1892), 83–5.
30. *Imperial calendar 1934*, HMSO, London, 1934, p. 190. PRO, Medical Research Committee and Medical Research Council: Files, FD 1/7110 MRC Statistics Committee: Report, including anaemia and measles, 1933–1936, extract of the Committee meeting of 26 October 1934; *Who was who, 1971–1980*, A. & C. Black, London, 1981, pp. 760–1; General Register Office, *Registrar general's statistical review for 1951 Text*, HMSO, London, 1954, p. 275.
31. General Register Office, *Registrar general's statistical review for 1934*, HMSO, London, 1936, pp. 4–8, 150–5; General Register Office, *Registrar general's statistical review for 1936*, HMSO, London, 1938, pp. 9–15.
32. For a general theory of systems development along these lines, see N. Luhmann, *Social systems*, Stanford University Press, Stanford, 1995.

Table XIV—Pentagamma Function

x	$\dfrac{d^4}{dx^4}\log_e(x!)$	δ^2 +	δ^4 +	δ^6 +	x	$\dfrac{d^4}{dx^4}\log_e(x!)$	δ^2 +	δ^4 +
0·00	6·49393 94023	1221 23348	5 06663	3636	0·50	1·40909 10340	111 15217	20034
·01	6·25106 18729	1151 06455	4 67960	3297	·51	1·37489 70527	106 90390	19000
·02	6·01969 49890	1085 57522	4 32554	2995	·52	1·34177 21104	102 84563	18034
·03	5·79918 38573	1024 41143	4 00143	2702	·53	1·30967 56244	98 96770	17119
·04	5·58891 68399	967 24907	3 70434	2467	·54	1·27856 88154	95 26096	16253
0·05	5·38832 23132	913 79105	3 43192	2231	0·55	1·24841 46160	91 71675	15440
·06	5·19686 56970	863 76495	3 18181	2039	·56	1·21917 75841	88 32694	14675
·07	5·01404 67303	816 92066	2 95209	1848	·57	1·19082 38216	85 08388	13946
·08	4·83939 69702	773 02846	2 74085	1687	·58	1·16332 08979	81 98028	13261
·09	4·67247 74947	731 87711	2 54648	1544	·59	1·13663 77770	79 00929	12618
0·10	4·51287 67903	693 27224	2 36755	1398	0·60	1·11074 47490	76 16448	11999
·11	4·36020 88083	657 03492	2 20260	1285	·61	1·08561 33658	73 43966	11425
·12	4·21411 11755	623 00020	2 05050	1177	·62	1·06121 63792	70 82909	10878
·13	4·07424 35447	591 01598	1 91017	1065	·63	1·03752 76835	68 32730	10358
·14	3·94028 60737	560 94193	1 78049	994	·64	1·01452 22608	65 92909	9867
0·15	3·81193 80220	532 64837	1 66075	886	0·65	0·99217 61290	63 62955	9408
·16	3·68891 64540	506 01556	1 54987	843	·66	·97046 62927	61 42409	8961
·17	3·57095 50416	480 93262	1 44742	742	·67	·94937 06973	59 30824	8550
·18	3·45780 29554	457 29710	1 35239	710	·68	·92886 81843	57 27789	8151
·19	3·34922 38402	435 01397	1 26446	626	·69	·90893 84502	55 32905	7782
0·20	3·24499 48647	413 99530	1 18279	603	0·70	0·88956 20066	53 45803	7421
·21	3·14490 58422	394 15942	1 10715	531	·71	·87072 01433	51 66122	7088
·22	3·04875 84139	375 43069	1 03682	503	·72	·85239 48922	49 93529	6766
·23	2·95636 52925	357 73878	97152	461	·73	·83456 89940	48 27702	6464
·24	2·86754 95589	341 01839	91083	422	·74	·81722 58660	46 68339	6175
0·25	2·78214 40092	325 20883	85436	392	0·75	0·80034 95719	45 15151	5903
·26	2·69999 05478	310 25363	80181	362	·76	·78392 47929	43 67866	5638
·27	2·62093 96227	296 10024	75288	334	·77	·76793 68005	42 26219	5397
·28	2·54484 97000	282 69973	70729	309	·78	·75237 14300	40 89969	5158
·29	2·47158 67746	270 00651	66479	285	·79	·73721 50564	39 58877	4934
0·30	2·40102 39143	257 97808	62514	263	0·80	0·72245 45705	38 32719	4724
·31	2·33304 08348	246 57479	58812	253	·81	·70807 73565	37 11285	4520
·32	2·26752 35032	235 75962	55363	213	·82	·69407 12710	35 94371	4328
·33	2·20436 37678	225 49808	52127	225	·83	·68042 46226	34 81785	4142
·34	2·14345 90132	215 75781	49116	189	·84	·66712 61527	33 73341	3972
0·35	2·08471 18367	206 50870	46294	177	0·85	0·65416 50169	32 68869	3801
·36	2·02802 97472	197 72253	43649	179	·86	·64153 07680	31 68198	3645
·37	1·97332 48830	189 37285	41183	149	·87	·62921 33389	30 71172	3496
·38	1·92051 37473	181 43500	38866	144	·88	·61720 30270	29 77642	3344
·39	1·86951 69616	173 88581	36693	148	·89	·60549 04793	28 87456	3217
0·40	1·82025 90340	166 70355	34668	108	0·90	0·59406 66772	28 00487	3077
·41	1·77266 81419	159 86797	32751	134	·91	·58292 29238	27 16595	2957
·42	1·72667 59295	153 35990	30968	100	·92	·57205 08299	26 35660	2838
·43	1·68221 73161	147 16151	29285	106	·93	·56144 23020	25 57563	2720
·44	1·63923 03178	141 25597	27708	91	·94	·55108 95304	24 82186	2615
0·45	1·59765 58792	135 62751	26222	98	0·95	0·54098 49774	24 09424	2509
·46	1·55743 77157	130 26127	24834	70	·96	·53112 13668	23 39171	2411
·47	1·51852 21649	125 14337	23516	91	·97	·52149 16733	22 71329	2316
·48	1·48085 80478	120 26063	22289	62	·98	·51208 91127	22 05803	2225
·49	1·44439 65370	115 60078	21124	75	·99	·50290 71324	21 42502	2137
0·50	1·40909 10340	111 15217	20034	56	1·00	0·49393 94023	20 81338	2058

(56)

9

Table making by committee: British table makers 1871–1965

MARY CROARKEN

The British Association for the Advancement of Science was set up in 1831 to promote the public understanding of science, an activity in which it is still has an important role today. By the 1870s it had come to play a central part

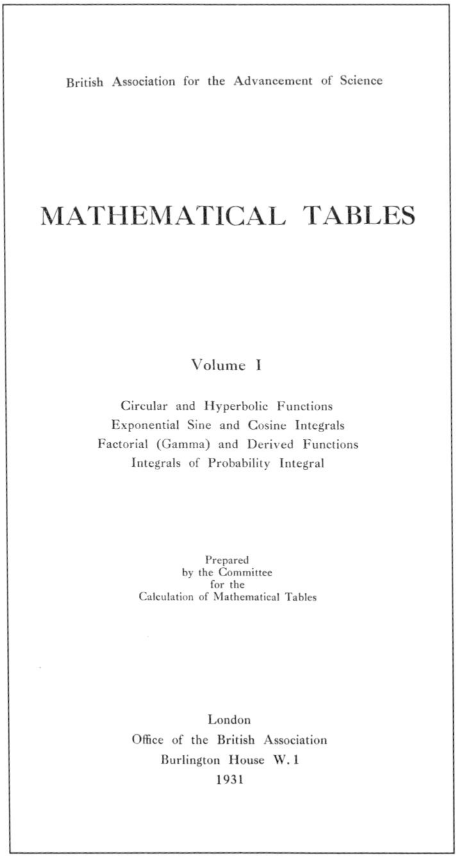

British Association for the Advancement of Science

MATHEMATICAL TABLES

Volume I

Circular and Hyperbolic Functions
Exponential Sine and Cosine Integrals
Factorial (Gamma) and Derived Functions
Integrals of Probability Integral

Prepared
by the Committee
for the
Calculation of Mathematical Tables

London
Office of the British Association
Burlington House W. 1
1931

Fig. 9.1 British Association for the Advancement of Science Mathematical Tables Volume 1. The first separate volume of mathematical tables produced by the British Association Mathematical Tables Committee was published in 1931. It set a standard of accuracy and ease of use which later volumes of tables strove to achieve. Although the tables were the result of a collaborative effort by members of the Mathematical Tables Committee, their typography and high standard of production were directly attributable to L. J. Comrie, the Committee's most influential Secretary. (By kind permission of the British Association for the Advancement of Science.)

in scientific life in England. The annual meetings, held in different cities around Britain and occasionally overseas, were regularly attended by eminent scientists and mathematicians as well as by their less well known colleagues. The meetings were an opportunity to listen to scientific papers, to engage in debate and to socialize. As part of its mission to promote the understanding of science, the British Association also supported new areas of scientific investigation through the creation of temporary research committees. The research committees were intended to undertake clearly defined tasks over a period of a few years and report their findings at the annual meetings. Some committees were supported by small grants to defray the costs of whatever investigation was being made. In 1870 the British Association allocated £1472 in research grants worth approximately £72 850 today.[1]

At the 1871 meeting of the British Association held in Edinburgh, a committee was set up for the purpose of reporting 'on Mathematical Tables, which it might be desirable to compute or reprint'.[2] The Committee consisted of Arthur Cayley, chairman, Sir William Thomson, Sir George Stokes, Henry Smith, and James Whitbread Lee Glaisher, secretary. It contained two of Britain's leading mathematicians, Cayley and Smith, and two leading British mathematical physicists, Thomson and Stokes. Glaisher, a student of Cayley's, was just beginning to establish his reputation as a mathematician and table maker.

The concerns which lay behind the creation of the Mathematical Tables Committee were based on the increasing amount of computation being required in scientific research especially in the physical sciences. Mathematical tables were the main computing tool for physicists, engineers, and mathematicians during the nineteenth century and had become increasingly numerous and diverse. While standard collections of 4 to 6 figure multiplication, logarithm, and trigonometric tables were commercially produced and easy to obtain, the more specialist mathematical tables were often published in mathematical and scientific journals, transactions of societies, pamphlets, etc. As a result it was becoming increasingly difficult to trace tables and to avoid the unnecessary labour of recalculating tables already in existence.

The lack of a method of identifying tables was a primary motivation for the creation of the Mathematical Tables Committee[3] but so too were the mathematical interests of some of the original committee members, namely James Whitbread Lee Glaisher and Arthur Cayley. Indeed Glaisher took the view that 'one of the most valuable uses of numerical tables is that they

Fig. 9.2 The British Association Mathematical Tables Committee 1871. Arthur Cayley (Chairman), J. W. L. Glaisher (Secretary), George Stokes, William Thomson, and Henry Smith (clockwise from top left) constituted the first Mathematical Tables Committee. The Committee represented two of Britain's leading mathematicians, Cayley and Smith, and two of Britain's leading mathematical physicists, Stokes and Thomson. Glaisher was a young man at the start of his long career as a Cambridge mathematician. (Thomson photograph with permission from the Institution of Electrical Engineers Archives and the others with permission from the London Mathematical Society Archives.)

connect mathematics and physics, and enable the extension of the former to bear fruit practically in aiding the advance of the latter'.[4]

Establishing the Mathematical Tables Committee, 1873–88

Glaisher, with guidance from Cayley, was the driving force behind the early work of the Mathematical Tables Committee. His first task was to draw up an extensive catalogue of general mathematical tables based on a classification system devised by Cayley. The catalogue was published in the British Association annual report for 1873.[5] It was a landmark in mathematical table bibliography and remains an authoritative source on tables available in the nineteenth century. It was supplemented two years later by a catalogue of number theory tables compiled by Cayley.

In addition to his bibliographic work, Glaisher soon began to make and publish tables both on his own and on behalf of the Mathematical Tables Committee. Glaisher had been introduced early to table making by his father, James Glaisher, head of the magnetic and meteorological department of the Royal Greenwich Observatory and an experienced and competent table maker. At the 1871 British Association meeting Glaisher read a paper on the calculation of e and expressed his intention of calculating tables of e^x and e^{-x} for which he hoped to gain financial support from the British Association. The project was never taken up by the Mathematical Tables Committee but may have been a contributing factor in arguing for the creation of such a committee. Glaisher published the tables independently in 1883.[6]

The first tables which Glaisher made on behalf of the Mathematical Tables Committee were Legendrian Functions, considered important because of their application to mechanics and to interpolation. In 1872 Glaisher designed the computation and supervised the calculation of the tables by a freelance computer paid from a British Association grant to the Committee. Another table of the same function was independently calculated and Glaisher himself compared and differenced the two tables to detect and eliminate as many errors as possible. Glaisher had hoped that the Mathematical Tables Committee would be able to publish any tables they produced in book form as he perceived one of the main difficulties with using mathematical tables was their dispersal throughout the periodical literature. However by 1879 it had become obvious that such a project was too ambitious and the Legendrian Function tables were published in the 1879 British Association report.[7]

As early as 1872 the Mathematical Tables Committee took the decision to systematically tabulate elliptic functions. Elliptic functions had many applications in mathematical physics and this project had the backing of the whole committee. Cayley was also in the process of preparing a book on the subject.[8] The project was a huge undertaking requiring the calculation and differencing of 64 800 tabular entries. Glaisher and his father supervised a team of eight computers who worked on the tables from October 1872 to, approximately, early 1876 producing 30 manuscript volumes at a cost of around £450 to the British Association. The British Association spent another £250 on printing proof copies of the tables. Glaisher, however, never took the final step of publishing the tables and, as far as can be established, was never held to account.

The next project undertaken by Glaisher on behalf of the Mathematical Tables Committee was the calculation of factor tables—the first of many number theory tables to be published by the Committee. Factor tables are the most universally useful of all number theory tables and Glaisher's 1873 catalogue, and his later work on the history of factor tables,[9] had identified factor tables covering numbers up to three million and numbers from six million to nine million. The Mathematical Tables Committee decided to fill the gap by publishing tables of the fourth, fifth, and sixth millions. James Glaisher Senior joined the Mathematical Tables Committee in 1877 specifically to oversee the work using two hired computers paid for by a grant of £250 from the British Association. The tables were published in three volumes in 1879, 1880, and 1883.[10] The publication of the factor tables effectively ended J. W. L. Glaisher's involvement with the Mathematical Tables Committee although he continued nominally to be a member until 1901.

Proceeding slowly: Bessel functions and number theory tables 1888–1928

In 1888, five years after the factor tables were published, the Mathematical Tables Committee was reappointed as a 'Committee appointed for the purpose of considering the possibility of calculating Tables of certain Mathematical Functions'.[11] The committee had five new members, Lord Rayleigh, Alfred Lodge, Alfred Greenhill, William Hicks, and Bartholomew Price, and retained Cayley, Glaisher, and Thomson. The committee's activities were strongly influenced by Rayleigh and Lodge who served as Chairman and Secretary respectively; the subject which dominated the work of the Mathematical Tables Committee during the late nineteenth century was Bessel function tables. Bessel functions were used in many branches of mathematical physics to describe physical phenomena and were perceived as an entirely appropriate subject for the Mathematical Tables Committee to address. Rayleigh had had first hand experience of publishing Bessel function tables in his *Theory of sound*[12] and knew that such tables were needed by mathematical physicists in many fields. Over the next few years the Mathematical Tables Committee began to calculate and publish Bessel function tables. The tables were published piecemeal in the British Association annual reports with the intention of gathering them all together to publish in book form at a later date—a goal which was not achieved until 1937.

Lodge who undertook most of the computing during this period, carried out the work on a much smaller scale than either Glaisher's elliptic function or factor table projects. Lodge did not ask for large sums from the British Association; rather he did much of the computation himself in his spare time with the help of two volunteer computers (one of whom was his sister Eleanor). Lodge was the first to apply a calculating machine to the Committee's work. In 1889 he used an Edmondson calculating machine to check the printed tables by continuous addition of the first differences. Initially the machine was borrowed from a colleague but later the Mathematical Tables Committee acquired a machine for its own use. Unfortunately the new machine proved mechanically unreliable which seriously slowed down the work and Lodge had to borrow a machine once again. In the early 1890s, to further speed up the computations, Lodge asked for, and received, £15 with which to employ a professional computer. Lodge remained a member of the Mathematical Tables Committee until his death in 1937. Throughout that time he proved to be an extremely diligent computer contributing many tables to the British Association Reports.

In 1895 Lt. Colonel Allan Cunningham turned the Committee's attention temporarily away from Bessel functions by suggesting that the Mathematical Tables Committee publish a table of residues of powers of 2 (useful for testing divisibility of numbers, factorization, and the solution of binomial congruences to base 2) which he had begun to compute. The Committee agreed to support the project and Cunningham took over from Lodge as Secretary in 1896 specifically to further the publication of his work. For the next four years Cunningham dominated the Committee, which paid for the computation of a second table to compare with the first and for the comparison of the two copies. By 1899 the work was complete and Cunningham asked the British Association to fund the printing costs but they granted only half of the required amount. The tables were published in 1900 with financial help from the Royal Society.[13] There is no doubt that Cunningham used the Mathematical Tables Committee to further his own reputation and to finance the publication of his tables.

Chairmen and Secretaries of the Mathematical Tables Committee

Chairman	Dates	Secretaries	Dates
A. Cayley	1871–1889★	J. W. L. Glaisher	1871–1889★
Lord Rayleigh	1888–1897	A. Lodge	1888–1896
W. Thomson	1897–1901	A. J. C. Cunningham	1896–1901
M. J. M. Hill	1906–1919	L. N. G. Filon	1906–1910
		J. W. Nicholson	1910–1920
J. W. Nicholson	1920–1931	J. R. Airey	1920–1929
E. H. Neville	1931–1947	L. J. Comrie	1929–1937
		J. Wishart	1937–1946
A. J. Thompson	1948	J. C. P. Miller	1946–1948
L. J. Mordell	1948–1950	Royal Society Administrators	1948–
C. G. Darwin	1951–1956		
D. R. Hartree	1957		
M. V. Wilkes	1958–1969		

★ For a short while there were two Mathematical Tables Committees but they were soon amalgamated.

Demand for the tables turned out to be low. Only 250 were printed and almost half of these were donated to libraries and mathematical institutions. Cunningham promptly left the Committee once his tables were published.

Following a six year period of inactivity the British Association Mathematical Tables Committee was again reconstituted in 1906—this time with a specific remit to tabulate Bessel functions. Initially the Committee consisted of just three members, Micaiah Hill as chairman, his student Louis Filon as Secretary and, the only link to the old Committee, Lodge. For the next twenty years the Committee gradually expanded to contain many of Britain's best mathematical table makers including John Airey, Arthur Doodson, and Ronald Fisher. In 1913 it reverted to the title Mathematical Tables Committee to reflect its members' growing interest in tables other than Bessel functions. The Committee published a steady stream of tables in the British Association annual reports but there was no strong direction or leadership of the Committee. All work was carried out on a voluntary basis by Committee members and sometimes unsystematically and, in the main,

Committee members made tables which were of specific interest to them. There appears to have been no cohesive policy to find out which tables were needed by the wider community in order to target Mathematical Table Committee activity.

Building a reputation, 1928–39

At the 1928 meeting of the British Association in Glasgow six new members were appointed to the Mathematical Tables Committee. One of the

Fig. 9.3 Selected Secretaries of the Mathematical Tables Committee 1888–1948. Until 1948, when the Committee was taken over by the Royal Society, the most influential person on the Committee was the Secretary. Of the eight people to hold the position following Glaisher's initial work, Allan Cunningham, Alfred Lodge, John Airey, Leslie Comrie, and Jeff Miller (clockwise from top left) were the most significant. (The photograph of Lodge is taken from *Mathematical Gazette* vol. 14, p. 394 with permission from the Mathematical Association. The other photographs are taken from R. C. Archibald, *Mathematical Table Makers*, Scripta Mathematica Studies No. 3, New York: Yeshiva University, 1948.)

six, Leslie John Comrie, had a profound effect on the work of the Committee. Comrie, then Deputy Superintendent at the Nautical Almanac Office, was establishing himself as Britain's leading table maker and expert on machine computation.[14] On a professional basis he supervised the production of the *Nautical almanac* and was engaged in a number of commercial table making projects. By 1929 Comrie had been elected Secretary of the Mathematical Tables Committee and brought with him vitality, vision, professionalism, and mechanical computing methods. His first task was to publish a collection of tables based on those published in the British Association annual reports. He soon found, however, that most of the tables needed recalculating to get them all into a consistent format and to allow for interpolation between the tabulated values. The Bessel function tables also needed systematic recomputation so these were left for a later volume.

Comrie very soon pruned the Committee of older members and organized other Committee members to be responsible for different aspects of the work. The first volume of the British Association Mathematical Tables Series was published in 1931, and set the standard for a series of tables which became a byword for accuracy and ease of use.[15] Although edited by John Henderson, the book was a true cooperative venture to which many Committee members directly contributed. Through it, and succeeding volumes, the British Association Mathematical Tables Committee built up a world wide reputation.[16]

The British Association had financially supported the publication of Volume 1 of the Mathematical Tables Series but could not commit itself to financing expensive publications on a regular basis. Comrie was fortunate in being able to gain sponsorship from other sources. The second volume of British Association Mathematical Tables was a joint venture between the Mathematical Tables Committee and the International Astronomical Union. In 1930, the International Astronomical Union asked the Mathematical Tables Committee to consider publishing a table of Emden functions, a specific form of differential equation used in the investigation of stellar structure. Comrie undertook to compute the tables of second order differential equations and used the opportunity to bring Donald Sadler and Jeffrey Miller onto the Committee. Sadler was Comrie's deputy at the Nautical Almanac Office where Comrie was now Superintendent and Miller was the man Comrie would have chosen for the post if ill health had not prevented it.[17] Comrie had great respect for both men's table making ability and trained them in his table making philosophy where attention to detail and a focus on usability dominated. While Miller was the better mathematician, Sadler had the ability to organize computing projects and to deal with typography issues. Both Miller and Sadler became dominant members of the Committee for the next 30 years and witnessed great changes in the computation and use of mathematical tables.

Volumes 3, 4, and 5 of the British Association Mathematical Tables Series were made financially possible by an earlier member of the Committee, Allan Cunningham. Cunningham died in 1928 and left £3272[18] to the Committee to be used to produce new number theory tables. After leaving the Mathematical Tables Committee Cunningham had continued to work on number theory tables and he also left money to the London Mathematical Society for a similar purpose. After extensive consultation with some of Britain's leading mathematicians[19] the Mathematical Tables Committee used the money to publish volumes entitled *Minimum decompositions into fifth powers, cycles of reduced ideals in quadratics fields*, and *Factor table giving the complete decomposition of all numbers less than 100,000*.[20]

Another important benefit of the Cunningham legacy was the financial freedom to purchase calculating machines for use by the Mathematical Tables Committee. Use of calculating machines for scientific computation was spreading and Comrie was one of the leading advocates of machine computation. Comrie purchased a Brunsviga-Dupla, a Nova Brunsviga, and a National Accounting machine with the money. Comrie had also been buying calculating machines for the Nautical Almanac Office but had been unable to persuade the Admiralty to finance the purchase of a National Accounting Machine. The British Association National was installed at the Nautical Almanac Office where Comrie and his staff soon developed very efficient differencing techniques[21] which he applied to British Association work. The other machines were held variously at the Nautical Almanac Office and at the Rothamsted Agricultural Experimental Station under the supervision of Ronald Fisher and were used by Committee members, or other computers, as needed for the work of the Committee.

The Cunningham bequest was also used to support the publication of Volumes 8 and 9 of the British Association series which owed their existence to another past member of the Mathematical Tables Committee, Glaisher. Manuscripts of number-divisor tables and a table of powers, prepared by Glaisher at the end of the nineteenth century, were donated to the Committee in the late 1930s. The Committee, which had already declined to publish the Glaisher elliptic function tables when they came to light, decided to use Cunningham's money to prepare the divisor and power tables for publication and both were issued in 1940.

The Mathematical Tables Committee had long been concerned with the tabulation of Bessel functions and Comrie ensured that the Committee remained committed to the project. A Bessel Function Subcommittee was created to deal specifically with the preparation of Bessel functions. The first collection of Bessel Function tables, Volume 6 of the British Association series, was published in 1937. Part two was published in 1952, parts three and four were issued in 1960 and 1964 respectively by the Royal Society. The Mathematical Tables Committee was not the only institution to devote considerable resources to the production of Bessel function tables. In the 1940s Howard Aiken used the Harvard Mark I to produce 12 volumes of Bessel function tables.[22] Volume 7 of the British Association Series, tables of the probability integral, was based on tables donated to the Committee by W. F. Sheppard shortly before his death and which the Committee took through to publication.

In 1936 Comrie was dismissed from the Nautical Almanac Office for carrying out external computations using Nautical Almanac staff and machines. For example, Comrie took responsibility for supervising most of the computations necessary for the first volume of Bessel Tables using Nautical Almanac Office facilities to do so. On leaving the Nautical Almanac Office Comrie set up a commercial computing business called the Scientific Computing Service and resigned from the Mathematical Tables Committee. Comrie's departure was a great loss to the Committee which had been shaped by his leadership and dynamism. The Committee also lost Comrie's practical table making skills and those of his staff. After Comrie's resignation from the Committee he offered to house the British Association National Accounting Machine at his new office and pay for any use he or his staff made of the machine at a rate of 9 pence an hour;[23] he also offered to carry out work on behalf of the Committee at the rate of 4 or 5 shillings per hour. This proved unacceptable to the Committee, who were concerned about the Council of the British Association's attitude towards the machines passing out of the direct control of Committee members[24] as well as about the cost differentials. It was arranged for the National Machine to be moved to the Galton Laboratory at University College, London where Fisher was professor of Eugenics.

Without Comrie's personal drive and his computing resources the work of the Committee slowed down immensely. Comrie had been privately

Members of the Mathematical Tables Committee in date order of joining the committee

J. W. L. Glaisher	1871	D. H. Sadler	1932
W. Thomson (Kelvin)	1871	E. L. Ince	1932
A. Cayley	1871	J. C. P. Miller	1933
G. G. Stokes	1871	W. G. Bickely	1934
H. J. S. Smith	1871	W. L. Stevens	1936
J. Glaisher, Sr.	1877	M. V. Wilkes	1938
A. Lodge	1888	F. Sandon	1938
Lord Rayleigh	1888	L. M. Milne-Thomson	1939
B. Price	1888	R. O. Cashen	1941
W. M. Hicks	1888	J. Todd	1944
A. G. Greenhill	1888	J. Womersley	1944
A. J. C. Cunningham	1895	E. T. Goodwin	1948
P. A. MacMahon	1895	L. Howarth	1949
L. N. G. Filon	1906	C. W. Jones	1948
M. J. M. Hill	1906	F. Yates	1948
J. W. Nicholson	1908	D. R. Hartree	1948
J. R. Airey	1911	L. Rosenhead	1948
A. G. Webster	1912	A. C. Aitken	1948
A. E. H. Love	1913	L. J. Mordell	1948
H. M. MacDonald	1913	M. H. A. Newman	1948
E. W. Hobson	1913	C. G. Darwin	1949
T. W. Chaundy	1914	D. C. Gilles	1949
A. T. Doodson	1915	Z. Copal	1949
H. G. Savidge	1915	H. Jeffreys	1949
G. N. Watson	1916	H. S. W. Massey	1949
G. Kennedy	1916	C. A. Coulson	1954
G. B. Matthews	1916	R. A. Scorer	1954
R. L. Hippisley	1919	A. Young	1956
D. M. Wrinch	1923	C. B. Hazelgrove	1957
R. A. Fisher	1925	L. Fox	1958
J. Henderson	1927	E. Knightling	1958
L. J. Comrie	1928	F. J. W. Olver	1959
A. J. Thompson	1928	E. S. Page	1959
J. Wishart	1928	D. R. Bates	1960
J. O. Irwin	1928	H. Swinnerton-Dyer	1960
J. F. Tocher	1928	M. S. Bartlett	1962
T. Whitwell	1928	C. W. Clenshaw	1962
E. H. Neville	1929	A. Delgarno	1962
F. Robbins	1930	I. M. Khabaza	1964
E. S. Pearson	1930	M. P. Barnett	1965

employing a computer to assist with the Bessel function computations but with Comrie's resignation her services were lost. While Committee members tried to get extra part-time computing support from colleagues and students[25] the task was becoming increasingly difficult and in January 1937 the Committee approved the employment of a full-time computer. Frank Cleaver, who had been put forward for the post by the National Cash Register Company (manufacturers of the National Accounting Machine), had a starting salary of £2 10s 0d for a 33 hour week and worked in the computing room of the Galton Laboratory at University College, London.[26] The Committee continued to employ a full-time computer until 1941 when war time labour restrictions meant it was no longer possible. Despite having access to a full-time computer Committee members, and some volunteers, continued to work on the Committee's behalf.

At the outbreak of the Second World War in 1939 the output of the Committee dropped dramatically. Many Committee members soon took up war time posts which meant that they had little or no time for Mathematical Tables Committee work or meetings. The Committee lost its full-time computer to war related work and, while Miller at Liverpool could occasionally employ temporary computers during the University summer vacations, the situation was far from satisfactory. In addition the Government leased the Committee's National Accounting Machine and moved it to Rothamsted where it was used for war related computations.

Post-war difficulties, 1945–8

The Second World War saw a huge increase in scientific computing and research into more efficient and effective computing methods. Electronic computers had their origins in ballistics and code-breaking work while the armed services sought to increase their computing power as much as they could using calculating machines, punched card machines, and analogue computing instruments. Mathematical tables were now only one of a range of computing tools but interest in them was still strong. For example the American journal *Mathematical tables and other aids to computation* began publication in 1943 and its main aim was to disseminate information about new published and unpublished mathematical tables.

In Britain, in the immediate post-war period, electronic computers began to be built at Cambridge University, Manchester University, and the

National Physical Laboratory. They promised a huge increase in computing power to the few people who would have access to them but initially would be of limited use to the majority of mathematicians, scientists, and engineers working in the rest of the country. This potential increase in computing power coupled with an increasing demand for mathematical tables led some members of the Mathematical Tables Committee to believe that the late 1940s and 1950s would be a golden era of mathematical table making. Their expectations were never realized.

By late 1945 members of the Mathematical Tables Committee had began to think about restarting their activities. Miller in particular had ambitious plans for the Mathematical Tables Committee and felt that the Committee should work in partnership with the new Mathematics Division of the National Physical Laboratory and the Cambridge Mathematical Laboratory. Both institutions had been set up to provide a range of computing services to their users and both were involved in building electronic computers. In Miller's view either or both of these institutions could be called upon to carry out some of the work of the Mathematical Tables Committee but that it was essential that the Committee itself remain separate from them in order that the Committee's role of independently producing high quality fundamental tables should not be lost. Miller and William Bickley, a committee member 1934–65, both proposed that the Committee hire a graduate who could supervise table making activity and undertake research and at least one ordinary computer at a cost of about £500 per annum. Publication and secretarial expenses would need to be met on top of this. Miller was of the view that the Committee should be expanding to meet modern needs and not drifting into an advisory role as had been suggested by some members of the Committee.[27]

The main obstacle to these plans was finance. After the war the British Association declared that it would fund research committee work for a maximum of five years only. The Mathematical Tables Committee was therefore facing a complete cut in funding. One option was to continue as a British Association Research Committee and try to exist on government grants but this would not give the Committee any long term stability. Another issue of concern at the time was that the British Association Mathematical Tables Committee had little contact with table users. Mathematical table users outside the Committee privately questioned the choices the committee had made in the tables they had produced. In many cases members had joined the Committee to make tables in which they had an interest and found the Mathematical Tables Committee a supportive place in which to do this. Senior figures in the British Association now sought to ensure that only tables with practical applications were sponsored by the Committee. Sir Charles Darwin, Director of the National Physical Laboratory and a member of the British Association Council, exerted a strong influence over decisions concerning the Mathematical Tables Committee's future. He expressed the opinion that the British Association should no longer be funding table making as the new Mathematics Division of the National Physical Laboratory (for which Darwin was directly responsible) could take over much of the computational work.[28] In Darwin's view an advisory body was what was needed. A further influence was the increasing awareness of computing issues within the Royal Society which was recommending to universities that computation and computational methods be top of the list of priority subjects to be taught to post-war undergraduates.[29] In response to these influences the Council of the British Association recommended that the Royal Society take over from the British Association as the sponsoring body.

Many members of the Committee fought vigorously against the British Association's decision to end its commitment to making mathematical tables, but the Committee was divided between members such as Miller, Sadler, and Bickley who wanted to push ahead for an active and expanding Mathematical Tables Committee and an older, tired generation of Committee members who were happy to become an advisory body. Once a decision had been taken by the Councils of the British Association and the Royal Society, the Committee membership was left with little say in the matter. They eventually bowed to the inevitable and entered into negotiations with the Royal Society. In June 1948 the Committee's assets and liabilities were transferred to the Royal Society and a new Committee constituted.

The Royal Society Mathematical Tables Committee 1948–65

The Royal Society Mathematical Tables Committee of June 1948 was a new Mathematical Tables Committee set up along traditional Royal Society committee lines. Its members, many of whom were Fellows, were selected by the Royal Society Council and the secretarial function performed by Royal Society administrators. The Committee was dominated by senior people in British computing. Representatives from Manchester, Cambridge, Liverpool, and the National Physical Laboratory sat on the Committee as did other senior figures. Less than half of the Committee was made up of members of the British Association Mathematical Tables Committee. Maurice Wilkes, Director of the Cambridge Mathematical Laboratory, committee member 1938–48 and chairman 1958–69, expressed the principle difference between the two committees. He wrote

> The [British Association] Committee consisted of a group of people who had long worked together and shared a similar enthusiasm, moreover, the committee itself decided on new members. It may thus be said to have had a personality of its own. The new... committee consisted of individuals arbitrarily selected by the Council of the [Royal Society] and appointed for limited periods. It had no personality. The take over made no-one happy. There was little enthusiasm for mathematical tables anywhere in the Royal Society... The new arrangements were unwelcome to members of the old committee who would have preferred to have been left alone to get on with the job as they had always done.[30]

Wilkes is being a little harsh in these comments in that the Royal Society Fellows do not appear to have been randomly selected to sit on the Committee but had all been involved in computing matters of one form or another. He was, however, correct in his assessment that there was 'little enthusiasm' within the Royal Society for mathematical tables. The new developments in electronic computers were exciting 'big' science, on a scale with nuclear technology, while mathematical tables were simple tools.

From the minutes of the Royal Society Mathematical Tables Committee and letters between Committee members it is clear that the first year of the Committee was a difficult one.[31] The Committee's terms of reference were to

(i) co-ordinate the activities of those engaged in the calculation of tables and decide priorities;
(ii) publish fundamental tables of the highest standards of accuracy and presentation;
(iii) give advice on the mathematical and computational aspects of tabulation and on the arrangement and presentation of tables;
(iv) administer special funds for the preparation of tables, e.g. the Cunningham Bequest.[32]

Initially the actual work of producing and tabulating mathematical tables was to be carried out by a General Sub-Committee which, in the main, drew its members from the British Association Committee. The role of the General Sub-Committee was to deal with actual table making decisions but it had no executive powers and could not initiate any work without first seeking approval from the main Committee. Members of the General Sub-Committee fought against their subordinate role, complained bitterly about the lack of progress being made and about the unrepresentative minutes being produced by Royal Society secretaries.[33] In May 1949 the main Mathematical Tables Committee bowed to pressure and the General Sub-Committee was dropped; the most active members of the British Association Committee taking places on the main Committee. In May 1949 Comrie was invited to serve on the Royal Society Mathematical Tables Committee but by this time he was ill and able to act only in a limited advisory capacity.

As things settled down the Royal Society Mathematical Tables Committee began to work on a series of mathematical tables many of which had their origins in the programme of work transferred from the British Association Committee. But progress was slow despite the Committee hiring a full-time computer from December 1949. The first volume of the Royal Society

THE ROYAL SOCIETY

MT/16(52) MATHEMATICAL TABLES COMMITTEE MT/16(52)

Sales of volumes of Mathematical Tables in 1951 (continuation of MT/8(5))

Volume No.	Stock on 1 January 1951	Sales Home	Sales Export	Stock on 31 December 1951
B.A.				
I	2000 printed	60	1	1929
II	-	-	-	-
III	-	-	-	-
IV	212	27	16	168
V	548	27	25	496
VI	713	30	36	647
VII	1	1	-	-
VIII	152	54	21	76
IX	990	30	37	923
Part Vol. A	639	30	13	596
Part Vol. B	638	30	11	597
Auxiliary I	848	17	3	828
Auxiliary II	848	17	5	826
R.S.				
I	371	25	19	327
Auxiliary I	490	14	1	475

Fig. 9.4 Mathematical Table Committee sales figures. A table showing the 1951 sales figures for published British Association and Royal Society Mathematical Tables Committee volumes. While this represents only a snap shot of the sales figures it does illustrate how low the number of tables sold was. (From Royal Greenwich Observatory Archives, Cambridge University Library, RGO 16 100C Folder VI by permission of the Syndics of Cambridge University Library and of the Particle Physics and Astronomy Research Council.)

Mathematical Tables Series, *The Farey series of order 1025*, was published in 1950, Volume 2 was issued in 1956, Volume 3 in 1954, and Volume 4 in 1958. There were several reasons why the production rate was so slow. First many of the Committee's principle table makers, Miller, Sadler, and E. T. Goodwin who joined the Royal Society Committee in May 1950 as a representative from the National Physical Laboratory Mathematics Division, were becoming extremely busy in their working lives and had only a limited amount of time to give to table making; all were involved in some way with the application of electronic computers. Secondly, and perhaps most importantly, the Royal Society Mathematical Tables Committee suffered from a lack of enthusiastic leadership and from having too many policy makers on the Committee and not enough workers.

While Wilkes observed that there was little enthusiasm for mathematical tables in the Royal Society; this was also true of many of the senior members of the Mathematical Tables Committee. For example, when Charles Darwin was chairman of the Royal Society Mathematical Tables Committee 1951–6 he would repeatedly question the value of the tables being computed which tended to have a dampening effect on the work.[34] The lack of activity was obvious also from the few meetings held (sometimes down to one or two a year) and the lack of business waiting to be executed. This feeling was summarized in 1953 by Alexander Thompson, a Mathematical Tables Committee member since 1929, who wrote 'So long as the management of a Committee has no real interest in the work, progress is bound to be slow and uncertain'.[35]

Despite the slow progress being made on preparing and publishing tables the Royal Society Mathematical Tables Committee developed other roles to serve the computing community. One such role was the development of a depository of unpublished tables. In January 1950 the Mathematical Tables Committee agreed in principle to collect unpublished mathematical tables and deposit them in the Royal Society Library.[36] The intention was to collect together those tables offered to the Committee but which proved to be too small or too specialized to realistically publish. The idea that the Committee should be actively seeking tables for deposit soon came to the fore and the depositary was formally announced in early 1951 in the journals *Nature* and *Mathematical tables and other aids to computation*[37] and anyone was invited to submit tables for deposit or use the depository. By late 1953 forty two tables had been accepted into the depository and the collection continued to grow so that by 1965 it held 86 tables.

Short biographies of selected members of the Mathematical Tables Committee

J. W. L. Glaisher (1848–1928)

James Whitbread Lee Glaisher was born in 1848 the son of James Glaisher astronomer, mathematician, and meteorologist. He went to Trinity College, Cambridge graduating second wrangler in 1871 when he was elected a Fellow of Trinity College. Glaisher spent the rest of his working life writing and lecturing in Cambridge publishing widely in many branches of mathematics including over 40 articles on mathematical tables. He was editor of the *Quarterly journal of mathematics* and the *Messenger of mathematics* for many years. Glaisher was an active member of the British Association, the Royal Astronomical Society, and the London Mathematical Society. He was elected Fellow of the Royal Society in 1878. Glaisher served on the Mathematical Tables Committee from 1871 to 1901 serving as the Committee's first Secretary.

James Glaisher (1809–1903)

James Glaisher began his career working on the ordnance survey of Ireland and was appointed First Assistant at the Cambridge Observatory in 1833 by George Airy. On Airy's appointment as Astronomer Royal, James Glaisher moved with him to Greenwich. Under Airy, James Glaisher gained extensive experience in mathematical computation and table making. In 1840 James Glaisher was appointed Superintendent of the Magnetic and Meteorological Department at Greenwich. He held the post for 40 years gaining a reputation as a pioneer in meteorology. He was also a well known balloonist. James Glaisher served on the Mathematical Tables Committee from 1877 to 1889.

Arthur Cayley (1821–1895)

In 1842 Arthur Cayley graduated from Cambridge as senior wrangler and soon began to publish original papers. In 1863, after a successful career in the law, Cayley was appointed Sadlerian professor of mathematics at Cambridge. Cayley published widely in many branches of pure mathematics and published several mathematical tables. Cayley was elected FRS in 1852 and was a very active member of the British Association. Cayley served on the Mathematical Tables Committee from 1871 to 1895.

Alfred Lodge (1854–1937)

After graduating from Magdalen College, Oxford, Lodge was appointed Professor of mathematics at the Royal Indian Engineering College. In 1904 he took up the post of assistant mathematical master at Charterhouse School where he stayed until his retirement in 1919. Lodge was the first President of

Short biographies of selected members of the Mathematical Tables Committee *cont.*

the Mathematical Association and Secretary of Section A (Mathematics and Physics) of the British Association from 1888 to 1892. In addition to teaching, Lodge was a tireless and extremely competent computer. He published nine mathematical tables and was a member of the Mathematical Tables Committee from 1888 to 1937 serving as Secretary from 1888 to 1896.

Allan Cunningham (1842–1928)

Allan Joseph Campneys Cunningham was born in India, educated in England and was commissioned into the Bengal Engineers in 1860 in which he both taught mathematics and saw active service. Cunningham retired from the Army as Lieutenant Colonel in 1891 and taught mathematics at the School of Military Engineering at Chatham from 1895 to 1900. Following his retirement from the Army Cunningham published extensively on mathematical tables. On his death in 1928 he left legacies to the British Association and the London Mathematical Society to further his work on number theory tables. Cunningham was a member of the Mathematical Tables Committee from 1895 to 1901 serving as Secretary from 1896.

John Airey (1868–1937)

Following a 1894 degree from London University, John Airey taught first at Porth County School in Glamorganshire before going to Cambridge where he took a second degree in 1906. Airey was principal of Morely Secondary School 1906–12, principal of West Ham Technical Institute 1912–18, and then principal of the City of Leeds Training College until his retirement in 1933 when he became a co-editor of *Philosophical magazine*. Throughout his life Airey was a prodigious calculator and table maker with a special interest in Bessel functions publishing over 40 mathematical tables during his lifetime. Airey was a member of the Mathematical Tables Committee from 1911 to 1937 serving as Secretary from 1920 to 1929.

William Bickley (1893–1969)

In 1913 William Gee Bickley graduated from University College, Reading and began a career teaching mathematics and engineering. In 1930 he was appointed Reader and Assistant Professor at Imperial College, London teaching mathematics to engineering students. In 1947 Bickley was appointed Professor of Mathematics at Imperial, a post he held until his retirement in 1953. Despite failing eyesight, Bickley computed and published mathematical tables during the late 1930s and 1940s. He went completely blind in 1949 but continued to lecture and write despite his disability. Bickely was a member of the Mathematical Tables Committee from 1934 to 1965.

L. J. Comrie (1893–1950)

Before serving in the New Zealand Expeditionary Forces during the First World War Leslie John Comrie graduated from University College, Auckland. After the war Comrie took a Ph.D. at Cambridge and taught in the United States before joining the Nautical Almanac Office in 1925 where he was soon appointed Deputy and then later Superintendent. While at the Nautical Almanac Office Comrie built up a reputation as Britain's leading mathematical table maker and expert on machine computation. Comrie left the Nautical Almanac Office in 1936 and set up the Scientific Computing Service, the first commercial scientific computing bureau of its kind. Comrie was a member of the Mathematical Tables Committee from 1928 to 1936 and from 1949 to 1950 serving as Secretary 1929 to 1936.

J. C. P. Miller (1906–1981)

Jeffrey Charles Percy Miller studied both mathematics and astronomy at Cambridge under A. S. Eddington before becoming a research assistant at Imperial College. In 1935 he became a lecturer at Liverpool University before joining Comrie's Scientific Computing Service as technical director from 1946 to 1950. In 1950 he was appointed to the Cambridge Mathematical Laboratory. Miller's main interests were mathematical tables and numerical analysis and he published widely. Miller was a member of the Mathematical Tables Committee from 1933 to 1965 serving as Secretary from 1946 to 1948.

D. H. Sadler (1908–1987)

Donald Sadler joined the Nautical Almanac Office in 1930 after graduating in mathematics at Cambridge earlier that year. Sadler was trained in mathematical table making and numerical computing by Comrie and he became first Deputy and, in 1936, Superintendent in Comrie's footsteps. Sadler remained Superintendent of the Nautical Almanac Office until his retirement in 1971. During the war he organized computations for the Admiralty Computing Service and served as General Secretary of the International Astronomical Union 1958 to 1964. Sadler was a member of the Mathematical Tables Committee from 1932 to 1965.

E. T. Goodwin (1913–)

E. T. Goodwin, 'Charles' to his friends, graduated from Cambridge in mathematics in 1934 and gained a Ph.D. in 1938. During the Second World War he worked for the Ministry of Supply and the Admiralty on a wide range of mathematical problems. In 1945 he joined the newly created Mathematics Division of the National Physical Laboratory being promoted to Superintendent in 1951.

> **Short biographies of selected members of the Mathematical Tables Committee** *cont.*
>
> He retired in 1974 as Deputy Director (C) of the National Physical Laboratory. Goodwin served as a member of the Mathematical Tables Committee in 1948 and from 1950 to 1965.
>
> **M.V. Wilkes (1913–)**
> Maurice Wilkes took a degree in mathematics at Cambridge in 1934 and a Ph.D. in 1938. In 1937 he was appointed as a demonstrator in the newly created Mathematical Laboratory at Cambridge. After war service Wilkes was appointed Director of the Cambridge Mathematical Laboratory where he designed and built EDSAC, one of the world's first electronic computers. Since his 1980 retirement from the renamed Cambridge Computer Laboratory, Wilkes has held senior consulting positions with the Digital Equipment Corporation, Olivetti and AT&T. He was elected FRS in 1956. Wilkes was a member the Mathematical Tables Committee from 1938 to 1948 and from 1958 to 1965 serving as Chairman for the latter period.

The tables came from a variety of sources. Tables made by government research establishments such as the National Physical Laboratory and the Admiralty Research Laboratory made up some of the collection. The Admiralty also deposited tables which had been made by the Admiralty Computing Service during the war. Some tables were submitted by members of the Committee and others came from industry and from universities. From time to time the Committee published lists of the tables held in journals such as *Mathematical tables and other aids to computation, Philosophical magazine*, and the *Journal of the London Mathematical Society*, and encouraged people to use the facility in person or request copies. Users of the tables are harder to locate. A list of early users issued in 1954[38] shows that institutions as diverse as the National Research Council of Canada, North American Aviation Inc., the US Navy, Ferranti Ltd, and the British Broadcasting Corporation all used the service.

Another role played by the Royal Society Mathematical Tables Committee was as a forum for technical debate on different aspects of table making or computation techniques. The most interesting of these discussions concerned the effect of the increasing number and sophistication of electronic computers on the production of mathematical tables.

The advent of electronic computers, 1952–65

In 1952 the Royal Society Mathematical Tables Committee held the first of its discussions on the effect of electronic computers on mathematical tables. It was prompted by an invitation from John Todd for a member of the Royal Society Mathematical Tables Committee to attend a symposium on *Mathematical tables in the age of electronic computers* to be held in Washington, DC. Todd, a member of the British Association Mathematical Tables Committee 1944–8, now ran the Computation Laboratory of the National Bureau of Standards in Washington. Todd was hosting the symposium, held on 15 May 1952, in response to suggestions that once 'high speed digital computing machines are readily available, the need for mathematical tables, in the conventional form, will disappear'.[39] Todd hoped that by bringing together as many table makers and table users as possible that it would be possible to outline a long term policy for the future needs of the table using community. Before the symposium Goodwin and Miller, in consultation with other active table makers on the Committee, reported on the matter to the Mathematical Tables Committee. In their view there were three stages to mathematical table production: planning, computation, and publication. Electronic computers, when they became widespread, would ease the burden of the computation stage by a factor of about 100 but the amount of planning that went into producing a table had remained fairly constant. The greatest part of the labour was now publication which was both very expensive and time consuming.

The Committee agreed to send Goodwin to the symposium and in preparation he canvassed opinions from other interested parties in Britain. The consensus of opinion, expressed in a Mathematical Tables Committee report,[40] was that the very small number of computers which were likely to be available in the foreseeable future was unlikely to meet the computing needs of the majority. In the Committee's opinion tables would continue to be needed and the number of machines available in England was too low to impact significantly on the computing of tables. The whole tone of the report is slightly dismissive of the notion that electronic computers would have a big impact on either the need for tables or their production. It differs subtly from Goodwin and Miller's earlier note and illustrates differences of understanding between those working at institutions which already had operational computers, such as Miller at Cambridge and Goodwin at the National Physical Laboratory, and those with less direct involvement. The following quote taken from the opening paragraph of the report illustrates the point.

> The need for extensive tables to moderate accuracy will continue especially for use in small organizations and individual workers. Tables of certain types of function (eg. those which machines would take a long time to compute) will be required to higher accuracy and the need will increase. The number of machines expected to be in use in the U.S.A. within the next year (50) is many times too small if tables were to be dispensed with. Tables are made by man and not machine.[41]

Goodwin attended the symposium along with C. W. Jones, who served on the Bessel Function Panel of the Mathematical Tables Committee from 1947 onwards and was a member of the full Committee 1948 and 1953–61. Goodwin had felt apprehensive about going to the Washington meeting as he felt their ideas were vague and that planning for the unknown was difficult but he need not have worried. The situation in America turned out to be even less clear than in Britain. Much of the meeting was taken up with discussions concerning the types of tables and problems which could be applied to electronic computers. The need to continue making tables to serve the thousands of scientists who would not have access to computers was discussed but the closure of the New York Mathematical Tables Project[42] in the late 1940s meant that there were few outlets for mathematical table publication in the United States and that this was felt to be a growing problem. No overall conclusions seem to have been reached and Goodwin felt that the symposium was 'premature'[43] and that it was too early to discuss fully what effect computers would have on the production of mathematical tables.

The debate about the future of mathematical tables was reopened in 1954 when the Massachusetts Institute of Technology hosted a 'Conference on Mathematical Tables: their publication and distribution, together with a consideration of their use in the light of the advent of high speed computing machines'. The Royal Society Mathematical Tables Committee did not send an official representative but Miller attended the conference in his own right. The driving force behind the conference seems to have been the difficulty table makers were experiencing getting tables published and only one session of the conference actually considered if tables still needed to be made; the answer was a resounding yes.

In 1958 Maurice Wilkes, Director of the Cambridge Mathematical Laboratory, was appointed Chairman of the Royal Society Mathematical Tables Committee. His appointment came at a time when it had become obvious to many that the use of electronic computers would become widespread. Wilkes was familiar with all kinds of computing techniques and had served on the

Fig. 9.5 Maurice Wilkes, the final chairman of the Mathematical Tables Committee, had the melancholy job of winding down the Committee's activities in response to the widespread development of electronic computers and calculators which heralded the end of the mathematical table era. (Photograph by kind permission of Sir Maurice Wilkes.)

British Association Mathematical Tables Committee 1938–48. On his appointment, which he took up rather reluctantly under pressure from Royal Society administrators, Wilkes realized that 'some of the members were still living in the past, or at any rate had not appreciated the change which had to come about'[44] and from the start Wilkes felt his job was to bring the Mathematical Tables Committee to an end. He took on this role in two ways.

On taking over as chairman Wilkes' first task was to initiate a discussion of the future of mathematical tables on October 1958. He produced a report[45] in which he stated clearly the changing situation. He noted that electronic computers had changed computational needs as well as making the computation of mathematical tables much easier. He pointed out that the good work of the Mathematical Tables Committee and others had ensured that most of the fundamental tables likely to be needed by desk machine operators had already been published and stated that even if electronic computers had not been developed, the Committee would have had to reconsider its role in response to that fact. He also suggested that the Committee consider tables for use with computers as a possible way forward. But Wilkes' true feelings can be detected in one sentence buried in the middle of the report. Wilkes wrote 'Much as I admire the work done by the Committee in the past I cannot feel enthusiastic about embarking on a future programme of tabulating functions of which individual values can be obtained by a digital computer in a few milliseconds'.[46] Wilkes invited comments on his report and over the next year

Committee members responded. Most of the responses suggest that the Committee look towards producing tables for use with computers such as regularly spaced values of functions with subroutines for interpolation.

It was slowly becoming inevitable, and obvious, that there was no longer any justification for producing traditional mathematical tables. While many older members of the Committee did not like the fact they did acknowledge it. Over the next few years Wilkes applied a gentle but persistent pressure to Committee members to either finish or abandon table making projects which had been on-going for several years. To this end seven of the eleven volumes of Royal Society Mathematical Tables Series were published after Wilkes took over the chairmanship of the Committee.

The Committee was finally dissolved in 1965 and replaced by a skeleton Committee set up to administer the sales income from existing Mathematical Tables Series publications and make reprinting decisions. It was the end of the mathematical table era.

Further reading

A history of the British Association for the Advancement of Science up to 1931 is given in O. J. R. Howarth, *The British Association for the Advancement of Science: A retrospective 1831–1931*, British Association for the Advancement of Science, London, 1931. The changing face of scientific computing in the UK over the period 1925 to 1955 is described in M. Croarken, *Early scientific computing in Britain*, Clarendon Press, Oxford, 1990. For the importance of mathematical tables in Victorian Britain see A. Warwick, 'The laboratory of theory' pp. 311–51 in M. N. Wise (ed), *The values of precision*, Princeton University Press, Princeton, 1995.

Biographical information on some table makers can be found in R. C. Archibald, *Mathematical table makers*, The Scripta Mathematica Studies Number Three, Yeshiva University, New York, 1948. See also the *Dictionary of national biography* for more biographical details of some the table makers mentioned.

The reports of the British Association Mathematical Tables Committee can be found in intermittent volumes of *Report of the British Association for the Advancement of Science* from 1873 to 1939. The final report is published in *The Advancement of Science*, vol. 5, 1948–9, 341–7 and reprinted in *Mathematical tables and other aids to computation*, vol. 3, 1948, 333–40. The reports of the Royal Society Mathematical Tables Committee are held at the Royal Society and in Sadler's papers held at Cambridge University Library RGO 16 100D. See also M. Croarken and M. Campbell-Kelly, 'Beautiful Numbers: The Rise and Decline of the British Association Mathematical Tables Committee, 1871–1965', *IEEE Annals of the history of computing*, 22:4 (2000), 44–61.

Notes

1. *Report of the British Association for the Advancement of Science 1870*, p. lxxi. Currency conversion data taken from <http://eh.net/hmit> accessed November 2001.
2. *Report of the British Association for the Advancement of Science 1871*, p. lxix.
3. 'Report of the Committee...on Mathematical Tables', *Report of the British Association for the Advancement of Science 1873*, 1–175, p. 2.
4. 'Report of the Committee...on Mathematical Tables', *Report of the British Association for the Advancement of Science 1873*, p. 172.
5. 'Report of the Committee...on Mathematical Tables', *Report of the British Association for the Advancement of Science 1873*, 1–175.
6. J. W. L. Glaisher, 'Tables of the Exponential Function', *Transactions of the Cambridge Philosophical Society*, 13 (1883), 243–372.
7. 'Report of the Committee...on Mathematical Tables', *Report of the British Association for the Advancement of Science 1879*, 46–57.
8. A. Cayley, *A treatise on elliptic functions*, Deighton Bell, Cambridge and London, 1876.
9. J. W. L. Glaisher 'On factor tables...', *Proceedings of the Cambridge Philosophical Society*, 3 (1878), 99–138.
10. J. Glaisher, *Factor table for the fourth million*, Taylor & Francis, London, 1879. *Factor table for the fifth million*, Taylor & Francis, London, 1880. *Factor table for the sixth million*. Taylor & Francis, London, 1883.
11. 'First Report of the Committee...', *Report of the British Association for the Advancement of Science 1889*, 28–9.
12. Lord Rayleigh, *The theory of sound*, 2 volumes, London, Cambridge, 1877 and 1878.
13. A. Cunningham, *A binary canon*, Taylor & Francis, London, 1900.
14. See M. Croarken, *Early scientific computing in Britain*, Clarendon Press, Oxford, 1990, pp. 22–46.
15. British Association for the Advancement of Science, *Mathematical Tables Volume 1: Circular and hyperbolic functions, exponential sine and cosine integrals, factorial (gamma) and derived functions, integrals of probability integral*, Office of the British Association, London, 1931.
16. See, for example, Preface, in E. Jahnke and F. Emde, *Tables of functions*, Leipzig, 1933.
17. For more information about Sadler's career see Chapter 11. Miller suffered a serious illness in the late 1920s the after effects of which precluded him from

passing the civil service medical examination which he would have had to have done to gain employment at the Nautical Almanac Office.
18. Approximately equivalent to £135 000 today. Data taken from <http://eh.net/hmit> accessed November 2001.
19. Comrie consulted G. H. Hardy and A. E. Weston among others. His letters are preserved in the Bodleian Library Oxford, Ms. Eng. Misc. 677 D1–D5.
20. L. E. Dickson, *Minimum decompositions into fifth powers,* British Association Mathematical Tables Series, Vol. 3, Cambridge 1933. E. L. Ince, *Cycles of reduced ideals in quadratics fields* (ed. L. J. Comrie), British Association Mathematical Tables Series Vol. 4, Cambridge 1934. *Factor table giving the complete decomposition of all numbers less than 100 000,* British Association Mathematical Tables Series Vol. 5, Cambridge, 1935.
21. L. J. Comrie, 'Inverse interpolation and scientific applications of the National Accounting Machine', *Supplement to the journal of the Royal Statistical Society,* 3:2 (1936), 87–114.
22. I. B. Cohen, *Howard Aiken: portrait of a computer pioneer,* Harvard University Press, Cambridge, Mass., 1999, Chapter 8.
23. Minutes of the Mathematical Tables Committee 5 November 1936, Bodleian Library, Ms. Eng. Misc. d. 1158.
24. Minutes of the Mathematical Tables Committee, 10 November 1936, Bodleian Library, Ms. Eng. Misc. d. 1158.
25. A. J. Thompson could supply computing labour from time to time but not on a regular basis, Minutes of the Mathematical Tables Committee 5 November 1936, Bodleian Library Ms. Eng. Misc. d. 1158. These probably came from his contacts in the General Register Office. Also students and facilities at the University of Liverpool were used (Report of the Mathematical Tables Committee, *Report of the British Association for the* Advancement *of Science 1937,* 272–3).
26. Minutes of the Mathematical Tables Committee, 21 January 1937, Bodleian Library Ms. Eng. Misc. d. 1158.
27. Letter from J. C. P. Miller to D. H. Sadler 16 December 1945, Cambridge University Library RGO 16 92/4 Folder II. Letter from W. H. Bickley to D. H. Sadler 22nd December 1945, Cambridge University Library RGO 16 92/4 Folder II.
28. Minutes of the British Association for the Advancement of Science Council 7 June 1946, Minute 7, Bodleian Library Dep BAAS 29 folio 283.
29. Minutes of the British Association Council 46/108. Cambridge University Library RGO 16 92/4 folder III.
30. M. V. Wilkes. Email to Martin Campbell-Kelly and Mary Croarken 3 August 1999.
31. The minutes of the Royal Society Mathematical Tables Committee are held at the Royal Society in London. Sadler's papers at Cambridge University Library contain copies of the minutes and letters between Committee members, RGO 16 100D.
32. Minutes of the Royal Society Mathematical Tables Committee 7 November 1947.

33. Letter J. C. P. Miller to D. H. Sadler 2 July 1948; D. H. Sadler to J. C. P. Miller 8 July 1948; W. G. Bickley to D. Brunt 1 February 1949. Cambridge University Library RGO 16 100D Folder I.
34. M.V. Wilkes. Email to Martin Campbell-Kelly and Mary Croarken 3 August 1999.
35. Letter from A. J. Thompson to D. H. Sadler 20 February 1953, Cambridge University Library RGO 16 100C Folder VII. Permission to quote has been kindly granted by the Syndics of Cambridge University and the Particle Physics and Astronomy Research Council.
36. Minutes Royal Society Mathematical Tables Committee 18 January 1950, minute 5.
37. 'Royal Society's depository of unpublished mathematical tables', *Nature*, Vol. 167, 24 February 1951, p. 303. 'Unpublished mathematical tables', *Mathematical tables and other aids to computation* 5 (1951), 83–4.
38. 'Tables supplied from the depository of unpublished mathematical tables (up to no. 20)', Royal Society Mathematical Tables Committee Report MT/10(54). Cambridge University Library RGO 16 100C Folder IX.
39. 'Symposium on "Mathematical Tables in the Age of Electronic Computers"', Royal Society Mathematical Tables Committee Report MT/12(52), Cambridge University Library RGO 16 100C Folder V. Permission to quote has been kindly granted by the Syndics of Cambridge University and the Particle Physics and Astronomy Research Council.
40. 'NBS Symposium on mathematical tables in the age of electronic computers', Royal Society Mathematical Tables Committee Report MT/15(52)(Amended). Cambridge University Library RGO 16 100C Folder VI.
41. 'NBS Symposium on mathematical tables in the age of electronic computers', Royal Society Mathematical Tables Committee Report MT/15(52)(Amended). Cambridge University Library RGO 16 100C Folder VI, p.1. Permission to quote has been kindly granted by the Syndics of Cambridge University and the Particle Physics and Astronomy Research Council.
42. See David Grier (Chapter 10) on the New York Mathematical Tables Project.
43. 'A conference on mathematical tables in the light of electronic computers', Typescript report Cambridge University Library RGO 16 100C Folder VI.
44. M.V. Wilkes, manuscript note dated 12 December 1978 attached to Royal Society Mathematical Tables Committee Correspondence file in Wilkes' possession.
45. M.V. Wilkes 'The future of mathematical tables', Royal Society Mathematical Tables Committee Report MT/15(58). Cambridge University Library RGO 16 100C Folder XIII.
46. M. V. Wilkes 'The future of mathematical tables', Royal Society Mathematical Tables Committee Report MT/15(58). Cambridge University Library RGO 16 100C Folder XIII, p.1. Permission to quote has been kindly granted by the Syndics of Cambridge University and the Particle Physics and Astronomy Research Council.

x	$\dfrac{2}{\sqrt{\pi}}\,e^{-x^2}$	$\dfrac{2}{\sqrt{\pi}}\displaystyle\int_0^x e^{-\alpha^2}\,d\alpha$
1.000	0.41510 74974 20595	0.84270 07929 49715
01	.41427 76978 09643	.84311 54854 78076
02	.41344 87300 69275	.84352 93486 22624
03	.41262 05958 47470	.84394 23832 16054
04	.41179 32967 83914	.84435 45900 92706
1.005	0.41096 68345 10002	0.84476 59700 88552
06	.41014 12106 48838	.84517 65240 41197
07	.40931 64268 15244	.84558 62527 89859
08	.40849 24846 15760	.84599 51571 75372
09	.40766 93856 48649	.84640 32380 40168
1.010	0.40684 71315 03900	0.84681 04962 28277
11	.40602 57237 63236	.84721 69325 85311
12	.40520 51640 00110	.84762 25479 58462
13	.40438 54537 79718	.84802 73431 96492
14	.40356 65946 58998	.84843 13191 49722
1.015	0.40274 85881 86635	0.84883 44766 70026
16	.40193 14359 03065	.84923 68166 10825
17	.40111 51393 40482	.84963 83398 27073
18	.40029 97000 22840	.85003 90471 75254
19	.39948 51194 65857	.85043 89395 13372
1.020	0.39867 13991 77022	0.85083 80177 00942
21	.39785 85406 55599	.85123 62825 98982
22	.39704 65453 92629	.85163 37350 70006
23	.39623 54148 70940	.85203 03759 78015
24	.39542 51505 65148	.85242 62061 88487
1.025	0.39461 57539 41661	0.85282 12265 68372
26	.39380 72264 58690	.85321 54379 86084
27	.39299 95695 66247	.85360 88413 11487
28	.39219 27847 06155	.85400 14374 15894
29	.39138 68733 12051	.85439 32271 72054
1.030	0.39058 18368 09392	0.85478 42114 54148
31	.38977 76766 15462	.85517 43911 37776
32	.38897 43941 39374	.85556 37670 99952
33	.38817 19907 82078	.85595 23402 19095
34	.38737 04679 36367	.85634 01113 75020
1.035	0.38656 98269 86882	0.85672 70814 48933
36	.38577 00693 10116	.85711 32513 23418
37	.38497 11962 74423	.85749 86218 82434
38	.38417 32092 40021	.85788 31940 11301
39	.38337 61095 59000	.85826 69685 96698
1.040	0.38257 98985 75329	0.85864 99465 26651
41	.38178 45776 24859	.85903 21286 90527
42	.38099 01480 35329	.85941 35159 79022
43	.38018 66111 26376	.85979 41092 84158
44	.37940 39682 09540	.86017 39094 99272
1.045	0.37861 22205 88268	0.86055 29175 19009
46	.37782 13695 57922	.86093 11342 39312
47	.37703 14164 05786	.86130 85605 57417
48	.37624 23624 11074	.86168 51973 71843
49	.37545 42088 44931	.86206 10455 82382
1.050	0.37466 69569 70448	0.86243 61060 90097

10

Table making for the relief of labour

DAVID ALAN GRIER

After the Mathematical Tables Project had shut its doors in New York and consigned its worksheets to the flames, its leaders liked to compare their efforts to those of Gaspard Riche de Prony (described by Ivor Grattan-Guinness in Chapter 4). Herbert Salzer, who was one of the assistant leaders of the group, would grow animated when he talked about the two organizations. 'We were like those French mathematicians', he would claim. 'We were modelled after them.'[1] In spite of his protestations, the best evidence suggests that the founders of the Mathematical Tables Project originally knew nothing about de Prony and his cadastral computers. Salzer, who served as

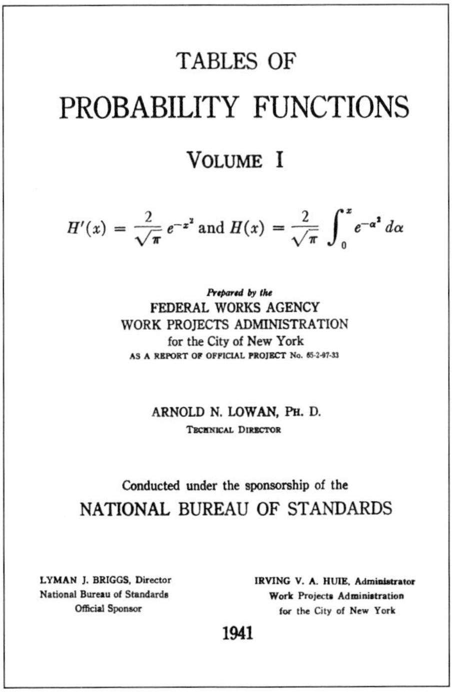

> TABLES OF
> PROBABILITY FUNCTIONS
> VOLUME I
>
> $H'(x) = \frac{2}{\sqrt{\pi}} e^{-x^2}$ and $H(x) = \frac{2}{\sqrt{\pi}} \int_0^x e^{-\alpha^2} d\alpha$
>
> *Prepared by the*
> FEDERAL WORKS AGENCY
> WORK PROJECTS ADMINISTRATION
> for the City of New York
> AS A REPORT OF OFFICIAL PROJECT No. 65-2-97-33
>
> ARNOLD N. LOWAN, Ph. D.
> Technical Director
>
> Conducted under the sponsorship of the
> NATIONAL BUREAU OF STANDARDS
>
> LYMAN J. BRIGGS, Director   IRVING V. A. HUIE, Administrator
> National Bureau of Standards   Work Projects Administration
> Official Sponsor   for the City of New York
>   1941

Fig. 10.1 *Tables of probability functions* published in 1941 was a typical example of the smaller interval high decimal figure tables which the WPA Mathematical Table Project published.

bibliographer for the group, probably made the connection while searching for old mathematical tables in the New York Public Library or the New York Engineering Society. Yet, even if he created the connection after the fact, Salzer was correct in drawing a parallel between the two groups. Both were born in times of social travail. Both employed large numbers of relatively unskilled computers. Both faced sceptical publics that were reluctant to accept the fruits of their work. The Mathematical Tables Project proved the more successful of the two computing organizations. Though their major product, twenty-eight volumes of mathematical tables, was at best lightly used, they produced a popular set of small 'Lilliputian' tables and hundreds of specialty computations for the Army, the Navy, the National Bureau of Standards, private corporations, and university professors. If their legacy had been nothing more than the Navy's LORAN navigation tables or Han Bethe's thermodynamic calculations, the group could count itself a success.

Over its ten year existence, the Mathematical Tables Project was constantly forced to argue for the value of its work. Each new calculation required the project leaders to claim that they knew what they were doing, that their computers were adequately trained, and that the final values could be trusted. For the most part, they suffered the damning criticism of faint praise from scientists. Such struggles are nothing new to science and are commonly found among scientific organizations that are trying to build a professional identity. George Daniels, in his study of the professionalization of American science, has argued that the process of building a professional discipline involves three stages. Scientists first assemble a body of knowledge and practices. Next, they build an institution to support that knowledge. Finally, they establish the legitimacy of their organization in the eyes of the public. The leaders of the Mathematical Tables Project followed much the same pattern, though with telling differences that underscore the extent to which computation has its roots not only in the intellectual discipline of mathematics but in the physical discipline of labour.[2]

The Mathematical Tables Project had to establish its legitimacy not among the general public, as Daniels suggests, but among scientists. When the project began operations in 1937, their human computers had no professional standing. None knew advanced mathematics. Few knew any algebra. Most understood only the rules of addition and even these didn't fully apprehend all the properties of basic arithmetic. The leaders of the project, Arnold Lowan and Gertrude Blanch, attempted to give their workers a professional appearance by imposing a rigorous, nearly mechanical form of discipline upon the group. They organized the computers by arithmetical operation and gave them little freedom in their work. They wrote strict instructions and prepared worksheets designed to identify mistakes, both accidental and deliberate. They even welcomed strict peer review of their own activities. In the end, the group was truly successful only when it was able to shed its work relief origins. As the Second World War engulfed the United States, Lowan and Blanch were able to release excess workers, adopt modern computing machinery, and distance themselves from the problems of work relief.

Origins of the Mathematical Tables Project

The Mathematical Tables Project was created by the Work Projects Administration (WPA), a government agency that attempted to create jobs for 'employable workers'.[3] It was organized in a hurried manner by Malcolm Morrow, an associate statistician in the Washington WPA office. Morrow's first task was to find a sponsor for the project, an organization that would take responsibility for daily operations. He approached New York University, which was initially interested until university officials discovered that they would be required to provide twenty per cent of the project's budget. Unable to find another sponsor from the academic community, he asked the New York mayor's office to sponsor the group. When they agreed, he turned to the National Academy of Sciences to provide the necessary mathematical expertise. The National Academy of Sciences was located just a few blocks from Morrow's WPA office. It was housed in a white marble building that appeared to be more of a Masonic temple to science than the home of an advisory body. The organization was formed during the civil war as a means of recognizing successful scientists and as a way to coordinate scientific advice for government projects. The academy also attempted to shape the scientific community by publishing monographs, bibliographies, and handbooks, financing graduate fellowships, and sponsoring meetings.

Morrow approached the leaders of the academy, told them of the proposed Mathematical Tables Project and asked if they might be willing to appoint a scientific advisory committee. The officers, who were represented by their permanent secretary, were generally pleased with the proposal. The president of the academy wrote to the WPA, 'I wish to assure you that we shall be most happy to render any service within our power.' Furthermore, 'the National Academy of Sciences in its part would be glad to appoint a committee of

Work Projects Administration

When Franklin Roosevelt became President of the United States in March 1933, he took the leadership of a country whose economy had been contracting for nearly three and one half years. His first act was to declare a bank holiday, in order to prevent a run on the money supply. He then set his new administration to implement a series of relief programs that would employ those who had lost their jobs and would rebuild the economy. During his first hundred days in office, his administration would initiate a collection of programs, which would later be known as the New Deal. They included the Agricultural Adjustment Administration, the first project; the National Recovery Administration, for relief of industry; and the Tennessee Valley Authority, to develop the resources of this key area.

The Work Projects Administration (WPA) was the third New Deal public works program. Originally called the Work Progress Administration, it was formed in 1935 to operate 'a nation-wide program of "small useful projects" designed to provide employment for needy employable workers'. It built on the lessons of its predecessors by providing funds to local agencies who would then use the moneys to construct public works.

The WPA undertook three kinds of projects. The most prevalent were construction projects. The WPA built roads, dams, parks, airports, and government offices. Equally prevalent were service projects: adult education classes, nursery schools, school lunch programmes, sewing and knitting projects, gardening, and canning efforts. Finally, the WPA sponsored cultural activities under a program known as Federal Project Number 1. This was a 'single nation-wide project which, with WPA sponsorship, provided a central administration for music, art, writer's and theaters projects and the historical records survey.' This project hired writers to prepare guidebooks for states, employed workers to review local government records and create historical indices, and organized library card catalogues. This group also sponsored the Federal Theatre Project, a group that WPA officials hoped would 'lead to the establishment of municipal theatres in a number of cities and even a nationally endowed theatre.' The project was criticized by conservatives in the press and the American congress. Orson Welles' play *Cradle will rock* particularly drew public anger. Congress terminated funds for theatrical productions in 1939.

The WPA funded about 4000 science projects. Most of them were small activities, the work of a single scientist or laboratory. Many dealt with social or economic research. By the nature of their charter, which required them to support labour intensive activities, they paid the salary of many computers, though most were engaged in statistical tabulation. They paid Iowa State University to conduct a large survey of rural life and, in the process, WPA workers constructed a new statistics building for the campus. They also funded a computing group in Philadelphia to do map computations for the Coast and Geodetic Survey.

TABLE MAKING FOR THE RELIEF OF LABOUR 269

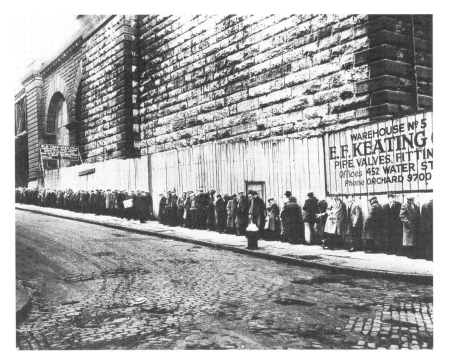

Fig. 10.2 A New York Breadline. Residents of New York City queuing for food underneath the Brooklyn Bridge, circa 1936. (Photograph courtesy of the Library of Congress.)

mathematicians to advise on specific undertakings.'[4] With this promise of advice for his project, Morrow left for New York. Before he departed, he asked the officers of the academy that 'no intimation of consideration of the project be permitted to be made publicly or find its way into the channels of the press.'[5] His experience in relief work had taught him the perils of announcing a project before it was ready. He still needed to file a formal application for the project and find adequate office space. Most importantly, he needed to find trained mathematicians to lead the group.

Waiting patiently for Morrow to return, the leaders of the academy discussed possible committee members. They considered well-regarded scientists, such as Vannevar Bush of MIT, and lesser known scientists who were considered skilled table makers, a leading example being Harold Davis of Indiana University. They concluded that the chair of the committee should be Oswald Veblen, a former president of the American Mathematical Society who had led the Army's mathematical ballistics efforts during the First World War.[6] By the time that they had decided upon Veblen, they had become anxious about Morrow. 'Mr. Morrow was in here saying we were to get a letter the next day', wrote one of the academy staff members. 'I have wondered what has happened to him. He seems to have disappeared off the horizon.'[7] In fact, Morrow had changed his plans without informing the academy. He had discovered a new sponsor for the project, Lyman Briggs, the director of the National Bureau of Standards. Briggs took immediate charge of the project. Rather than wait for the appointment of a new advisory group, he recruited an existing committee from the National Academy of Sciences, the Committee on the Bibliography for Mathematical Tables and Other Aids to Computation (MTAC). This committee had been founded in 1934. Its chair, A. A. Bennett of Brown University, had been one of Veblen's assistants during the war.

The leaders of the National Academy of Sciences reacted politely when they learned of the meeting. However, other members of the academy were less sanguine about the unannounced change of plans. The scientists who oversaw the MTAC committee felt that their authority had been circumvented. They requested a full report on the subject from Professor Bennett and advised him that he should not undertake such efforts. 'It was the consensus,' wrote the chair of the group, 'that the Committee work should be limited to a bibliography.'[8]

Starting operations

When Malcolm Morrow proposed the project to the National Academy, he had spoken lavishly of the resources that would be available to compute mathematical tables. He mentioned that the white collar relief budget for New York City was $2 300 000, and intimated that a large part of it 'could be devoted to this project.' He said that the WPA could pay the salaries of the directors, would provide furniture and office machines. He even speculated that the project could get an extra grant of $100 000 for the purchase of computing machines.[9] The reality of the project proved to be quite different.

The budget proved to be a small fraction of what Morrow suggested. The offices were an old industrial loft and the only computing machines for the groups were those that had been abandoned by other government agencies. The WPA recruited two underemployed scientists to direct the project,

Fig. 10.3 Arnold Lowan. When asked to lead the programme, Arnold Lowen (photograph circa 1938) was a professor at Yeshiva University, New York. (Picture taken from R. C. Archibald, *Mathematical Table Makers*, Scripta Mathematica, New York, 1948. Courtesy of Yeshiva University.)

neither of whom had ever done this type of calculation. The director of the project would be Arnold Lowan. He was an immigrant from Romania, where he had taken a degree in chemical engineering. Upon arriving in the United States, he had worked as a combustion engineer for a New Jersey utility. Step by step he had mastered English and started to study physics. After becoming a naturalized citizen in 1929, he had enrolled in the PhD program at Columbia University. He shunned the new and revolutionary topics of quantum physics and turned to a more conventional subject, the cooling of the Earth. After completing the degree in 1934, he had briefly tasted success when he was named one of the first post-doctoral students at the Institute for Advanced Study in Princeton, New Jersey. Lyman Briggs found Lowan after he had left the institute and was shuttling between two jobs in New York. By day, he was a professor at Yeshiva University, a Jewish school on the north end of Manhattan. At night he augmented his salary by teaching physics courses at Brooklyn College. Though he had never overseen a large laboratory, he was interested in calculation and willing to take the job when Briggs offered it.[10]

Fig. 10.4 Gertrude Blanch (photograph mid-1930s) brought to the program a Ph.D. in mathematics from Cornell University and nearly twenty years of office experience. (Courtesy Deborah Stern.)

Lowan recruited Gertrude Blanch to be the mathematical leader of the project. In the fall of 1937, Blanch was a mathematics PhD who worked as an office manager for a photographical company. Like Lowan, she was an immigrant, and the late recipient of a doctorate. She had completed her degree in 1936, six months after her fortieth birthday. Her advisor had been Virgil Snyder, a former president of the American Mathematical Society. After graduation, the only job she could find was a one-year visiting position at Hunter College for women in New York City. When that job ended, she took an office job in a well appointed building on East 23rd Street. The firm, Devin Photographic, was flourishing during the depression through the unlikely strategy of selling a luxury good to the wealthy. The work was pleasant and easy. She kept accounts, handled correspondence and tracked orders. It gave her a enough free time to read mathematics on the side and so she registered for a course on relativity at Brooklyn College.[11] Lowan was the instructor. As the term progressed, he became increasingly curious how Blanch, who showed little inclination during the lectures she attended, could do such fine work on the homework sets. One day he sat next to her on the bus and casually asked why she didn't attend class more regularly. Blanch, often a private individual, deflected his query. A few days later, he was more direct. He complemented her work and asked about her scientific background. At that point, she admitted that she held a PhD in mathematics.

The story of her background intrigued Lowan. After the next class, he told Blanch about the WPA project and asked if she would be interested in being the chief mathematician. He took her to visit the offices for the new project. They were located in what was then Manhattan's seedy west side, a neighbourhood just a few blocks from the Hudson River piers. It was the west side of 'West Side Story', of gang conflict, tenements, and industrial buildings. The offices for the Mathematical Tables Project were located on the tenth floor of an old structure; they could only be reached via a freight elevator. To less disciplined eyes, the prospects were not promising. The floor was covered with dust. The furniture was broken and worn. The WPA offered few funds for supplies. Blanch may have hesitated for a moment. She knew only the rules of commercial calculation, little about numerical analysis and nothing about table making. Nonetheless, she concluded that this was her best opportunity to become a scientist again, 'I resigned from a beautiful office downtown' and took the position with the Mathematical Tables Project.[12]

Blanch had the task of preparing the work for the computers. Ever the optimist, she later recalled that some of the relief workers did fairly well. 'The New York high schools were pretty good in those days,' she commented. 'We had a few very competent youngsters.'[13] However, most of the computers were unable to do the work of the competent few. These individuals were not merely out of work. They had faced the full brunt of the depression and had learned the lesson, as had much of the American middle class, that it was something more than a shrinkage of the money supply or a loss of confidence in the stock market. Blanch did not let her computers work independently. She dictated every step they took. She outlined how they would compute their functions, how they would check the results and even how they would perform basic arithmetic. First, she restricted each computer to a single operation. The largest group of the computers would do only addition. The next group would do subtraction. A considerably smaller number would undertake multiplication and they would only multiply a single digit against a larger number. Only a handful would undertake the hardest operation, long division. The most skilled computers were generally assigned to a special group, which did preliminary computations for Blanch and the other mathematicians.

Computing the error function: steps the WPA Mathematical Tables Project used to compute the error function tables published in their *Tables of Probability Functions Volume I* published in 1941

The error function is derived from the famous bell curve (see Fig. 10.5). The bell curve describes the behaviour of certain kinds of statistical data. Data which follow the bell curve are likely to take values near their mean (the central hump of the curve) and are less likely to fall at a distance (the sloping curves to the right and left that form the bell). The curve had applications in statistics, economics, electrical engineering, psychology, and medicine. The Mathematical Tables Project tabulated a version of the function which was derived from the right half of the bell curve, a function given by the expression:

$$F(x) = \frac{2}{\sqrt{\pi}} \int_0^x e^{-a^2} da$$

The tabulation had six steps:

1. *Initial values* The special computing group calculated 60 key starting values of the function. $f(x) = \frac{2}{\sqrt{\pi}} \int_0^x e^{-a^2} da$. For this step they used an old table of the exponential function that had been published by the National Academy of Sciences.
2. *First derivatives* For each of these sixty values, they computed the first derivative $f^{(1)}(x)$ with the formula $f^{(1)}(x) = -2xf(x)$,
3. *Higher derivatives* The special computers calculated higher derivatives using the formula $f^{(n)}(x) = -2xf^{(n-1)}(x) - 2(n-2) f^{(n-2)}(x)$. The number of derivatives varied at each point and was dependent upon the need to get 25 digits of precision. They always did at least seven derivatives and as many as twenty. All calculations were done to 25 digits.
4. *Crude tabulation* The computing floor began to tabulate the function $F(x)$ using two formulae derived from a Taylor expansion of $F(x)$:

$$F(x+h) = F(x) + hf(x) + h^2/2! \, f^{(1)}(x) + \cdots h^n/n! f^{(n-1)}(x) \quad (1)$$

and

$$F(x-h) = F(x) - hf(x) + h^2/2! \, f^{(1)}(x) - \cdots + (-h)^n/n! f^{(n-1)}(x). \quad (2)$$

Starting with $F(0)$, they would use formula (1) to create $F(.1)$. Then, using both formula, they would produce pairs of values at increments of .1. The second step would produce $F(.2)$ and a new calculation of $F(0)$, which they would compare to the original $F(0)$ to see if the calculations were

accurate. The third step would produce a $F(.3)$ and a new calculation of $F(.1)$ to compare with the old values. This procedure continued until they had computed $F(6)$.
5. *Final tabulation* This step repeated the computations of step 4 with increasingly smaller increments. They repeated the calculations first at steps of .01, then at steps of .001 and finally at steps of .0001.
6. *Checking for errors* The manuscript was checked with a fifth difference test. Computers took differences of adjacent numbers (first differences) then differences of those differences (second differences), then continued the process until they reached the fifth difference. Any unusual values identified potential errors. Once they were corrected, the numbers were rounded to fifteen decimals and typed. The final typescript was subjected to a difference test using third and fourth differences.

Source: Mathematical Tables Project, *Tables of probability functions volume 1*, New York, Columbia University Press, 1941, pp. xxi–xxiii.

Many of the first computers did not fully understand the basic laws of arithmetic. Even those that did were not allowed to use their own ways of implementing those rules. Blanch isolated each group of computers in a different part of the office and posted instructions for them to follow. For example, she required all computers to write negative numbers in red and positive numbers in black. Her instructions for addition told the computers that they could add two red numbers or two black numbers but not a mixture of red and black numbers. When faced with numbers of different colours, they had to send the calculation to the subtraction group.

For each computation, Blanch prepared worksheets for the computers. These sheets divided the calculation into its fundamental operations. A computer assigned to the addition group, for example, would take a worksheet, complete all the additions and then pass it to another group. In the course of a day, a sheet would pass between several groups as the calculations were done. The worksheets required the computers to match some of their values to pre-determined results. Most of the tables were computed through the method of subtabulation. Subtabulation is similar to interpolation. It starts with two known values and computes the numbers that fall between them. Because of this, Blanch and her staff knew which values needed to appear in the final column of the worksheet. She would print those values on the paper and would instruct the computers to compare

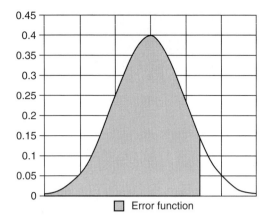

Fig. 10.5 The error function is derived from the famous bell curve and describes the behaviour of certain kinds of statistical data. With applications in statistics, economics, electrical engineering, psychology, and medicine it was a common function and easily justified its inclusion in the Mathematical Tables Project *Tables of probability functions*.

Fig. 10.6 Computing floor of the Mathematical Tables Project (photograph circa 1940). Mathematical Tables Project computers worked at long tables in an industrial building in New York. Their worksheets were kept in large binders. (Courtesy of the U.S. National Records Administration.)

them with their results. If the figures did not match, the computers would be responsible for correcting them.

Correct values in the final column did not guarantee the accuracy of the entire sheet. Errors could appear in intermediate values. To search for such mistakes, Blanch and her assistants would test all the values on each worksheet. They would derive a series of numerical tests in order to check for errors. A common test was the difference test. They would check the worksheet by taking differences of adjacent numbers on the sheet. Certain patterns would indicate the presence of an error. The members of the Mathematical Tables Project were proud of their techniques for checking their work. In reviewing the first two volumes of tables, the mathematician John Curtiss, wrote 'It is claimed that each volume is entirely free from error, and the precautions taken seem to give considerable weight to the claim.'[14] The claim that the tables were 'entirely free from error' proved to be over-confident, as errors did creep into the process from time to time. Nonetheless, the known errors in Mathematical Tables Project calculations is remarkably small.[15]

Selling to the scientists

Not only did Lowan and Blanch have to present the Mathematical Tables Project as a professional, scientific institution, they also had to keep their workers busy. A staff with more than a hundred computers is a surprisingly powerful computing organization, even when all calculations have to be checked and rechecked multiple times. Keeping the computers active was a crucial strategy for any project of the WPA. Rightly or wrongly, the WPA had a reputation of supporting malingerers, labourers who actively avoided work. Commentators joked that its acronym stood not for 'Work Projects Administration' but for 'We Poke Along.'[16] The job of keeping the computers busy was complicated by the procedure for approving projects, a procedure designed to give an unquestionable pedigree to the calculations. The approval process attempted to establish that the computations would be of general use and benefit more than a single individual or organization. It also examined the mathematics behind the computation and verified that it produced results as accurate as any other procedure. Finally, it examined the computing procedure itself to ensure that the human computers could actually do the work.

The computing plans were prepared by Gertrude Blanch and a small staff of mathematicians. They were approved first by Lowan, who would solicit comments from mathematicians and scientists outside the project. Generally, Lowan didn't worry about the amount of effort the calculations required. Though Blanch rarely chose the simplest methods, she tended to favour methods that minimized the number of calculations. Excessive calculations simply provided opportunities for the computers to make mistakes. She regularly rejected methods that were commonly used by skilled computers. These methods might require the subtle use of another table or a complicated re-use of intermediate values.

After Lowan received comments, he would then submit the plan to Lyman Briggs at the National Bureau of Standards. Bureau scientists would give the project a second review and would try to assess the number of scientists who might make use of the calculations. After receiving his staff's evaluation, Briggs would send the plan to the New York Office of the WPA. The WPA employed a member of the mathematics group at Bell Laboratories to review all scientific plans. As before, another round of mathematicians would be asked to comment on the plan. At this stage it was often difficult to find knowledgeable scientists who had not already offered their comments. Once the New York Office had approved the plan, they would send it to the WPA headquarters in Washington, DC for final approval.

The approval process appears to be burdensome and ungainly. Indeed, Lowan occasionally thought it so. More than once he wired the Washington WPA office reporting that the project was running out of work and urging them to approve a computing plan.[17] Yet, the review process was really an extensive, highly ritualized version of peer review. Each of the reviewers commented on the plans and each added something to it. Briggs and his staff often had the best sense of the scientists that might use the work. The outside reviewers regularly identified literature that had been overlooked by the project staff. The staff in the New York Office of the WPA commonly identified inconsistencies in plans and corrected mistakes.[18]

The large tables, which were published in distinctive tan colored volumes, included lengthy prefaces, written by outside mathematicians, physicists, or engineers. Lowan spent substantial effort in trying to recruit the most prominent scientists possible. These prefaces defined the functions to be tabled, described the computing plan, and presented an analysis of underlying algorithms. They shifted attention from how the tables were prepared to the mathematics behind them. The books never mention the human computers, the worksheets or Blanch's instructions. They list only the members of the committee that prepared the computing plan.[19] Through their efforts, Lowan, Blanch, Briggs and the staff of the WPA established a reputation for quality work. The published reviews of the books uniformly praised the Mathematical Tables Project. However, this reputation could not stifle a lingering private criticism of the group. A few scientists felt that the project was misguided. For all of their success, the human computers were relief workers, not trained mathematicians. In their mind, the project was a charitable organization, not a scientific laboratory.

The most vocal critic of the project was Wallace J. Eckert. Eckert was the director of the T. J. Watson computing laboratory at Columbia University, a laboratory barely two miles from the offices of the Mathematical Tables Project. This laboratory used punched card tabulators to compute mathematical functions. In his private correspondence, Eckert would openly question the methods of the project. 'In discussing a large project of table making one must consider whether the idea is to avoid work or to make it' he wrote with almost a mocking tone. In framing his criticism in terms of the amount of work, he was often promoting his own laboratory at the expense of the WPA. Punched card machines, he wrote, 'are not well suited to the latter and hence are not recommended as a solution of the unemployment problem during a depression.'[20]

Eckert was a strong conservative who liked little about the Roosevelt administration. He once attacked a proposed government laboratory by asking 'How much government can we stand?' He suggested that it would be best if the 'government undertakes nothing that can be done well otherwise.'[21] Such opinions were occasionally echoed by L. J. Comrie, who normally a strong supporter of the project. Comrie was a member of the MTAC committee, a former secretary of the British Association Mathematical Tables Committee (described by Mary Croarken in Chapter 9), and a former superintendent of the British Nautical Almanac Office. He was an enthusiastic advocate for computation. His support for the Mathematical Tables Project was so strong that many of the project leaders came to believe, quite mistakenly, that he single-handedly saved the project at the start of the Second World War.[22] Like Eckert, Comrie preferred computing machines to human computers. He also believed that some of the projects had little value and had been undertaken only as a means to keep the computers busy. In a letter that must have surprised Lowan and Blanch, he wrote that a proposed calculation seemed to be 'extravagant, and to savour of computing gone amok.' In a charge that would resonate for the remaining life of the project, he continued, 'In fact this criticism applies, although often in a lesser measure, to other work done by your project.'[23]

Lowan and Blanch were remarkably immune to the complaints of Comrie and Eckert. Lowan, in particular, was unflappable. His letters were filled with happy quotes from satisfied scientists, testimonials which he carefully edited in order to present the Mathematical Tables Project in the best possible light. At times, these letters present an unrealistically positive assessment of the group and ignore the real problems they faced. The historian George Daniels, has noted that scientists often feel that it is necessary to exaggerate the nature of their work in their attempt to gain the approval of the public.[24] Lowan clearly felt that he needed to do this in order to gain the attention of scientists.

In Lowan's mind, the scientists didn't need to know about the social problems among the computing staff. He never mentioned epilepsy, or criminal records, or the fact that many of the computers did not know how to subtract a bigger number from a smaller one. Occasionally, he actually misled his potential collaborators. During the first months of the project's operation, he wrote to a scientist, 'we are operating with a staff of 110 workers under the supervision of a planning section of which I am in charge. Most of the work is of course done with the aid of calculating machines.'[25] At the time, the project owned exactly three adding Monroe machines, all of which were used by the planning committee.

It is easy to castigate Lowan for his unwillingness to admit that most of the calculations were done with red and black drafting pencils on graph paper and that few of his computers knew enough mathematics to balance a cheque book. Yet, we must remember that he received little response for his efforts. The few scientists who replied usually gave cautious, non-committal answers. John von Neumann, who was just starting to become interested in computing sent a short note: 'Many thanks for the announcement of your project. I am much interested in your program and should like to get your material,' he wrote. 'I may have some remarks and suggestions in connection with these things and will write to you concerning them before long.'[26] However, despite regular prodding from Lowan, von Neumann never offered any comments and never proposed any work for the project.

Lowan's saviour was not von Neumann but Phil Morse, a professor at the Massachusetts Institute of Technology. Morse learned of the project in September 1938 through a circular that had been forwarded to him by a friend. 'I have been in charge of a number of similar projects here at Technology' he wrote to Lowan and he had been advised by 'Dr. Compton, President of MIT that I should get in touch with you.'[27] Morse helped find work for the project. His contacts brought work that Lowan considered highly prestigious, the first of these was a computation for Hans Bethe and his student, Robert E. Marshak.

In November 1938, Bethe requested a computation for a paper he and Marshak were preparing on the internal structure of the Sun. The request had to go through the usual approval process. Bethe was quite patient as each office looked at his plans, assessed its use to a broader public and judged the correctness of his proposal. The review found inconsistencies in certain constants, which Bethe adjusted. Such errors could be a major problem. A computing group at Indiana University had undertaken a large computation for a physicist ten years earlier and had used a constant later found to be incorrect. Once the work was over, the physicist and the director of the computing laboratory decided that they had no choice but to re-do the experiment so that it matched the calculation.[28]

In the end, Gertrude Blanch decided to do the calculation herself, rather than prepare it for the entire computing floor. Working with a few assistants, she completed the task in three weeks. The resulting paper, which is now regarded as a seminal contribution to the field, is usually cited as Bethe and Marshak but in fact, Blanch is listed as the lead author and Lowan is included as a co-author.[29] Lowan insisted that his name appear on every publication from the project, a policy that apparently produced a confrontation with Blanch. Blanch prevailed in the conflict and Lowan was not mentioned on the next paper to draw upon these computations, which was credited only to Blanch and Marshak.[30]

While Blanch was working on the Bethe computations, Lowan was reaching out to the military. In anticipation of a European war, the army was preparing a new series of maps. They required a detailed series of grid points to be computed for each map and they contracted the job to the Mathematical Tables Project. The contract brought not only new work to the project, but also new equipment. It paid for three new Monroe calculators, several dozen used Sunstrand calculators, a punched card tabulator and a sorter. It expanded the project to its largest size. In July 1941, the group had 100 computers working by hand, another 100 using Sunstrand calculations, a group of 12 using Monroe calculators to do special projects, a second small group preparing tables, and a final group calculating grid points with the IBM tabulator. This expansion also brought a new layer of control. All the military projects had to be approved not only by the National Bureau of Standards and the WPA but by the Army as well.[31]

Shrinking for war

As the country began mobilizing for war, the Mathematical Tables Project began to shrink. Computers left to take better paying war jobs in factories. A few of the most skilled computers found work doing computations for other governmental offices. After a flurry of letter writing and the intervention of Phil Morse, Lowan succeeded in gaining the designation 'critical war project' for the Mathematical Tables Project. This designation kept the project alive within the WPA but it did not guarantee the resources, human or financial, to keep the project at full strength. After US entry into the war on 8 December 1941, many of the male computers left to join the military. The Army reclaimed their IBM equipment in July 1942. The greatest threat to the project came on 3 December 1942, when President Roosevelt announced that he was liquidating the WPA.[32]

Lowan again worked to find other sources of funding for his project but once more, the group was saved by Philip Morse. Morse helped to arrange for the group to be divided between two organizations, the Navy's Hydrographic office and the Applied Mathematics Panel, the mathematical branch of the Office for Scientific Research and Development. It required Lowan to reduce his staff and dispose of some of his equipment. As difficult as these changes may have been, they actually improved the stature of the project in the scientific community.

The Navy took the largest group of computers, initially sixty-five in number, and established them in a separate office.[33] For the duration of the war, Lowan had no control over this group, which was preparing navigation tables for the LORAN radio navigation system. LORAN was a joint project of MIT and Bell Laboratories. It relied on pairs of radio stations to broadcast carefully timed signals. Navigators would record differences in these times and use them to identify their position on the globe. Graphically the technique was simple. From a single pair of stations, navigators could determine that their position fell on a certain curve. From a second pair of stations, they would find a second curve. Their true position fell where the two curves intersected. In practice, the system required navigators to solve a complicated series of equations to find that intersection. The LORAN tables reduced the work to the simple process of looking up values in a book of tables.[34]

The preparation of the LORAN tables were one of the bottlenecks of the war. The military found it difficult to deliver men and material until it had the navigation system in place. Milton Abramowitz, the leader of the LORAN computing group spent much of his time working to improve the efficiency of his organization. He acquired calculating machines for each of his computers, reorganized the way in which the computers operated and identified a more efficient algorithm for preparing the tables. Abramowitz ran a race against time, for the members of his computing staff slowly left for better paying war jobs. As it's work came to a close in the spring of 1945, the LORAN group was producing a new table every three weeks with a staff less than half the size of the initial group.

Lowan and the remaining Mathematical Tables Project computers became a general service organization for the divisions of the Office for Scientific Research and Development. They computed trajectories of bombs, the dispersion of shock waves, the transmission of heat, and the distances on maps. Lowan had hoped to preserve a group of fifty computers for this work but, after negotiations with the chair of the Applied Math Panel, he agreed to reduce his staff to a mere twenty-five computers.[35] The twenty-five were the best trained in the project and all worked with their own adding machine. Additionally, Blanch tried to increase the professionalism of the group by teaching them applied mathematics. During the lunch hour, she and the senior staff offered a series of six mathematics courses. The most elementary course dealt with the basic properties of arithmetic. The most advanced covered the theories of calculus, table making, and numerical analysis. For this last course, Blanch prepared a textbook on advanced subjects of applied mathematics. Published only in mimeographed form, it circulated widely among the war offices involved in computation.[36]

The problem of security reminded the staff of the Mathematical Tables Project that they occupied a subservient role in the scientific infrastructure. No one in Lowan's group held security clearances. Both Lowan and Blanch had applied for clearances but were denied. Were it not for the general nature of mathematics, the Applied Math Panel might have been forced to disband the group and assemble a computing organization with a more trusted background. 'In mathematics you didn't need clearances,' argued Blanch 'The only thing that is classified is dimensions, which you didn't need.'[37] Her words reflected a stoic way of viewing the circumstances. In practice, it meant that the Mathematical Tables Project did the work of others but never initiated projects of their own. For Blanch, the lack of a clearance was more personal and frustrating. In 1944, she was asked to join the Manhattan Project in Los Alamos. The offer was withdrawn after her security review.[38]

The mightiest computing organization in the world

The war ended with a flush of glory and an anxiety over the future. It re-unified the two parts of the Mathematical Tables Project and encouraged the computers to think that they had earned the respect of the scientific community and deserved to be a permanent organization. They enjoyed thinking of themselves as the 'Mightiest Computing Team the world has ever known,' a title that had come from their supporter and occasional critic, L. J. Comrie.[39] Certain members of the Office for Scientific Research and Development agreed with that assessment, but their agreement had its limits. Most were more interested in supporting the development of electronic computers than in supporting table making. Oswald Veblen, who was organizing post-war mathematical research, felt that the group might be helpful in developing new computing machines and that there would be a dozen members 'who will be hard to replace once scattered.'[40] That dozen included Gertrude Blanch and her assistants: Milton Abramowitz, Irene Stegun, Herb Salzer, Ida Rhodes, Jack Laderman, and a few others. It would also embrace a few of the computers who had completed Blanch's course of study. It would not include a majority of the staff and nor would it include Arnold Lowan.

The fate of Arnold Lowan summarized the struggle over the role of the Mathematical Tables Project. After the war, he paid the price for his aggressive efforts in promoting the project and for his inflated claims. Freed from the constraints of the WPA and boasting a solid record of accomplishments from the war, Lowan should have been able to find a good position for himself and his computers. However, he clung to a personal vision for the project, a vision in which the Mathematical Tables Project would remain an independent scientific laboratory. As the debate over the future of the project progressed, Lowan returned to the rhetoric that had helped him in the 1930s. He argued vigorously for his ideas, invoked the aid of his allies, and exaggerated the accomplishments of the group. To his surprise, he discovered that the post-war scientific leaders valued his project more than they valued him.

In November 1945, Lowan lost his strongest champion when Lyman Briggs retired from the National Bureau of Standards. Briggs's successor, Edward Condon, wanted to continue the group as a service organization, located in Washington, DC. Almost immediately, he indicated his plans by putting it under the authority of one of his assistants, the mathematician

John Curtiss. In the early years of the war, Curtiss had been a friend to programme. He had written reviews of the first tables. Curtiss had lost none of his respect for the project but he clearly conceived the group as a unit within a new national mathematics research group. The Mathematical Tables Project would be the Computation Laboratory of this group, the smallest of its four major divisions. It would serve the mathematical researchers and the scientists at the National Bureau of Standards.[41]

Lowan immediately began lobbying Curtiss to keep his independence and to leave the Mathematical Tables Project in New York. His efforts continued for nearly three years, becoming increasingly shrill, angry, and accusatory. Witnesses to the conflict tended to cast the conflict in personal terms, a struggle for control between Lowan and Curtiss.[42] Lowan, indeed, may have had some personal motivation for remaining in New York. His wife and her family were from New York. His appointment at Yeshiva University was, by then, a full-time position that he did not want to leave. Curtiss, equally, may have concluded that he would have had limited control if the group remained in Lowan's hands. The computers worked only a 32 hour week, as they had when they were relief workers. WPA rules limited the hours of operation in order to let the workers search for a full time job in the private economy. Many of the workers, including Lowan himself, exploited these rules to take a second job.[43]

Even if it was motivated by personal reasons, the argument between Lowan and Curtiss was ultimately about the independence of computing organizations. Lowan lost this battle. His vision of an independent computing laboratory succumbed to Curtiss' idea that computing should be subservient to groups doing scientific research. It would be a battle that would be repeated in the coming decades on university campuses, in the National Science Foundation, in corporate laboratories, and at the national research facilities. Though many organizations would establish computing centres, they generally saw scientific computing drift back to the laboratories, where scientists could directly oversee it.

Lowan made his case by trying to establish the professional nature of the Mathematical Tables Project. He first attempted to do so by arguing that the Mathematical Tables Project was the creation of existing scientific organizations. Ignoring the role of the WPA, he stated that it had been organized by the National Bureau of Standards and that he had been chosen to lead it by two well-known scientists, C. E. Van Orstrand and John von Neumann. Like many of his claims about the group, this statement was based in truth and yet misrepresented the way in which the organization began operations. Lowan soon abandoned this claim, as few had forgotten that the laboratory had begun as a work relief effort.

The more powerful argument was to catalogue the quality of the project's work, though Lowan weakened its impact by exaggerating his claims. He stated that the project had worked for the Navy Bureau of Ordnance, the Army Signal Corps, the Fire Control Division of the Office for Scientific Research and Development. Indeed, the project had actually contributed computations to each of these groups but they had tended to work with individual researchers rather than the directors of these organizations. Furthermore, they had usually provided general mathematical tables which might be used in many kinds of scientific research. For the Army Signal Corps, for example, the project provided tables of the binomial distribution, a common statistical model.

Lowan tried to invoke the prestige of the Manhattan Project, which was then considered the epitome of war research. Again, the Mathematics Tables Project had made real contributions to the atomic bomb effort, but they were far from the centre of research. Their largest task had been to supply tables of Bessel functions to a Canadian metallurgist who was working on the problem of separating uranium isotopes. The work had provoked a certain amount of irritation between Lowan and his client. Both Warren Weaver, head of the Applied Math Panel, and James Conant, director of the National Research and Defence Council, had intervened in the correspondence to assure that the computations were done quickly and with little public attention. Late in the war, the Mathematical Tables Project did some calculations for the plutonium bomb. As they began this work just three weeks before the first test of the bomb, they were probably double-checking the work of computers at Los Alamos.[44]

As before, Lowen turned to his friends in the scientific community, including Phil Morse and John von Neumann. Morse was the more helpful. He tried to find alternative sources of funding for the project. The most promising seemed to be the new Brookhaven National Laboratory on Long Island. He enlisted von Neumann in his attempts to incorporate the Mathematical Tables Project in the laboratory organization. Initially, the leaders of the new laboratory seemed cautiously interested in working with the project. In July 1947, Lowan happily reported that 'The most fortunate solution to the problem of locating the Computational

Laboratory would be the scientific atmosphere of the Brookhaven National Laboratory.'[45] This hope was short lived. By early fall, the administrators at Brookhaven were raising concerns about the idea and soon thereafter the project stalled.

In August 1947, John Curtiss informed Lowan that he would move the computing office in June of the following year. Lowan, growing increasingly desperate, began looking for support outside of the scientific community. It was a step that alienated him from other scientists. He contacted the New York Congressional delegation and asked them to pressure Curtiss. He also asked for support from the United Public Workers, the union that represented the human computers, and fed them information about the move. The leader of the union, Samuel Finkelstein, asked for a meeting with Curtiss and tried to bully him into changing his decision. 'His own words indicate that the proposed move is based on public relations considerations and the desire to make the Laboratory a Washington showpiece for Congressmen and other officials.'[46]

As winter progressed into spring, the rhetoric became heated and personal. Perhaps the most vitriolic words came from R. C. Archibald, who had succeeded A. A. Bennett as the chair of the National Academy of Sciences MTAC committee. Archibald, who had once a good working relationship with Curtiss, advocated his removal on the grounds of 'his brutal treatment of Lowan, who is certainly a gentleman, which John is not.'[47] The crisis lasted through early summer of 1948, when the two sides reached a compromise. Lowan would have a year to find outside funding for a New York office, while Curtiss would establish a laboratory in Washington. It was a compromise that offered little hope to Lowan. Blanch left the project that spring to join the Institute for Numerical Analysis in Los Angeles. Other senior staff members began packing their books and moving to Washington. Few believed that Lowan could maintain his base in New York. By winter, even Lowan himself conceded that the game was lost. He wrote to the faithful Morse, 'The fact that you have not replied to my letter of November 5th, 1948 would seem to substantiate my feeling that you are seemingly unable to do anything which would change the trend of events.'[48] In replying, Morse confirmed that the New York office would shortly close: 'the situation you are up against is well nigh unbreakable. I don't like it but there it is.'[49] Lowan shut the door of the New York office for the final time on Thursday 30 June 1949.

Epilogue

The demise of the Mathematical Tables Project provides a striking example of how science can continue, at least temporarily, during social and organizational upheavals. At the height of the debate over the future of the project, the human computers were working on the first major trial of the simplex algorithm for linear programming. This trial represented a major change for the group as it was a problem that could not be tabulated and re-used later. Though he had not initiated the calculations, John von Neumann followed the trial closely and tried to estimate the time that the ENIAC would require for such a problem.[50]

The Mathematical Tables Project finally reached a professional status with the publication of *The handbook of mathematical functions*, a compendium of applied mathematics.[51] The handbook is perhaps the single most widely circulated scientific work. The National Bureau of Standards estimates that about one million copies have been printed. The project outlined several versions of such a book during the 1940s, but could not find the time or money to complete it. The idea was revived at a 1952 conference on table-making. The two editors of the book were veterans of the Mathematical Tables Project, as were a third of the contributing authors. Most of the chapters are based directly on tables created by the Mathematical Tables Project.

In spite of its origins in the WPA, the book hides its lineage. Conceivably, the National Bureau of Standards may still have been sensitive to the idea that the group was nothing but an unqualified collection of relief workers, but this is unlikely. The book was published in 1964, twenty years after the liquidation of the WPA and fifteen years after the formal end of the Mathematical Tables Project. In all likelihood, the project had become what Lowan had wanted it to be, a research program accepted by all in the scientific community. It did so with little time to spare. One year after the publication of the *Handbook*, the Bureau of Standards incorporated the remnants of the project into its Center for Computer Sciences and Technology, thus ending the last organizational trace of the Mathematical Tables Project of the WPA.

Further reading

More detailed descriptions of the table making work of the Mathematical Tables Project and the National Bureau of Standards can be found in

D. Grier, 'Gertrude Blanch of the Math Tables Project', *IEEE Annals of the History of Computing* 19:4 (1997), 1–10; D. Grier, 'The Math Tables Project: the reluctant start of the computing era', *IEEE Annals of the History of Computing* 20: 3 (1998), 33–50; J. Todd, 'A prehistory and early history of Computation in the U.S. National Bureau of Standards' in S. G. Nash (ed.), *A history of scientific computing*, ACM Press, New York, 1990, pp. 237–50; A. N. Lowen, 'The Computational Laboratory at the National Bureau of Standards', *Scripta Mathematica* 15 (1949), 33–63; and G. Blanch and I. Rhodes, 'Table making at NBS', in B. R. Scaife (ed.), *Studies in numerical analysis: papers in honour of Cornelius Lanczos*, Academic Press, London, 1974.

More wide reaching discussions of table making projects in the United States can be found in M. Rees, 'The mathematical sciences and World War II', *American mathematical monthly* 87:8 (1980), 607–21 and D. Grier, *When computers were human*, Princeton University Press, Princeton, NJ, forthcoming. For a general history of Roosevelt's employment policy during the depression, see A. J. Schlesinger, *The coming of the New Deal*, Houghton Mifflin, New York, 1959.

Notes

1. Interview with author, February 1996.
2. G. Daniels, 'The process of professionalization in American science: The emergent period, 1820–1860', *Isis* 58 (1967), 151–66.
3. *Final Report of the WPA*, Government Printing Office, Washington, DC, 1946, p. 1.
4. F. Lillie to E. Ross, 29 October, 1937, Scientific Advisory Board Files, National Research Council Records, Archives of the National Bureau of Sciences.
5. Office memorandum No. 433, Scientific Advisory Board Files, National Research Council Records, Archives of the National Bureau of Sciences.
6. D. A. Grier, 'Dr. Veblen Goes to War,' *American mathematical monthly* 108: 10 (Dec. 2001), 922–931. See also L. B. Feffer, 'Oswald Veblen and the capitalization of American mathematics', *Isis* 89 (1999), 474–97.
7. Paul Brockett to Frank Lillie, 16 November, 1937, Scientific Advisory Board Files, National Research Council Records, Archives of the National Bureau of Sciences.
8. Minutes of the Executive Council for the Physical Sciences Division of the National Research Council for 1934, p. 39, Archives of the National Academy of Sciences.
9. Office memorandum No. 433, *op. cit.*

10. D. A. Grier, 'The Math Tables Project: The reluctant start of the computing era', *IEEE annals of the history of computing* 20:3 (1998), 33–50. Ida Rhodes to Ira Merzbach, 4 November, 1969, Smithsonian National Museum of American History Computing Machinery Collection Supporting documents; Arnold Lowan to Phil Morse, 20 July, 1946, Box 10, Papers of Phil Morse, Morse papers, Special Collections, MIT Libraries.
11. D. A. Grier, 'Gertrude Blanch of the Math Tables Project', *IEEE Annals of the history of computing* 19:4 (1997), 1–10.
12. Interview with G. Blanch and H. Thatcher in San Diego, 17 March 1989, Stern Family Collection.
13. Interview with G. Blanch by H. Tropp, 16 May, 1973, Smithsonian Interviews with Computer Scientists, Record Group 196, Archives Center, Smithsonian National Museum of American History.
14. J. H. Curtiss, 'Table of the First Ten Powers of the Integers 1 to 1000; Tables of the Exponential Function', *American mathematical monthly* 48:1 (1941), 56–7.
15. The definitive list of errata is found in A. Fletcher, *et al.*, *An index of mathematical tables*, second edition, Reading Massachusetts, Addison-Wesley, 1962. For example, the authors list (page 881) only ten errors in the *Tables of the exponential function*, a book with 535 pages. In comparison, H. T. Davis' 1933 book, *Tables of higher mathematical Functions*, has 60 known errors in 250 pages of tables. At the far extreme, K. Hayashi's 1926 book, *Sieben- und mehrstellige taflen der kreis- und hyperbelfunktionen*, has about 2000 errors in 200 pages of tables.
16. R. Wallace, personal communication.
17. Night Telegram from A. Lowan to L. Briggs, 1 August, 1940, Lyman Briggs Correspondence for 1940, National Bureau of Standards Records on Electronic Computers, National Archives and Records Administration.
18. The approval process is best illustrated by a series of letters between Arnold Lowan, L. Briggs, A. Rogers, I. Gordy and L. M. Milne-Thompson in June 1941, Administrative Records of WPA Project 765-97-3-10, New York City Office of the WPA, 1940–42, Microfilm Reel 8659, Record Group 69, National Archives and Records Administration.
19. The second volume, *Tables of the exponential function*, gives the most complete narrative of the project but it only describes the committee meeting on January 28. Subsequent volumes list the names of the senior mathematicians.
20. W. J. Eckert to D. W. Rubidge, Commercial Research Department of IBM, 11 January, 1941, Papers of Wallace J Eckert, CBI9, Charles Babbage Institute, University of Minnesota.
21. W. J. Eckert to R. C. Archibald, 16 September, 1941, Papers of Wallace J Eckert, CBI9, Charles Babbage Institute, University of Minnesota.
22. H. S. Tropp, 'Leslie John Comrie', *Annals of the history of computing*, 4:4 (1982), 371–2, This version of the story, conveyed by I. Rhodes, is

contradicted by the correspondence of the Applied Mathematics Panel. Lowan promoted Comrie's support as a reason to keep the panel (see A. Lowan to P. Morse 31 December 1942, Morse Papers). However, at its inception on 12 November 1942, the members of the Applied Mathematics Panel discussed the importance of utilizing the Mathematical Tables Project (see report of W. Weaver, 12 November 1942, Inception of the Applied Math Panel, Box 17, General AMP Records 1942–46, Record Group 227). Furthermore, W. Weaver seems to have kept his own council in regards to the Mathematical Tables Project (see W. Weaver to J. Conant, 8 February 1943, Correspondence of the Applied Math Panel 1942–43, Applied Maths Panel Records, National Archives and Records Administration).

23. L. J. Comrie to J. L. Ginniff, 6 May 1942, Administrative Records of the Mathematical Tables Project, 1940–1942, Project Number 765-97-3-10, Records of the Work Projects Administration for New York City, Records of the Federal Emergency Relief Agency (FERA), Record Group 69, National Archive and Records Administration.
24. G. Daniels, 'The process of professionalization in American science: The emergent period, 1820–1860', *Isis* 58 (1967), 151–66.
25. A. Lowan to P. Morse, 29 September 1938, W.P.A. Math Table Files, Morse Papers, Special Collections, MIT Libraries.
26. J. von Neumann to A. Lowan, September 1940, von Neuman Papers, Library of Congress.
27. P. Morse to A. Lowan, 21 September 1938, Administrative Records of the Mathematical Tables Project, 1940–1942, Project Number 765-97-3-10, Records of the Work Projects Administration for New York City, Records of the Federal Emergency Relief Agency (FERA), Record Group 69, National Archives and Records Administration.
28. H. T. Davis, *Adventures of an ultra-crepidarian*, San Antonio, private printing, 1962, p. 273.
29. G. Blanch, A. N. Lowan, R. E. Marshak, and H. A. Bethe, 'The Internal Temperature Density Distribution of the Sun', *Journal of astrophysics* (1942), 37–45.
30. G. Blanch and R. E. Marshak, 'The internal temperature density distribution of main sequence stars', *Journal of astrophysics*, 1946.
31. "Annual Report for 1941 "Administrative Records of the Maths Tables Project 1940–1942, Project 765–97–3–10, Records of Works Project Administration for New York City, Records of FERA, Record Group 69, National Archives and Records Administration.
32. D. A. Grier, 1998 *op. cit.*
33. Memo from Assistant Chief of Navy Bureau of Docks and Yards, 3 April 1942, Mathematics and Statistics, Box 46, Records of Office of Naval Research.
34. J. A. Pierce, *et al.*, *Loran, long range navigation*, McGraw Hill, New York, 1948, pp. 19–20.

35. W. Weaver to E. L. Moreland, 15 February 1943, Correspondence of the Applied Math Panel 1942–43, Box 16, General Records. AMP Records 1942–46, Record Group 227, National Archives and Records Administration.
36. G. Blanch, *Notes for a class on numerical analysis, taught 1943–45 at the New York Mathematical Tables Project*, Blanch Papers, Stern Family Collection.
37. Interview with G. Blanch and H. Thatcher in San Diego, 17 March 1989, Stern Family Collection.
38. Memo to Director, FBI, 20 September 1955, FBI File of Gertrude Blanch.
39. L. J. Comrie, 'The Math Tables Project', mimeograph note attached to A. Lowan to W. Weaver, 9 March 1944, AMP Records. Quoted by I. Rhodes in interview with H. Tropp, Record Group 196, Smithsonian Archives Center.
40. O. Veblen to L. Jordan, 13 June 1945, Executive Offices of Research Board for National Security, Oswald Veblen Papers.
41. J. Curtiss, 'The National Applied Mathematics Laboratories—A prospectus', *Annals of the history of computing* 11:1 (1989), 13–30.
42. J. Todd, A. Hillman, H. Salzer interviews with the author.
43. J. Todd, interview with the author, June 1998.
44. M. Rees to A. Lowan, July 19, 1945. Applied Maths Panel Records, ZG 227, National Archieves and Records Administration.
45. A. Lowan to P. Morse, 26 July 1947, Morse Papers, Special Collections, MIT Libraries.
46. Circular attached to S. Finkelstein to R. C. Archibald, 9 April 1948, MTAC Records; similar language appears in a 11 May, 1948 letter to Morse, Morse Papers, Special Collections, MIT Libraries.
47. R. C. Archibald to R. C. Gibbs, 27 April 1948, MTAC Files, Physical Sciences Division, National Research Council Records, Archives of the National Academy of Sciences.
48. A. Lowan to P. Morse, 5 January 1949, Morse Papers, Special Collections, MIT Libraries.
49. P. Morse, to A. Lowan, 24 January 1949, Morse Papers, Special Collections, MIT Libraries.
50. G. Dantzig to J. von Neuman, 28 April 1948, Correspondence 'D', von Neumann Papers, Library of Congress.
51. The book is uncopyrighted and it is difficult to give an exact total. The Bureau of Standards has sold about 150 000 copies. They believe that Dover publications has sold 600 000 to 900 000 copies. In addition, there are editions published in foreign languages by foreign presses. D. Lozier, National Bureau of Standards, communication with author, 18 July 2001.

SUN, 1960
FOR 0ʰ EPHEMERIS TIME

Date	Longitude Mean Equinox of 1960·0		Red. to App. Long.	Latitude Ecliptic of			Hor. Par.	Prec. in Long.	Nutation in Long.	Obl. of Ecliptic
				1960·0	1950·0	Date				23° 26′
Jan. 0	278 33 36·9	3670·8	−20·8	+0·65	+5·21	+0·65	8·95	− 0·185	+0·178	30·254
1	279 34 47·7	3670·8	20·6	·58	5·12	·58	8·95	− 0·048	0·261	30·294
2	280 35 58·5	3670·7	20·4	·49	5·01	·49	8·95	+ 0·090	0·295	30·337
3	281 37 09·2	3670·4	20·3	·37	4·87	·37	8·95	0·228	0·285	30·372
4	282 38 19·6	3670·1	20·2	·24	4·72	·24	8·95	0·365	0·246	30·397
5	283 39 29·7	3669·7	−20·1	+0·10	+4·55	+0·10	8·95	+ 0·503	+0·197	30·408
6	284 40 39·4	3669·4	20·0	− ·02	4·39	− ·03	8·95	0·641	0·152	30·405
7	285 41 48·8	3669·0	19·9	·14	4·25	·15	8·95	0·778	0·121	30·393
8	286 42 57·8	3668·5	19·8	·24	4·12	·25	8·95	0·916	0·113	30·376
9	287 44 06·3	3668·2	19·6	·33	3·99	·34	8·95	1·054	0·131	30·358
10	288 45 14·5	3667·7	−19·5	−0·40	+3·89	−0·41	8·95	+ 1·191	+0·172	30·344
11	289 46 22·2	3667·3	19·3	·44	3·82	·45	8·95	1·329	0·235	30·339
12	290 47 29·5	3666·9	19·0	·45	3·77	·46	8·95	1·466	0·308	30·345
13	291 48 36·4	3666·4	18·8	·44	3·75	·45	8·95	1·604	0·382	30·360
14	292 49 42·8	3666·1	18·6	·40	3·75	·41	8·95	1·742	0·447	30·389
15	293 50 48·9	3665·7	−18·4	−0·32	+3·78	−0·34	8·95	+ 1·879	+0·493	30·426
16	294 51 54·6	3665·3	18·3	·24	3·82	·26	8·95	2·017	0·513	30·469
17	295 52 59·9	3665·0	18·2	·13	3·89	·15	8·95	2·155	0·501	30·514
18	296 54 04·9	3664·6	18·1	·01	3·96	− ·03	8·95	2·292	0·461	30·552
19	297 55 09·5	3664·3	18·0	+ ·12	4·05	+ ·10	8·94	2·430	0·403	30·580
20	298 56 13·8	3663·9	−17·9	+0·25	+4·13	+0·23	8·94	+ 2·567	+0·337	30·593
21	299 57 17·7	3663·6	17·8	·37	4·21	·35	8·94	2·705	0·280	30·593
22	300 58 21·3	3663·2	17·7	·47	4·26	·45	8·94	2·843	0·250	30·581
23	301 59 24·5	3662·8	17·6	·56	4·30	·54	8·94	2·980	0·256	30·561
24	303 00 27·3	3662·3	17·4	·61	4·30	·59	8·94	3·118	0·307	30·545
25	304 01 29·6	3661·8	−17·1	+0·64	+4·28	+0·62	8·94	+ 3·256	+0·395	30·540
26	305 02 31·4	3661·0	16·9	·63	4·21	·61	8·94	3·393	0·503	30·553
27	306 03 32·4	3660·3	16·7	·59	4·12	·57	8·94	3·531	0·605	30·585
28	307 04 32·7	3659·5	16·4	·53	3·99	·50	8·94	3·669	0·675	30·633
29	308 05 32·2	3658·4	16·3	·43	3·84	·40	8·93	3·806	0·701	30·689
30	309 06 30·6	3657·4	16·2	+0·31	+3·66	+0·28	8·93	+ 3·944	+0·678	30·741
31	310 07 28·0	3656·1	16·1	·18	3·47	·15	8·93	4·081	0·617	30·783
Feb. 1	311 08 24·1	3654·9	16·0	+ ·05	3·28	+ ·02	8·93	4·219	0·536	30·812
2	312 09 19·0	3653·6	16·0	− ·08	3·09	− ·11	8·93	4·357	0·451	30·825
3	313 10 12·6	3652·2	15·9	·21	2·90	·24	8·93	4·494	0·377	30·826
4	314 11 04·8	3650·8	15·8	0·32	+2·73	−0·35	8·93	+ 4·632	+0·325	30·819
5	315 11 55·6	3649·4	15·7	·41	2·57	·44	8·93	4·770	0·297	30·811
6	316 12 45·0	3647·9	15·6	·48	2·44	·51	8·92	4·907	0·293	30·803
7	317 13 32·9	3646·4	15·4	·53	2·32	·56	8·92	5·045	0·310	30·802
8	318 14 19·3	3645·0	15·2	·55	2·23	·58	8·92	5·183	0·341	30·810
9	319 15 04·3	3643·6	−15·1	−0·54	+2·18	−0·57	8·92	+ 5·320	+0·376	30·830
10	320 15 47·9	3642·0	14·9	·50	2·15	·53	8·92	5·458	0·404	30·860
11	321 16 29·9	3640·7	14·7	·44	2·14	·47	8·92	5·595	0·415	30·900
12	322 17 10·6	3639·2	14·6	·36	2·15	·39	8·92	5·733	0·402	30·947
13	323 17 49·8	3637·8	14·5	·25	2·19	·28	8·91	5·871	0·358	30·996
14	324 18 27·6	3636·5	−14·4	−0·13	+2·24	−0·16	8·91	+ 6·008	+0·285	31·040
15	325 19 04·1		−14·4	+0·01	+2·31	−0·02	8·91	+ 6·146	+0·188	31·074

To obtain the longitude referred to the mean equinox of 1950·0, subtract 8′ 22″·7.

11

The making of astronomical tables in HM Nautical Almanac Office

GEORGE A. WILKINS

Introduction

A table of the computed values of the position of a moving astronomical body, such as the Moon or a planet, is usually called an 'ephemeris'. This word is also used in the title of *The Nautical Almanac and Astronomical Ephemeris*, which was first issued for the year 1767 to give astronomical data for navigation at sea. At this time the Astronomer Royal, Nevil Maskelyne,

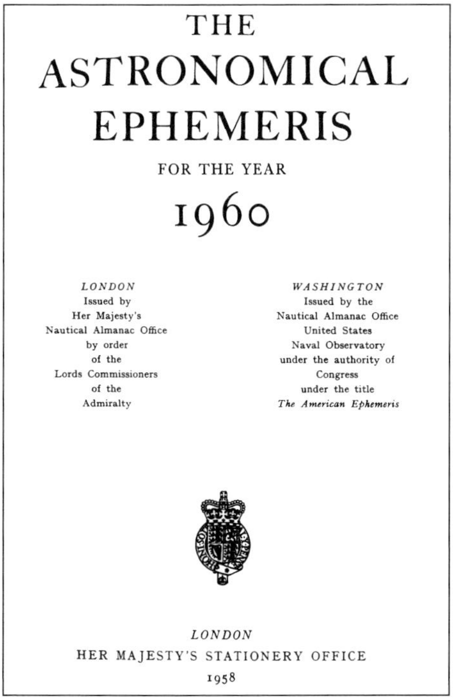
THE
ASTRONOMICAL
EPHEMERIS
FOR THE YEAR
1960

LONDON
Issued by
Her Majesty's
Nautical Almanac Office
by order
of the
Lords Commissioners
of the
Admiralty

WASHINGTON
Issued by the
Nautical Almanac Office
United States
Naval Observatory
under the authority of
Congress
under the title
The American Ephemeris

LONDON
HER MAJESTY'S STATIONERY OFFICE
1958

Fig. 11.1 *The Astronomical ephemeris for 1960.* The original title of *The Nautical almanac and astronomical ephemeris* was replaced in 1960 by the shorter title of *The Astronomical ephemeris* when the British and American almanacs for astronomy and navigation were unified and the work of computation and typesetting was shared. The table on the facing page was computed with the use of punched card machines and was typeset by hand. (By kind permission of HM Nautical Almanac Office.)

organized the computations in such a way that each ephemeris was computed independently by two human computers and the results were then examined by a comparer whose task it was to resolve any discrepancies between the two. The computers and comparers were scattered around the country, with a concentration in Cornwall, while Maskelyne was at the Royal Observatory at Greenwich. Consequently, the process was very time-consuming. This system was replaced in 1831 when the Nautical Almanac Office (NAO) was set up as a separate establishment in London. This led to changes in the distribution of work amongst the staff and it allowed the expansion of the scope of the work to match the developments in dynamical and positional astronomy.

The principal tabulations in the *Nautical almanac* give the celestial coordinates of the Sun, Moon, and planets for each day of the year. These coordinates change slowly and regularly from day to day in accordance with the adopted theory of the motion of the body concerned. For an analytic theory, each coordinate is obtained by summing the contributions from the terms of a trigonometric series whose amplitudes and arguments are given by the theory. In the past each theory was represented by a set of tables that was designed to reduce the amount of calculation required to compute an ephemeris; in particular, every effort was made to reduce the number of multiplications required. This technique continued in use even when logarithms were superseded by desk calculating machines and when commercial accounting and punched-card machines were adapted for scientific purposes. The other principal technique that was used to reduce the amount of calculation was that of systematic interpolation, or subtabulation as it is now known. The coordinates were calculated from the precomputed tables at a wide interval and then interpolation by the method of finite differences was used to obtain the coordinates at the interval to be given in the ephemeris. Flexible calculating machines and new formulae had to be developed before this process could be successfully mechanized. When powerful electronic computers became available the use of precomputed tables and subtabulation were both replaced by direct evaluation of the trigonometric series of the theory of each body or by direct numerical integration of the equations of motion of a system of bodies.

The techniques of computation used in the NAO and the design of the almanac changed only slowly during the nineteenth century. The International Meridian Conference of 1884, which adopted the Greenwich meridian for the zero of longitude, was followed by other conferences that led to international agreements in 1896 on the bases of the ephemerides and star positions and then in 1911 on arrangements for the sharing of the work of their computation.[1] The pace of change in the NAO accelerated sharply after 1925 following the appointment of L. J. Comrie, who was its Superintendent from 1930 to 1936. Consequently, this article contains only a brief account of the techniques used for table making for the *Nautical almanac* from 1767 until 1925. Not only did Comrie introduce new machines for use in the computations, but he also redesigned the *Nautical almanac* to make it more user-friendly and introduced new publications, such as the first in a series of volumes of *Planetary co-ordinates*. Comrie developed new techniques so that National accounting machines and punched-card machines could be used effectively to produce tables for general use in mathematics as well for astronomical purposes. More new publications, especially *The Air almanac* and auxiliary tables for use in navigation, and new methods were introduced by the next Superintendent, D. H. Sadler.

The changes in the aids to computation from logarithms to desk calculating machines, then to 'difference engines' and, finally from 1959 onwards, to electronic computers and calculators entailed changes in the way that the work was organized and in the mathematical techniques and formulae that were used. New techniques for printing and data distribution introduced during the second half of the twentieth century affected both the methods of production and the design of the tabulations. Consequently, there were changes in the extent and nature of the quantities that were published and in the forms in which the tables were made available to those who wished to use them. The printed page has been superseded for some applications by other media, while for other applications programs and data are provided to enable the user to generate directly the numerical data that are required. These four aspects of table making—computation, presentation, printing, and publication—are reviewed in this article.

Early procedures for preparing the Nautical Almanac

The arrangements for the computation and printing of the early Nautical Almanacs have been described but as far as I am aware no examples of the original calculations for the *Nautical almanac* exist.[2] Edwin Dunkin, who was a Chief Assistant at the Royal Observatory at Greenwich, states that he could find no record of the early computations.[3] He does, however, give background

The publications of HM Nautical Almanac Office

The Nautical almanac and astronomical ephemeris was published for the years 1767 to 1959. The title was changed to *The Astronomical ephemeris* in 1960. It was produced in cooperation with the Nautical Almanac Office in the United States Naval Observatory and was identical in content to *The American ephemeris and nautical almanac*. In 1981 both titles were changed to *The Astronomical Almanac*.

From 1896 the tabulations for navigators were given in Part I of the almanac and this was published separately. From 1914 the navigational ephemerides were redesigned and published only in *The Nautical almanac, abridged for the use of seamen*; this title was changed to *The abridged nautical almanac* in 1952. Confusingly, the main almanac, which gave high-precision data for astronomers, retained its original title and the abbreviation (*NA*) until 1959 even though it was no longer intended for use by navigators. The title *The Nautical almanac* has been used for the almanac for marine navigation since 1960. It was unified with the American almanac in 1958.

The Air almanac was first published for October—December 1937 and was continued until 1997. It was unified with *The American air almanac* from 1953. From 1998 the daily pages were omitted from the British edition and its name was changed to *The UK air almanac*. The full tabulations are still given in the American almanac.

The NAO has prepared *The Star almanac for land surveyors* for the years 1951 onwards. It also prepared annual volumes of *Apparent places of fundamental stars* for the years 1941 to 1959. This was intended for general international use and it eliminated the need to include such data in each of the national almanacs that were then published. The responsibility for the publication was taken over for 1960 onwards by the Astronomisches Rechen-Institut in Heidelberg, which then ceased to publish a separate almanac in German.

A series of volumes of *Planetary co-ordinates* for the years 1900–1940, 1940–1960 and 1960–1980 were published for the primary purpose of tabulating rectangular coordinates of the planets with respect to the Sun for use in the calculation of the orbits of comets by amateur astronomers. The scope was widened in the succeeding volumes of *Planetary and lunar coordinates* for 1980–1984 and 1984–2000. (The overlap was due to the introduction of new fundamental ephemerides for the almanacs from 1984 onwards.)

In addition to the navigational almanacs, the NAO prepared a variety of sets of auxiliary tables for marine and air navigation; most of these gave the transformation from celestial coordinates to altitude and azimuth. These tables were also later unified with the American editions so that design, computation, and printing could be shared.

The NAO also produced various mathematical tables for general use. They included *Interpolation and allied tables* (1936 and 1956), *Subtabulation* (1958), *Seven-figure trigonometrical tables for every second of time* (1939) and *Five-figure tables of natural trigonometric functions* (1947). Many different investigations were carried out and their results usually took the form of numerical tables. The most important were published formally, but many were written up in the form of *NAO technical notes* that had only a limited distribution.

Sets of the NAO almanacs and occasional publications are held in the RGO archives in the Cambridge University Library.

information about the computing arrangements that Maskelyne[4] put in place by which two persons independently computed each ephemeris and a third compared them and resolved any discrepancies. In particular, Dunkin describes the work of the Rev. Malachy Hitchins, who was one of the computers for the early editions and who then became a comparer until his death in 1809.

The most important ephemeris in the early editions of the *Nautical almanac* was that giving the lunar distances (angles) of the Sun and bright stars from the limb (nearest edge) of the Moon. In addition, in order to determine the Greenwich time of his observations, the navigator needed a set of *Requisite tables*, which had been written by Maskelyne, and trigonometric and logarithmic tables, which were included in later editions of the *Requisite tables*. The almanac also included an ephemeris of the Sun so that the navigator could determine his local time and hence his longitude.[5] The *Nautical almanac* was produced several years in advance as many ships, especially those on voyages of exploration, were away from their home ports for long periods.

Maskelyne's successor, John Pond, was less meticulous in his oversight of the work of the computers and comparers and so this task was taken over by Thomas Young, who was appointed Superintendent of the Nautical Almanac in 1818. The accuracy of the almanac was restored, but Young ignored the requests for the inclusion of more data for use by astronomers. The almanac continued to give lunar distances up to the edition for 1906 even though accurate chronometers gradually came into widespread use during the nineteenth century.

The computational work of the Nautical Almanac Office, 1831–1925

Lieutenant W. S. Stratford was appointed Superintendent of the Nautical Almanac after the death of Young. He was then the secretary of the Royal Astronomical Society which made many recommendations to the Admiralty about the content of the almanac in order to make it more useful for astronomical purposes. Most of the recommendations of the Society were accepted by the Admiralty and were introduced in the almanac for 1834. Stratford changed the system of calculation by setting up the Nautical Almanac Office in London towards the end of 1831. Edwin Dunkin's father, William Dunkin, who had assisted Hitchins from about 1804, moved from Truro to Camden Town in order to join the mainly new staff of the Office and he continued to work in the Office until his death in 1838.[6]

The methods for the computation of the almanac remained largely unchanged for another century. Printed multiplication tables and logarithm tables were used. Some of the tables gave directly the logarithms of the trigonometrical functions in sexagesimal measure. As far as I am aware, no examples of the computations during this period have survived, although some printed sheets give the formulae and precepts used for subtabulation. These make use of fourth-order interpolation formulae and include pre-calculated tables for the contributions of the differences.[7]

Stratford was succeeded by J. R. Hind, who is best known for his discoveries of asteroids, and then by A. W. M. Downing and P. H. Cowell, who were transferred from the Royal Observatory where the computational methods were similar to those in the NAO. Cowell was refused funding for a research group in celestial mechanics and from then on was content to allow the NAO to continue its long-established practices. One member of the staff, T. C. Hudson, was, however, keen to introduce new techniques, but he received no support from Cowell. Hudson published very little and so we do not know the full extent of his contributions to the making of mathematical tables. Two members of the staff of the NAO who had known him wrote down their recollections of his work for me.[8] Hudson is credited with the installation of a Burroughs adding machine in 1911, a Leyton arithmometer, and an early Brunsviga. He also made a 'star-correction facilitator' to simplify the computation of the apparent places of stars and devised a double entry table for subtabulation using Everett's formula using a cork bathroom mat.[9] Comrie paid tribute to Hudson in a letter to

The Superintendents of the *Nautical Almanac*	
Thomas Young	1818–1829
John Pond	1829–1831
W. S. Stratford	1831–1853
J. R. Hind	1853–1891
A. W. M. Downing	1891–1910
P. H. Cowell	1910–1930
L. J. Comrie	1930–1936
D. H. Sadler	1936–1971
G. A. Wilkins	1970–1989
B. D. Yallop	1989–1996
A. T. Sinclair	1996–1998
P. A. Wallace	1999–present

Sinclair used the title 'Head', rather than 'Superintendent', and Wallace has continued this practice.

Mrs. Hudson dated 1928 February 23. He wrote 'Ever since I joined the office I have felt most grateful to your husband for the successes which attended his efforts to get [the Burroughs] machine. The ideas which he applied to it were original and I have derived much inspiration from them, and have been able to save many hours of work by the application of ideas that have been evolved from these inspirations'.[10] There are few records remaining of the methods used prior to 1926 as all past work and records were scrapped in March 1930.[11]

Twentieth century procedures for computation: electromechanical machines

L. J. Comrie joined the staff of the NAO in 1925 and became Deputy Superintendent in the following year. He immediately proceeded to revolutionize the computing procedures of the Office and to redesign the *Nautical almanac* to make it more useful to astronomers. He obtained more desk calculating machines in order to replace completely the use of logarithms and he made corresponding changes in the choice of formulae.

He obtained the latest models of the Burroughs accounting machines[12] and later he discovered the more flexible Ellis accounting machines, which developed into the National accounting machines. He also introduced the use of punched card machines, which were designed for commercial work, by obtaining the use of machines in the Admiralty or by hiring them commercially as he was able to demonstrate the savings in cost.

Comrie held strongly to the view that it was best to make use of commercially available machines, rather than to attempt to design and build special machines for scientific purposes. One of his first major achievements was to use punched card machines for the summation of the hundreds of harmonic terms in the theory of the motion of the Moon.[13] His expertise became widely recognized and so he was able to obtain many different types of machines for use in the NAO on trial. It was found that the manual Brunsviga calculating machines were better for general use than electromechanical machines, although the latter were faster for some jobs. The Brunsviga remained the standard desk machine in the NAO until it was eventually replaced by electronic computers.

The Ellis machine, first seen at the Business Efficiency Exhibition in 1932, had four registers and could be used for differencing to, and summation from, fourth differences. The later National machines had 6 registers and a keyboard, each with a capacity of 12 digits. The Office acquired two machines, one of which used sexagesimal arithmetic. These machines were used for checking by differencing and also to provide the differences that were required for subtabulation by the method of bridging differences. In addition to their basic use for the summations in subtabulation, they were also used for summations in numerical integration. They had other uses and the collection of 'National set-ups' contained the necessary operating instructions for about 100 jobs.[14] The machines were operated on a rota system by scientific assistants, who would operate a machine for, say, two hours at a time. Each set-up specified the arrangement and functions of the 'stops' on the control bar, the sequence of the operations to be carried by the operator and the quantities to be printed. Each 'stop' determined the way in which the contents of one or more registers were to be added to or subtracted from another register.[15]

Comrie was also involved in the design, computation, and publication of new mathematical tables with extremely high standards of accuracy and typography.[16] The interval and interpolation aids were chosen to match the number of figures in the table so that the table would be easy to use, while the typeface and the spacing of the figures and columns were chosen so as to minimize the risk of error in extracting data from the table. Unfortunately, Comrie did not make a clear separation between official work for the NAO and private work for new general-purpose mathematical tables, so that he was suspended from duty in 1936.[17] His young successor, D. H. Sadler, continued to expand the work of the NAO while maintaining the high standards set by Comrie.[18] During the Second World War the staff of the NAO was expanded so that it could act as the computing centre for the Admiralty Computing Service.[19] After the war, many of the new staff were transferred to the National Physical Laboratory to form the nucleus of the new Mathematics Division and in 1949 the NAO moved to the new home of the Royal Greenwich Observatory at Herstmonceux Castle in Sussex.

The NAO continued to use National accounting machines but it also installed a powerful set of punched card machines. These were of the Hollerith type with 80-column rectangular-hole cards, with the 11 and 12 rows being used for minus and plus signs. The basic installation consisted of a sorter, reproducer, collator, and multi-register tabulator, which could read numbers from the cards into its registers, add and subtract them, and print the results. It could be linked to the reproducer in such a way that it was possible to punch the results on cards for input to the next stage of the computation. Each machine was controlled by a plugboard that was used to determine the ways in which the data were read from the input card and processed and in which the results were punched (on the same or another card) or printed.

The system was completed (apart from the card-controlled typewriter) in 1951 by the addition of an IBM 602A calculating punch, which was able to multiply and so could be used for tasks that up to that time had to be

Fig. 11.2 Herstmonceux Castle. The Nautical Almanac Office was based at Herstmonceux Castle in Sussex from 1949 to 1990 as a department of the Royal Greenwich Observatory. The Office is now part of the Rutherford Appleton Laboratory at Chilton, Oxfordshire.

carried out manually. It was very slow by current standards as an operation of the form $a + b.c$ took about 2.5 seconds. It had a very large plugboard with nearly 1500 sockets and it was possible to carry out a complex programme of operations in one run.[20]

The staff of the Machine Section of the NAO provided the technical expertise in the use of the machines, while the staff in the other sections were responsible for the overall planning of each job, for providing the data that were punched on the input cards, for monitoring progress, and for checking the results. There was a gradual increase in the range of the tasks for which they were used. Scheduling of the work was a major task as many different jobs, some large, some small, would be in progress at any one time. One job with which I was associated—the evaluation of the series for the nutation of the Earth for a period of 100 years—took almost a year from inception to completion. This was probably the last major use of the 'method of cyclic packs', which was based on the principle used by Brown in his *Tables of the motion of the Moon* and which was developed by Comrie for punched-card machines.[21]

The use of these machines led to enormous increases in the productivity of the staff. In particular, the use of cards to hold intermediate results obviated the need for writing them down and then setting them for the next stage. This saved time, but more importantly it reduced the number of errors that were made and so much less time was required to find and correct errors. It was still necessary to apply independent checks to verify that the work had been planned and carried through correctly and it was still necessary to check by differencing that the successive values in an ephemeris were free from isolated errors. For example, an operator might make an error in replacing a damaged card. It was also possible to use high-order differences for subtabulation with many figures, so that fewer pivotal values had to be calculated. 'Throw-back' methods developed by Comrie were used, as these allowed the effects of high-order differences to be taken partially into account by modifying the lower-order differences.[22]

A major advance was made in the USA in 1953 when the IBM Selective Sequence Electronic Computer (SSEC) was used to carry out a long-term numerical integration of the coordinates of the outer planets.[23] It was agreed internationally that this should be the basis of the ephemerides for 1960 onwards and the NAO had the task of computing the apparent geocentric spherical coordinates from the rectangular heliocentric coordinates. This was within the capacity of our punched card machines.

The Office also made use of an 'occultation machine', which may be regarded as an analogue computer that simulated the passage over the Earth of the shadow of the Moon as cast by a star.[24] The input settings represented the angular position and motion of the Moon with respect to the Earth. The shadow was represented by a cylinder of light that moved over the surface of a rotating terrestrial globe on which the positions of a list of observers had been marked. The operator turned a handle to move the shadow, rotate the Earth and move a needle around a circular dial that was read to give the times when the edge of the shadow passed over each of the observers. Apart from providing approximate predictions of the times of the occultations, the machine showed clearly the positions where occultations would not be observable and so it obviated an enormous amount of exploratory calculation that would otherwise have been necessary. The machine remained in use even when electronic computers became available.

Twentieth century procedures for computation: electronic computers

I was given the task of leading the programming effort when the NAO obtained its first electronic computer in 1959. My introduction to computers had been at Imperial College post-graduate lectures in 1950 and I later attended a short course on EDSAC at Cambridge in 1954.[25] I gained valuable experience in the use of IBM 650 computers in the USA in 1957–8, firstly at the US Naval Observatory and then at the Yale University Observatory in New Haven, Connecticut. My main project was to make a new determination of the orbits of the satellites of Mars. Sputnik 1 was launched while I was in the USA, and for a short while the NAO provided a satellite prediction service for the UK.[26]

Before I left for the USA we had reviewed the commercial computers that were becoming available in the UK and we had chosen an English Electric DEUCE computer, which was based on the ACE Pilot Model at the National Physical Laboratory. I had taken DEUCE programming manuals to the USA, but I returned to find that the Admiralty had decided that that the NAO should have a HEC 4 computer. This computer, which was renamed ICT 1201 by the time of its delivery in 1959, had so little capacity and speed that it did not allow scope for the development of new techniques for table making. Nevertheless, it was used for a wide variety of jobs

Fig. 11.3 The ICT 1201 computer was used in the Nautical Almanac Office from 1959 to 1964; it had a magnetic drum store of only 1024 words. The machine is pictured here in the West Building of Herstmonceux Castle with operators Valerie Cann (on the left) and Lynn Ellis.

by staff throughout the Royal Greenwich Observatory. It was operated by NAO staff.[27]

From 1963 onwards, while waiting for a decision on our bid for a better computer, we hired time on an IBM 7090 (later 7094) computer in London.[28] I learnt how to program in Fortran by adapting a program written at the NASA-funded Jet Propulsion Laboratory in Pasadena, California, for computing the ephemeris of the Moon.[29] Nevertheless we still used differencing to verify that there were no random errors in the final results. We found, but could not explain, one isolated error.

Again we did not get our computer of choice—we took delivery in 1966 of an ICT 1909 computer instead of an IBM 360 computer that would have given us compatibility with the computer at the US Naval Observatory. The ICT 1909 proved to be a powerful system but the computing demands of the Observatory rose rapidly. A link to the ICL 1906A computer at the Atlas Computer Laboratory was installed in 1972. The ICT 1909 was upgraded in 1974 to an ICL 1903T. These computers were used by the NAO for work on the publications, for various occultation programmes, and for some research in dynamical astronomy involving numerical integrations of the orbits of minor planets and satellites. The NAO did not, however, have the resources to carry out major projects in the dynamics of the solar system. Instead the NAO concentrated on the improvement of the presentation of the tables and of the techniques for printing and publication.

The Computer Section of the NAO became the Computer Department of the Royal Greenwich Observatory in 1974 and the NAO became a minor user of the computing facilities after a VAX 11/780 system was obtained in 1980 as a node of the national STARLINK network for image processing.

Fig. 11.4 The ICT 1909 computer was in use in the Nautical Almanac Office from 1966 to 1974, when its central processing unit was replaced by an ICL 1903T unit.

Presentation of astronomical and mathematical tables

There are several factors that determine the quality of a numerical table apart from the prime requirement that the values be free from errors. The factors that make a user-friendly table include:

(1) the choice of the quantities—functions, arguments, and interpolation aids—that are tabulated on each page or pair of facing pages;
(2) the choice of the interval and number of figures for each function;
(3) the overall design—layout, headings, typeface, type size, spacing, and rules—of the page; and
(4) the availability of explanatory notes about the basis and use of the quantities in the table.

These factors are especially important in an almanac that is intended for practical use in situations in which speed and accuracy of use are both crucial. Thus in an almanac for navigation an attempt is made to give in each opening all the quantities that vary from day to day, either for a particular day or for a group of days, while essential auxiliary data are given on end-flaps that can be seen at the same time as the daily data.

The mathematical tables produced by Comrie showed very clearly the benefits of using white space rather than rules to separate the columns and rows of the numbers. Unfortunately, the demands of economics and the desirability of making all the required data available at one opening meant that this could not be applied throughout the almanacs produced by the NAO. A further factor was that the main almanacs were produced jointly with the Nautical Almanac Office of the US Naval Observatory, which in turn had to satisfy the US Navy. We found that the Americans were resistant to such changes and so rules were often retained unnecessarily. Comrie also split the numbers into small groups and changed from level figures (in which the top and bottom edges of the numbers are level) to head-and-tail figures (in which the tops of the sixes and eights are higher than those of the other numbers and the bottom edges of the threes, fours, fives, sevens and nines are lower than the other numbers) in order to reduce the risk of errors in the reading from the tables. The differences in approach of the offices are clearly visible in the *Astronomical ephemeris* from 1960 onwards, as the first half was typeset in the UK while the second half was typeset in the USA. The mixture of styles was perpetuated in the *Astronomical almanac* from 1981 onwards even though it was printed only in the USA. Level figures were, however, used throughout. In general, publications that were designed and printed in the UK were printed without rules.

Another important design factor that affected the ease of use of a table of a smoothly varying function was the choice of aids to interpolation. In general, differences were not published if the table could be interpolated linearly; the user was expected to form the first difference mentally and then either to use a first-difference correction table or to calculate and apply the correction by using a desk machine. On the other hand, if second differences were significant, then the first differences would be printed. Only rarely did the almanacs give functions that required the use of higher orders of differences.

It was until recently the practice to publish separately the instructions and auxiliary data for calculating a time or position from the sextant observations. Several commercial publications were available, while the Royal Navy had its own procedures. Similarly, the almanac contained only brief descriptions of the quantities used by astronomers until 1931 when an extended 'Explanation' was introduced by Comrie. Some sections, such as those on the calendar and interpolation tables, were published as separate booklets. The scope of the auxiliary tables was expanded to include, for example, formulae and tables for computing derivatives from differences and for the numerical integration of differential equations. This material was collected together in 1936 in the booklet *Interpolation and allied tables*.[30] It included formulae and tables for the use of sixth differences as well as throw-back formulae. It also gave formulae for 'differences in subdivided intervals', but it did not discuss the procedures used for subtabulation. A completely revised and extended version of *Interpolation and allied tables* was published in 1956; it was reprinted many times.[31]

SUN, 1931.

Date.	Apparent Right Ascension.	Var. per Hour.	Apparent Declination.	Var. per Hour.	Semi-diameter.	Equation of Time. App.–Mean	Var. per Hour.	Sidereal Time.
	h m s	s	° ′ ″	″	′ ″	m s	s	h m s
Jan. 1	18 41 49.81	11.050	−23 06 07.2	+11.04	16 17.54	− 3 06.22	−1.193	06 38 43.59
2	18 46 14.85	11.037	23 01 28.3	12.19	16 17.55	3 34.70	1.180	06 42 40.15
3	18 50 39.57	11.023	22 56 21.9	13.34	16 17.56	4 02.86	1.166	06 46 36.71
4	18 55 03.93	11.007	22 50 48.1	14.47	16 17.56	4 30.66	1.151	06 50 33.27
5	18 59 27.91	10.991	22 44 47.1	15.60	16 17.55	4 58.08	1.134	06 54 29.83
6	19 03 51.48	10.974	−22 38 19.1	+16.73	16 17.54	− 5 25.10	−1.117	06 58 26.39

SUN, 1960
FOR 0ʰ EPHEMERIS TIME

Date	Apparent Right Ascension	Apparent Declination	Radius Vector	Semi-diameter	Equation of Time Apparent − Mean
	h m s	° ′ ″		′ ″	m s
Jan. 0	18 37 14.65 265.52	−23 09 55.6 +250.1	0.983 3048 −183	16 17.50	− 2 32.88 −28.96
1	18 41 40.17 265.22	23 05 45.5 277.7	.983 2865 144	16 17.52	3 01.84 28.66
2	18 46 05.39 264.88	23 01 07.8 305.3	.983 2721 104	16 17.53	3 30.50 28.33
3	18 50 30.27 264.52	22 56 02.5 332.7	.983 2617 60	16 17.54	3 58.83 27.96
4	18 54 54.79 264.13	22 50 29.8 360.0	.983 2557 14	16 17.55	4 26.79 27.58
5	18 59 18.92 263.69	−22 44 29.8 +387.0	0.983 2543 +35	16 17.55	− 4 54.37 −27.14

SUN, 1973
FOR 0ʰ EPHEMERIS TIME

Date	Apparent Right Ascension	Apparent Declination	True Distance from the Earth	Semi-Diameter	Ephemeris Transit
	h m s	° ′ ″		′ ″	h m s
Jan. 0	18 40 58.87 265.24	−23 06 33.6 +273.8	0.983 2884 −82	16 17.52	12 03 09.11 +28.54
1	18 45 24.11 264.95	23 01 59.8 301.3	.983 2802 41	16 17.52	12 03 37.65 28.24
2	18 49 49.06 264.62	22 56 58.5 328.8	.983 2761 0	16 17.53	12 04 05.89 27.89
3	18 54 13.68 264.25	22 51 29.7 356.1	.983 2761 39	16 17.53	12 04 33.78 27.50
4	18 58 37.93 263.86	22 45 33.6 383.2	.983 2800 79	16 17.52	12 05 01.28 27.09
5	19 03 01.79 263.43	−22 39 10.4 +410.2	0.983 2879 +116	16 17.52	12 05 28.37 +26.65

SUN, 1981
FOR 0ʰ EPHEMERIS TIME

Date	Julian Date	Ecliptic Long. for Mean Equinox of Date	Ecliptic Lat.	Apparent Right Ascension	Apparent Declination	True Geocentric Distance
	244	° ′ ″	″	h m s	° ′ ″	
Jan. 0	4604.5	279 28 37.08	+0.51	18 41 12.42	−23 06 06.5	0.983 3281
1	4605.5	280 29 46.88	.53	18 45 37.58	23 01 31.6	.983 3210
2	4606.5	281 30 56.98	.52	18 50 02.44	22 56 29.1	.983 3185
3	4607.5	282 32 07.33	.48	18 54 26.99	22 50 59.1	.983 3203
4	4608.5	283 33 17.85	.42	18 58 51.18	22 45 01.9	.983 3262
5	4609.5	284 34 28.46	+0.33	19 03 14.98	−22 38 37.6	0.983 3361

Fig. 11.5 Changing type faces in the *Nautical almanac*. The typographic style as well as the contents of the tables in the *Nautical almanac and astronomical ephemeris* changed over time. The examples shown here are taken from the Sun's ephemeris for 1931, 1960, 1973, and 1981 and show changes in the typeface for the figures and in the extent of the rules used to separate headings and columns. (By kind permission of HM Nautical Almanac Office.)

It appears that subtabulation by the method of 'bridging differences' was first developed by Comrie.[32] It was developed further by Sadler and others in the NAO to meet particular circumstances. The first systematic account of the method was written much later by Wilkins for the booklet *Subtabulation*, which gives descriptions, formulae and tables for several methods.[33] These procedures soon became obsolete as fast electronic computers made it feasible to compute the required values *ab initio*.[34]

At the time of the revision of *Interpolation and allied tables* there was a growing realisation that Chebyshev polynomials could be used to provide more strongly convergent interpolation formulae than those based on Gaussian polynomials at the expense of giving much higher errors outside the central region. In turn a truncated Chebyshev series could be replaced by an economized polynomial that could be evaluated by the use of a simple recurrence relation without the need to look up interpolation coefficients. This technique was first introduced in the *Astronomical ephemeris* for 1972 and it was used in the *Astronomical almanac* for 1981 in order to replace the lengthy Moon's hourly ephemeris. It allowed the printing of equivalent data, given in decimals of a degree instead of in sexagesimal measure, on 23 pages instead of 122 pages.

The annual reprinting of lengthy sections of the Explanation of the *Nautical almanac* was stopped during the war as an economy measure and it was not until 1961 that the long-promised *Explanatory supplement*[35] was published. This gave the basis and method of calculation of the tabulations, as well as information about many related matters; it included auxiliary tables and other reference data. The various sections of the book were drafted by members of both the UK and US NAOs, and were then subjected to exhaustive criticism, checking and editing. It was published in the UK and was reprinted several times. Parts of it became obsolete before a completely revised edition was published in the USA in 1992.[36]

Copy preparation, proofreading, and printing procedures

Until the 1930s the printer's copy was prepared manually by copying the figures from the computation sheets. This process, and the subsequent manual typesetting, were prolific sources of errors. Errors on the copy were still made even when it became possible to make up the copy by a scissors-and-paste technique. Columns of figures that had been printed by a National machine or a punched card tabulator were cut out and stuck on pieces of card.[37] Headings and auxiliary data were usually inserted manually. Moreover, the computed figures could themselves be in error, and so the NAO used a combination of methods of proofreading.

There were several stages in the proofreading of each publication and at each stage at least four sets of proofs would be examined independently. The code number on each set of proofs indicated the stage and type of reading, so that P indicated a first proof, R a revised proof and S a stereo proof. One member of staff acted as the editor for each publication.[38]

The 'main readers', who were usually the more senior staff, mentally differenced the figures on the proofs in small groups looking for unexpected discontinuities. The leading figures were examined separately as it was otherwise easy to miss an error. The printed differences were checked for consistency with the function values and were themselves differenced. The headings and minor items of auxiliary data were also examined separately. The main reader also verified that the tabulations were continuous from page to page and from year to year and he or she applied occasional independent checks that had been computed by someone other than the persons responsible for producing the original computations. The proofreading of text was also broken down so that spelling, typography, grammar, and technical meaning were verified separately. Examples were checked for appropriateness as well as for consistency with the formulae.

The 'secondary readers' compared their proof with the copy by eye—not by listening to another person reading from the copy—and carried out such tasks as had been specified in the instructions for the page concerned. In addition the Superintendent would scan a set of proofs and would sometimes find errors that had not been found by either the main or secondary readers.

The editor of the publication was responsible for collating the errors that were found, for returning to the printer one set of proofs on which the required corrections were marked and, later, for examining the next set of proofs to verify that the corrections had been made. Sometimes a further round of correction and verification was needed. The standards of proofreading were monitored by noting when each reader missed an error that had been found by another reader. Users occasionally reported errors in the printed volume and these were listed in a later edition.

For many years the almanacs were printed from the moveable type—the printers used the Monotype system, rather than the Linotype system used

Fig. 11.6 A group of NAO staff after a tennis match 1953. Preparing the *Nautical almanac* was a collaborative process. In addition the staff also came together for recreation. Clockwise from top left are Gordon Taylor, George Wilkins, Donald Sadler, Mavis Gibson, Evelyn Grove, and Aileen Grogan. (By kind permission of the *Sussex Express and County Herald*, Lewes, East Sussex.)

for newspapers, because of the complexity of the material. This had the disadvantage that new errors could be introduced during the printing processes after the proofs had been read, during the initial correction phase, or even after the corrections had been verified.

Considerable gains were made by using stereomoulds, which were made from the moveable type after it had been corrected. A second round of main reading was carried out by a different reader on the stereo proofs in order to pick up any errors missed on the first proofs or any new errors introduced by the printer, perhaps in places where no errors had previously been found. If necessary, new stereo moulds would be made.

Increases in productivity came, firstly, from the use of better-quality printed output for printer's copy and, secondly, from the use of IBM card-controlled typewriters to provide copy that could be reproduced directly by photolithography. The first such system was used at the US Naval Observatory in 1945 for the *American air almanac for 1946*.[39] The first NAO system was installed in 1953 for use for the *Abridged nautical almanac* and the *Apparent places of fundamental stars*. It was also used for some special publications such as

Nutation 1900–1959 and *Planetary co-ordinates*. It was necessary to use preprinted forms that gave the page and column headings and the rules. Special procedures and great care by the operator were required to ensure that figures remained centred between the rules at the end of the page. It was sometimes necessary to paste on symbols, such as those for the phase of the Moon. The pages were photographed and reduced in size to, usually, 70% for the production of the photolithographic plates.

A new system based on the IBM 870 document writing system was installed in 1963.[40] At that time other typewriters that were controlled by punched-tape did not give such high-quality output. This system was replaced in 1971 by a UDS 6000 automatic typewriter, with paper-tape readers and punch, which was used for the editing of text as well as for tabular output.

The final step was the introduction of automatic composition by computer techniques. The first experiment was actually carried out in 1964 by Arthur H. Phillips of HM Stationery Office. He arranged for a student at the National Physical Laboratory to program in Algol the setting of one page from the *Astronomical ephemeris*. The computer output was on paper-tape and was used to drive a Monophoto filmsetter. I then developed a Fortran program that could be used more generally to set any required page layout. For our first run in 1968 we used paper tape to drive a Monophoto filmsetter for the *Astronomical ephemeris 1972*.[41] The problems we encountered confirmed our belief in the advantages of punched cards over punched paper tape. From then on, we used a Linotron phototypesetter that was being used by HM Stationery Office to print telephone directories for the Post Office and we were able to supply the data on 9-track magnetic tape. This system was used for the *Star almanac* as well as for the *Astronomical ephemeris*.[42] It was able to set headings and footnotes, but was not suitable for text. At the insistence of the Superintendent, I built into the program all the rules that were then in use for manual composition, such as those for the suppression of common leading figures after the first line of a block of 5 lines.

This may have been the first use of such a system in the UK. I believe that other systems were being developed in USA for printing tables, but none of them aimed to provide the sophisticated editorial control that was expected for the NAO almanacs. We made no attempt to publicise our development of this system as we were too busy to write it up for publication.[43] The system proved to be so reliable that it was used for about 15 years before a new system was introduced.[44]

Publication and distribution media

As far as I am aware the printed page was the only medium for the distribution of the ephemerides and other tables of astronomical data until after the Second World War. The first examples of the exchange of data on punched cards were probably for star catalogues. The increasing level of cooperation between the almanac offices led to exchange of almanac data with the US Naval Observatory and later with other offices. For example, the heliocentric coordinates of the outer planets that were computed in 1953 in the USA by numerical integration were supplied on cards so that we could compute the geocentric coordinates to be published in the almanacs. By 1962, the NAO had built up a library of ephemerides, star catalogues, and observational data on punched cards and supplied copies on request.

The next step was the use of 7-track magnetic tape and then 9-track magnetic tape for the interchange of star catalogues and astronomical ephemerides. This was partly prompted by the space research programmes, which needed existing astronomical data and which also supplied new data. The Jet Propulsion Laboratory, for example, became the major generator of high-precision orbits of the Moon and planets. International cooperation in this matter was required since it was necessary, for example, to agree on standards for the interchange of such data. At its first meeting in Prague in 1969 a joint working group of the International Astronomical Union (IAU) and COSPAR (Committee on Space Research of the International Council of Scientific Unions, ICSU) set the pattern for later cooperative activities.

The importance of international cooperation in the production and distribution of both computed and experimental or observed data was recognized by other scientific unions and was organized by CODATA (ICSU Committee on Data for Science and Technology). In 1970, the IAU appointed me as chairman of its Working Group on Numerical Data and its delegate to CODATA. During the following nine years while I held those positions, there were major advances in the capacity of storage devices for data and in the software for the organization of databases and for the retrieval of information from them. Astronomers shared in the development of these techniques.

In 1978, in an attempt to reduce costs, microfiches were used to publish a catalogue of observations of occultations of stars by the Moon.[45] Each 15 cm × 10 cm microfiche held 98 pages. Nevertheless, the data were also made available in machine-readable form for those who wished to analyse the data in depth.

The increasing use of small programmable electronic calculators led to the production of a new type of publication that allowed the user to bypass the ordinary printed almanac for the data needed for astronavigation. The first booklet giving data for 5 years in compact form was published by the RGO in 1981 for trial purposes,[46] but later volumes were published by HMSO.[47] This booklet made use of economized polynomials, but the users were expected to key the data into their hand-held programmable calculators. This was both a chore and a source of error, but the technique was clearly preferred to the manual use of the more expensive main almanac by many navigators.

The development of small computers, such as the BBC microcomputer (which was tried in the NAO in 1985), and later of the ubiquitous PC, led to the use of cassette tapes, floppy discs, and now CD-ROMs for the distribution of astronomical and astronavigational data and of software that can be used with the data. For example, the current issue of *Compact data* contains a CD-ROM with the *NavPac* program for astronavigation and much more information than the printed text and tables in the booklet itself. It was prepared by HM Nautical Almanac Office and the dedication reads:

This book is dedicated to the Royal Greenwich Observatory (RGO) which closed during the preparation of this fifth edition. It is poignant to note that the RGO was originally established in 1675 in order to '... find out the so-much-desired longitude of places for the perfecting of the art of navigation'.[48]

A further development by the American NAO is represented by the *Multiyear interactive computer almanac (MICA)*, which gives data and both Mac and PC software on CD-ROM for longer periods for most of the ephemerides in the *Astronomical almanac* itself.[49] Moreover, the user may obtain directly the required data for any particular time or place without the need for interpolation. Although the user now has the means to prepare his own almanac it nevertheless seems likely that there will still be a role for the old-fashioned printed almanac, which is a convenient, reliable source of much of the information that is required.

Further reading

A biography of Maskelyne along with an authoritative account of the early years of the *Nautical almanac* is given by Derek Howse, *Nevil Maskelyne: the seaman's astronomer*, Cambridge University Press, 1989. Howse also gives explanations of how lunar distances were used for the determination of longitude in *Greenwich time and the discovery of longitude*, Oxford University Press, 1980. (New edition 1997, with many additional illustrations, published by Philip Wilson, London.) An introductory account of the development of the method of lunar distances and of later methods for position finding at sea and in the air is given in *Man is not lost: a record of two hundred years of astronomical navigation with the Nautical Almanac 1767–1967*, HMSO, London, 1968.

An overview of the history of the Nautical Almanac Office is given by G. A. Wilkins, 'The expanding role of H M Nautical Almanac Office, 1818–1975', *Vistas in Astronomy*, 20, 1976, 239–43, while M. Croarken's *Early scientific computing in Britain*, Oxford University Press, 1990, provides an overview of Comrie's work at the Nautical Almanac Office and elsewhere. Wilkins has also given a general account of the history and work of the Office in a paper with many references and a chronological list of the main events between 1767 and 1998 in Alan D. Fiala & Steven J. Dick (ed.), *Proceedings: Nautical almanac office sesquicentennial symposium: U.S. Naval Observatory, March 3–4, 1999*, US Naval Observatory, Washington, DC, 1999, pages 54–81. This volume contains many papers on the history of the US Nautical Almanac Office and the methods employed there.

A survey by P. Kenneth Seidelmann, who was Director of the US Nautical Almanac Office, of the basic formulae of celestial mechanics and of the procedures for the numerical integration of the equations of motion precedes a historical review of the use of computers (especially in the US Naval Observatory and related organizations) for celestial mechanics and for typesetting is given in J. Belzer, A. G. Holzman and A. Kent, (eds.), *Encyclopaedia of computer science and technology*, vol. 4, (1976) 243–67, Marcel Dekker Inc, New York and Basel.

The availability of computers and new methods for the acquisition, distribution and storage of data led to new problems for astronomers who had previously relied on large printed volumes for their catalogues of numerical data on stars and other astronomical objects and phenomena. The proceedings of the early conferences to discuss these problems highlight the importance of (a) measures to ensure the quality of the data and (b) techniques to retrieve data from the databases that were then being developed. Although primarily concerned with observational data, many of the new techniques were relevant to the storage and distribution of ephemerides and the results of computations on models of stars and stellar systems. The most relevant are C. Jaschek & G. A. Wilkins (eds.), *Compilation, critical evaluation and distribution of stellar data*, Proceedings of IAU Colloquium No. 35, Dordrecht-Holland: Reidel Publishing Company, 1977 and C. Jaschek & W. Heintz (eds.), *Automated data retrieval in astronomy*. Proceedings of IAU Colloquium No. 64. Dordrecht-Holland: Reidel Publishing Company, 1981.

Notes

1. An account of international cooperation in the production and publication of astronomical ephemerides is given in the introductory chapter of the *Explanatory supplement to the astronomical ephemeris and the American ephemeris and nautical almanac*, HMSO, London, 1961.
2. E. G. Forbes, 'The foundation and early development of the Nautical Almanac', *Journal of the Institute of Navigation* (London), 18:4 (1965), 391–401. The papers of Joshua Moore (one of Maskelyne's computers) in the Manuscript Room of the Library of Congress in Washington contain some examples of the prepared noon and midnight lunar distances for the computer to interpolate between but do not show details of the workings. See also M. Croarken, 'Tabulating the heavens: Computing the nautical almanac in eighteenth century England; *IEEE Annals of the history of computing*, 25:3 (2003), July–Sept, forthcoming, for information about the first computers of the Nautical Almanac.
3. E. Dunkin, 'Notes on some points connected with the early history of the "Nautical Almanac" ', *Journal of the Royal Institution of Cornwall*, 9 (1886), 7–18, p. 10.
4. See Maskelyne's notebook in Royal Greenwich Observatory Archives, Cambridge University Library RGO 4/34.
5. See, for example, D. H. Sadler, 'The bicentenary of the Nautical Almanac', *Journal of the Institute of Navigation*, 21 (1968) 6–18, or the books listed for further reading.
6. See Chapter 1 of P. D. Hingley & T. C. Daniel (eds.), *A far off vision. A Cornishman at Greenwich Observatory*. Royal Institution of Cornwall, Truro, 1999. This spiral-bound book (218 pp.) contains a transcription of the autobiographical notes of Edwin Dunkin (pp. 29–172), a foreword by Allan Chapman, an introduction by the editors and 6 appendices.
7. Royal Greenwich Observatory Archives, Cambridge University Library RGO 16, box 54.
8. Personal communications by W. A. Scott on 21 February 1966 and by E. Smith on 6 April 1966.
9. There is an example of Hudson's 'cork mat' in the Royal Greenwich Observatory Archives, Cambridge University Library RGO 16/54, together with an explanation by D. H. Sadler.

10. Royal Greenwich Observatory Archives, Cambridge University Library RGO 16/15.
11. Personal communication by E. Smith.
12. L. J. Comrie, 'The Nautical Almanac Office Burroughs machine', *Monthly notices of the Royal Astronomical Society*, 92 (1932), 523–41.
13. E. W. Brown, *Tables of the motion of the Moon*. Yale University Press, New Haven, 1919. L. J. Comrie, 'The application of the Hollerith tabulating machine to Brown's tables of the Moon', *Monthly notices of the Royal Astronomical Society*, 92 (1932), 694–707.
14. Royal Greenwich Observatory Archives, Cambridge University Library, RGO 16, box 63.
15. L. J. Comrie, 'Inverse interpolation and scientific applications of the National accounting machine', *Journal of the Royal Statistical Society* (Supplement) 3:2 (1936), 87–114. A. E. Carter & D. H. Sadler, 'The application of the National accounting machines to the solution of first-order differential equations', *Quarterly Journal of Applied Mathematics* 1:4 (1948), 433–41.
16. R. C. Archibald, *Mathematical table makers*, Scripta Mathematica, New York, 1948. See pp. 16–18 for notes on L. J. Comrie, with a list of his principal tables. M. Croarken, 'L.J. Comrie: a forgotten figure in the history of numerical computation', *Mathematics Today*, 36:4 (April 2000), 114–18.
17. The correspondence relating to Comrie's suspension from duty is held in Royal Greenwich Observatory Archives, Cambridge University Library RGO 16/60.
18. G. A. Wilkins, 'Donald Harry Sadler, O.B.E. (1908–1987)', *Quarterly journal of the Royal Astronomical Society*, 32 (1991), 67–8.
19. J. Todd and D. H. Sadler, 'Admiralty Computing Service', *Mathematical tables and other aids to computation*, 2:19 (1947), 289–97.
20. J. G. Porter, 'Punched card machines at Herstmonceux', *Journal of the British Astronomical Association*, 61:7 (1951), 185–9.
21. G. A. Wilkins, 'Calculation of the nutation from the new series', in *Improved lunar ephemeris 1951–1959*, US Government Printing Office, 1954, 420–2.
22. L. J. Comrie, 'On the construction of tables by interpolation', *Monthly Notices of the Royal Astronomical Society*, 88 (1928), 506–23.
23. W. J. Eckert, D. Brouwer, and G. M. Clemence, 'Coordinates of the five outer planets, 1653–2060', *Astronomical papers prepared for the use of the American ephemeris and nautical almanac*, vol. XII, 1953, U.S. Naval Observatory, Washington, DC.
24. NAO, *The prediction and reduction of occultations. Supplement to the nautical almanac for 1938*, HMSO, London, 1937. This booklet includes a photograph and a description of the occultation machine and its use.
25. Royal Greenwich Observatory Archives, Cambridge University Library RGO 16, box 40.

26. D. H. Sadler, 'The prediction service of HM Nautical Almanac Office', *Proceedings of the Royal Society of London*, Series A, 248 (1958), 45–8.
27. The HEC 4 computer was manufactured by the British Tabulating Machine Company (BTMC), which had supplied most of the punched-card machines in the NAO system. It merged with another company to form International Computers and Tabulators (ICT). Some years later it merged with English Electric to form International Computers Limited (ICL). The HEC 4 was aimed at the commercial market and had no scientific software. Its main store was a magnetic drum with a capacity of 1024 words. Guidance and information about its use was given by Wilkins and later by others in a series of lectures and in *NAO computer circulars*. See Royal Greenwich Observatory Archives, Cambridge University Library RGO 16, boxes 40 and 411.
28. Royal Greenwich Observatory Archives, Cambridge University Library, RGO 16, boxes 61 and 222 (file 13R).
29. Royal Greenwich Observatory Archives, Cambridge University Library, RGO 16, box 35.
30. NAO, *Interpolation and allied tables*, HMSO, London, 1936. Printed from the stereo plates for the *Nautical almanac* for 1937.
31. NAO, *Interpolation and allied tables*, HMSO, London, 1956. This was a completely rewritten version of the 1936 edition. It was considerably extended in its scope.
32. L. J. Comrie, 'Inverse interpolation and scientific applications of the National accounting machine', *Supplement to the journal of the Royal Statistical Society*, 3 (1936), 87–114.
33. NAO, *Subtabulation: a companion booklet to interpolation and allied tables*, HMSO, London, 1958. The introductory section includes bibliographic notes on methods of subtabulation. Part II of this booklet gives tables for some 'Direct methods'; Part III (by D. H. Sadler) describes 'The method of precalculated second differences'; while Part IV (by G. A. Wilkins) describes 'The method of bridging differences'.
34. A generalized technique for the use of bridging differences was given by Enid R. Woollett, 'Subtabulation with special reference to a high-speed computer', *Quarterly journal of mechanics and applied mathematics*, 11 (1958), 185–95.
35. *Explanatory supplement to the astronomical ephemeris and the American ephemeris and nautical almanac*, HMSO, London, 1961.
36. P. K. Seidelmann (ed.), *Explanatory supplement to the astronomical almanac*, University Science Books, Mill Valley, California, 1992.
37. The white card was cut from obsolete Admiralty charts while the glue was sold under the name of 'Cow Gum'! It was desirable to open the windows while it was being used.
38. See Royal Greenwich Observatory Archives, Cambridge University Library, RGO 16/52 for collections of proofreading notes.

39. W. J. Eckert and R. F. Haupt, 'The printing of mathematical tables', *Mathematical tables and other aids to computation*, 2:17 (1947), 197–202.
40. Royal Greenwich Observatory Archives, Cambridge University Library, RGO 16, box 8.
41. Royal Greenwich Observatory Archives, Cambridge University Press, RGO 16, boxes 5, 22.
42. Royal Greenwich Observatory Archives, Cambridge University Archives RGO 16, boxes 22, 85.
43. Royal Greenwich Observatory Archives, Cambridge University Library, RGO 16, files 40P and 101/8 (IV) contain further notes and correspondence about the development and use of the system.
44. The new system is briefly described in the report on the work of HM Nautical Almanac Office in: *Royal Greenwich Observatory: Telescopes, instruments, research and services: October 1 1980–September 30 1985*, RGO, undated, pages 92–3. This report also includes paragraphs on the changes in the *Astronomical ephemeris for 1984* and on new publications and techniques.
45. L. V. Morrison, 'Catalogue of observations of occultations of stars by the Moon 1943–1971', *Royal Greenwich Observatory bulletins* No. 183, 1978. 16 pages and 5 microfiches.
46. B. D. Yallop, 'Compact data for navigation and astronomy for the years 1981 to 1985', *Royal Greenwich Observatory bulletins* No 185, 1981.
47. B. D. Yallop and C. Y. Hohenkerk, *Compact data for navigation and astronomy for the years 1986 to 1990*, HMSO, London, 1985.
48. C. Y. Hohenkerk and B. D. Yallop, *NavPac and compact data 2001–2005*, The Stationery Office, London, 2000. 132 pages and a CD-ROM. The quotation is from the royal warrant for the appointment by Charles II of John Flamsteed, the first Astronomer Royal.
49. Astronomical Applications Department, US Naval Observatory, *Multiyear interactive computer almanac 1990–2005 (MICA)*, Willmann-Bell, Inc., Richmond, Virginia, 1998.

VISICALC® USER'S GUIDE | **IBM PERSONAL COMPUTER**
LESSON FOUR | TUTORIAL

The screen should look like the following photograph:

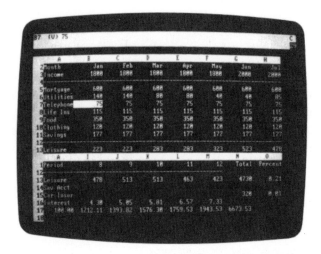

If you would like to use this version of the budget worksheet for your own use, save it by typing /**SSbudget** ↵.

ORDER OF RECALCULATION

So far, you've seen that the VisiCalc program recalculates the values of all the formulas on the worksheet, but you haven't been told much about how this is done. Some characteristics of recalculation can affect results on a complex worksheet.

The VisiCalc program recalculates by starting at the upper left corner of the worksheet, working its way down and to the right until it reaches the lower right corner. Each formula is evaluated only once unless you ask for an extra recalculation by typing !.

As a rule, this means that formulas that reference other entries must be located below and to the right of the referenced entries. An entry at position A1 cannot be a formula that references other positions.

12

The rise and rise of the spreadsheet

MARTIN CAMPBELL-KELLY

The mathematical table, as a paper-based artefact, is close to the end of its technological life. Tables had two main uses—as a calculating aid (such as a logarithm table) and as a data storage device (such as an actuarial or census table). Logarithmic tables have now been made obsolete by the electronic calculator, while data tables are increasingly being replaced by online databases. Examples of these uses of tables have been given in several of the previous chapters. The decline in the use of tables was much in evidence before the arrival of the electronic spreadsheet on the scene, and by and

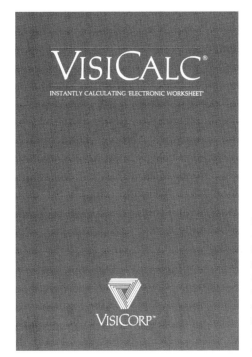

Fig. 12.1 *User's guide* for VisiCalc, c. 1981. The *User's guide* was a necessary complement to the early spreadsheet program. It explained an unfamiliar technology and provided user assistance at a time when computer memories were too small to support on-line help. (Reproduced with permission of International Business Machines.)

large the spreadsheet has not taken over these functions of the table. Yet, the perception of the spreadsheet as an historical successor to the table is intuitive and appealing. In what sense can this said to be true?

Spreadsheets vis-à-vis tables

I believe one answer to the above question is offered by the psychological concept of 'affordance' which has been much used by architects and human computer interaction designers in recent years. Affordance can be defined as:

A bundle of properties about some specific object that provides us with an opportunity to perceive something specific or to move through the world in a characteristic way.[1]

Thus well-trained architects now design doors with either handles that afford *pulling* or flat plates that afford *pushing*, depending on which way the door opens. Similarly designers of civic structures now understand that a horizontal surface at about knee height affords sitting—which they may or may not want to encourage; while a smooth vertical surface encourages graffiti—which they probably do not want to encourage.

Seen in this light, a writing surface affords the property of organizing information in a two-dimensional grid, and therefore tables can be viewed almost as an historical inevitability. Conversely, a two-dimensional writing surface does not afford easy three-dimensional representations of data, and these have not evolved to any significant extent. Even before the invention of paper, it is clear from Eleanor Robson's chapter on Mesopotamian tables that the clay tablet lent itself to two-dimensional expression. It seems plausible that the two-dimensional table would arise spontaneously in any civilization where a writing surface was used.

The screen of a personal-computer shares the two-dimensional character of a writing surface, but has two additional properties—easy erasure, and the ability to act as a 'window' onto a much larger virtual surface. Hence the spreadsheet can be seen to make use of two-dimensional affordance in exactly the same way as a table, but to have the additional properties of dynamic alteration, and quasi-infinite extent in the plane. So, while one cannot assert that the mathematical table is the direct historical ancestor of the spreadsheet, one can argue that they sprang from similar historical and technological forces.

Because tables and spreadsheets share these fundamental properties, in this chapter the table-centric view of *Sumer to spreadsheets* will be used as a lens through which to examine the spreadsheet, by asking the same kind of questions that one might ask of tables. How were spreadsheets produced and sold? Who were the users of spreadsheets? Within these larger themes, some secondary issues will be explored, such as ease of use, errors, and the problem of plagiarism. Babbage, and many of his successors, were fascinated by the visual presentation of tables, and how this could facilitate their use and minimize errors. How have spreadsheet designers married form and function to assist the user? Babbage, above all, was obsessed with eliminating the possibility of errors occurring in tables, whether made by human computers or printers. In a similar way, spreadsheet makers have gone to considerable lengths to make their products easy to use, and to minimize the potential for error. As Mary Croarken has noted elsewhere, Comrie was famous for the fact that he introduced small rounding errors in his edition of *Chambers six-figure mathematical tables*, as a trap for plagiarists.[2] How have spreadsheet makers sought to deter would-be copyists?

Prehistory

The Library of Congress catalogue lists over a thousand texts on spreadsheets published during the last 20 years, an astonishing testament to the extent to which they have entered the commercial and scientific culture. Very few of these books suggest that the spreadsheet has a history.

There is no record of who introduced the word spreadsheet into the computer lexicon, but it was not the inventor of VisiCalc, which was originally called 'an electronic worksheet.' Both of the terms *worksheet* and *spreadsheet* have their origins in accounting, although the former is much more prevalent. Accountancy textbooks published up to the 1970s typically have an index entry for worksheet (or, rather, work sheet), but not for spreadsheet. A worksheet was (and is) a standard tool for constructing a trial balance. According to a textbook first published in 1956:

The information needed for a formal balance sheet and income statement can be obtained quickly by means of a work sheet. As the work sheet can be completed before the adjusting and closing entries are made in the books, its use avoids delays in presenting these financial statements to management. Even though the worksheet is not part of the formal accounts, all worksheets should be preserved.

These are generally required by outside auditors in any examination of the books, not only in audits by public accounting firms but also in those made by government tax examiners in connection with income tax returns.[3]

Hence the worksheet was both an aid to systematic calculation and an auditing device. The technique became so standardized that pre-printed worksheets, typically in 6 or 10 column format, could be purchased from stationery suppliers.

By the 1970s, the term *spreadsheet* had acquired a meaning that can best be explained by a simile: as a [human] computer was to a tabulation, so an accountant was to a spreadsheet. And just as the human computer had large sheets of paper feint-ruled into quarter-inch squares, the accountant had spreadsheet paper ruled into a grid for recording financial data.

In the 1960s and 1970s many software packages for financial analysis were designed for use on mainframe computers and commercial time-sharing systems. Such packages were primarily used by middle-management for financial forecasting, strategic planning, and budgeting. These early tools originated in the discipline of 1960s management science, which itself has a deep historical connection with post-war operational research and long-range planning. One example of such a system was the Basic Business Language (BBL) offered by the popular Tymshare system. This package was a dialect of the BASIC programming language for financial analysis:

> One of the most useful applications of BBL is for report generation: preparing and formatting reports. Reports of this type, sometimes called spread sheets, are often used to summarize historic information or forecast future performance.[4]

The term 'what-if analysis', much used in the spreadsheet literature, also appears to date from the early 1970s. Thus a writer of 1972 noted:

> After years of merely processing clerical transactions, it now appears that the computer has progressed to the point where it provides management with useful tools for the planning process. Two of the most powerful tools available are financial modeling and "what if" budgeting.[5]

There were three main problems with these financial analysis packages. First, they were expensive. For example the median price in 1974 of a mainframe financial analysis package was $2800; alternatively, running a program on a time-sharing system would have cost at least $10 an hour.[6] Second, the programs had poor interactivity so that results would at best be available after a minute or two using a teletype terminal attached to a time-sharing computer, or at worst after several hours when using a conventional mainframe. Third, the packages were difficult to use, and usually required more than a modicum of computer programming knowledge.

The acknowledged inventor of the electronic worksheet was Dan Bricklin, and the program he and his partner Bob Frankston produced was sold as VisiCalc from late 1979.[7] Bricklin was a 26 year-old Harvard MBA student who had at one time been a software designer with the Digital Equipment Corporation, where he worked on, among other things, word processing software and newspaper typesetting systems. Bricklin enrolled for his MBA as a springboard to a career in the financial services industry. He was exposed to financial analysis techniques using tedious pencil and paper methods, and the idea for a 'visible calculator' was born of that experience.

The detailed history of how the spreadsheet was invented and kick-started the personal computer revolution is a standard item in every account of the rise of the personal computer. In short, Bricklin and Frankston incorporated a development company Software Arts, Cambridge, Mass., to write the program, and used a software publisher Personal Software to manufacture and distribute it. VisiCalc went on sale at a retail price of $99 in October 1979. Sales were only modest at first, but following favourable press reviews and word-of-mouth recommendations sales took off during the second half of 1980, reaching 12 000 copies a month by the end of the year. At the same time, the price was increased to $249, a level the market was evidently willing to bear.

Fig. 12.2 Dan Bricklin inventor of the personal-computer spreadsheet demonstrating VisiCalc at the West Coast Computer Faire, 5 December 1979. (Reproduced with permission of Bob Frankston.)

Whether or not VisiCalc merely provided features already available on existing financial analysis packages was of less importance than its cheapness, interactivity, and unprecedented ease of use. Although personal computers were still relatively expensive in 1979, they were very cheap compared with corporate mainframes and mini-computers. VisiCalc's 'instantly calculating' nature was a revelation to seasoned computer users, a point well articulated by an analyst of 1985:

> Although there have been financial modeling and planning programs on mainframes, none had exactly the properties of VisiCalc. ... It was able to recalculate quickly the rows and columns of a spreadsheet every time a single number was changed. Because the response time was virtually instantaneous, the machine was able to keep up with the thinking speed of the user. Mainframe financial modelling packages were often on-line systems, but they were not real-time. The real-time interactive nature of VisiCalc was not just an improvement on mainframe financial planning software, it was a new paradigm.[8]

VisiCalc's ease of use arose from the fact that the user did not need any programming knowledge. In this respect VisiCalc was analogous to a word processor, where the user directly arranges the print on the page, as opposed to a typesetting system where a set of instructions is used to lay out the text. Indeed, Bricklin sometimes described VisiCalc as a word processor for numbers. In financial analysis packages such as BBL the user had to code the problem in traditional programming terms, but in VisiCalc the user simply entered values into the worksheet and specified the mathematical relationships between them; VisiCalc took care of the rest. The difference between these two problem solving styles is a recurring theme in computer science, known as the procedural and declarative programming paradigms. Jerrold Kaplan, the Chief Technology Officer of the Lotus Development Corporation put this very well:

> While some spreadsheet users may not recognize it immediately, this process is actually a method of programming. Rather than write a program as a procedure to be executed (a series of well-defined steps to be performed in a particular sequence), you write the program as a description to be maintained (a set of relationships between data elements). It turns out that some problems are easier to solve using the procedural approach, such as processing orders or updating a database, while others are easier to solve using the declarative approach, such as projecting profit and loss or performing a lease-versus-buy analysis.[9]

While it took hours or days to learn even rudimentary computer programming, anyone with a basic mathematical knowledge could use a spreadsheet straight out of the box.

Anatomy of a spreadsheet

	A	B	C	D	E	F
1		Reading	Writing	TOTAL		
2	Janet	22	21	43		
3	James	14	18	32		
4	John	19	17	36		
5	AVERAGE	18.33	18.67			
6						

The basic form of the personal-computer spreadsheet has remained essentially unchanged since it was first introduced in VisiCalc in 1979.

A worksheet consists of a matrix of cells, of which the top left corner is shown above. Most spreadsheets have 256 columns indexed by a one or two letter combination: A, B, C, ... X, Y, Z, AA, AB, AC, ... AZ, BA, BB, ... IV. The rows are numbered 1, 2, 3 .... The number of rows was as few as 256 in early spreadsheet programs, up to many thousands on a modern program. It is difficult to conceive of applications that would be able to make use of such a large worksheet. At any time, the user can see only a 'window' on the spreadsheet of perhaps 8 or 10 cells wide by 20 deep.

Individual cells are denoted by the concatenation of the column and row indices: thus in the figure the name Janet is in cell A2, while John's TOTAL is in cell D4. A range of cells is denoted by the start and end cells of the range, separated by a double period. Thus the AVERAGE of the range of cells B2..B4 is shown as 18.33 in cell B5, while the table as a whole occupies the rectangular range of cells A1..D5. The contents of a cell can be formatted appropriately by the user: in the example above, some values are shown as integers, while others are given with two places after the decimal point; some labels are centred while others are aligned left or right.

Cells can contain one of three entities: a *label*, a *value*, or a *formula*:

> A *label* consists of a textual item, which cannot enter into a calculation and is usually used to name neighbouring cells. Thus in the figure the labels Janet, James, John, and AVERAGE are row labels, while Reading, Writing, and TOTAL represent column headings.

> A *value* cell contains quantitative information that can be used in a computation. Values are usually decimal numbers, but they can also include currency values, percentages, or even dates. In the figure, the six central cells B2..C4 are all integer values entered by the user.

> **Anatomy of a spreadsheet** *cont.*
>
> A *formula* cell defines a computation involving other cells. Thus in the figure, cell D2 evaluates the sum of Janet's Reading and Writing scores. In VisiCalc this formula would have been expressed as +B2+C2; subsequent spreadsheets used similar (though not always identical) notations. It is possible to use pre-defined arithmetic and commercial functions (sometimes known as @functions in early spreadsheet programs because of the syntactic @-sign that preceded their name). In the figure above, the average score for Reading is evaluated in cell B5 by the formula @AVERAGE(B2..B4). Note that the spreadsheet displays the result of the formula, not the formula itself. A formula can use the results of other formula cells, allowing for very complex analysis. Much of the analytical power of modern spreadsheets is achieved by incorporating dozens of functions for statistical calculations and financial analysis.
>
> The single most important feature of the spreadsheet program was its ability to immediately show the effect of changes. For example, if John's score for Writing, in cell C4, was changed from 17 to 19, cells D4 and C5 would *instantly* change to 38 and 19.33 respectively.

The production and diffusion of spreadsheets

Spreadsheets are a huge business. The diffusion of spreadsheets was dramatic from the very beginning. For example, VisiCalc had sold 100 000 copies by May 1981, and an estimated 700 000 copies by late 1984, when it had been on the market 5 years.[10] By the mid-1990s the leading spreadsheet, Microsoft Excel, was selling more than 20 million copies a year.[11]

The market for spreadsheets in highly concentrated, and only a handful of products have ever achieved real prominence. VisiCalc had an early-start advantage, being the first program of its kind for the then market-leading Apple II computer. VisiCalc was subsequently re-written for the Radio Shack TRS-80, and several other early computers. During the early 1980s there were many competing personal computers, which were largely incompatible, each of which had to be supplied with a spreadsheet. Hence within two or three years of the launch of VisiCalc there were dozens of what the trade press termed 'Other Calcs', 'Visi-Clones', or 'Calcalikes'. Some of these were quite successful, such as Sorcim's SuperCalc and Microsoft's Multiplan. On any single platform, however, just one spreadsheet tended to dominate with 70 per cent or more of sales.

In August 1981, IBM announced its personal computer, subsequently known simply as the 'PC'. This created a standard personal computer platform for the first time, and soon IBM 'clones' accounted for 80 per cent of new computer sales. As with the earlier platforms, a single spreadsheet dominated the IBM-compatible PC market, although this changed over time. Thus VisiCalc dominated from 1981 to 1983, Lotus 1-2-3 from 1984 to 1989, and Microsoft Excel from 1990 to the present. Economists have a number of explanations for the existence of these 'serial monopolies': *the theory of increasing returns, user lock-in*, and *network effects*.

The theory of increasing returns—sometimes called 'Microsoft Economics' in business magazines—explains the tendency for a single product to dominate many high-tech markets. The leading protagonist of the theory of increasing returns is the Stanford School, led by the economist Brian Arthur.[12] The theory applies to many high-tech products, but especially to software in which increasing-returns behaviour is most pronounced. The theory runs along the following lines.

Suppose there are two equally priced and equally desirable information products (say spreadsheet A and spreadsheet B). Then the market will not stabilize at a 50:50 split as one might have expected, but one product will eventually dominate strongly. The reason is that a 50:50 split is unstable and the market will tend to tilt slightly in favour of one product or the other (spreadsheet A, say). Once product A has achieved a small sales advantage, the magic of Microsoft economics kicks in. Because the marginal cost of software production is trivial compared with the original development cost, incremental sales generate very high profits. These profits can then be reinvested in product improvement, advertising, and lower prices; this makes product A more desirable and get still further ahead of B. Today software marketers have become very adept at 'tipping' an unstable market in their favour when introducing a new product. Tactics include free trials, competitive upgrades, bundling lacklustre products with market leaders, and so on.

Once market dominance has been gained, this advantage is sustained by user lock-in and network effects. Both of these make it very difficult for users to switch from their current spreadsheet program to a competitive one. These 'switching costs', besides the cost of the new package itself, include the lock-in factor of the intellectual investments a user has made in learning to use the program and in creating application files. Network effects refers to the fact that a successful package creates a community of users who share files, and their ability to co-operate depends on them using the same package, or at least the same file format. Hence, it is even more difficult for a group of workers to switch products than an individual. Needless to say manufacturers of the dominant software packages have become accomplished at making the switching costs as high as possible, while competitors work hard to make the change as painless and cheap as possible. Because the costs of switching from one product to another are so high, users usually only make a change when there is a 'platform discontinuity'. In the case of the IBM-compatible PC there have been two such discontinuities.

In August 1981, when the IBM PC was introduced, VisiCalc was the first and dominant spreadsheet for it. However, people came to the IBM PC from many different non-compatible computer models, or no computer at all, and so there was no lock-in effect on their purchasing decisions. Users simply selected what they perceived to be the best spreadsheet on the market. In January 1983, when people were beginning to buy the IBM PC *en masse*, the Lotus Development Corporation introduced its 1-2-3 spreadsheet. Lotus 1-2-3 was a major improvement over VisiCalc. For example, it included graphics and database capabilities in addition to the spreadsheet function (hence the name 1-2-3). Thus 1-2-3 users could present their results in compelling graphical forms such as bar and pie charts, and they could use information permanently stored in a database. In addition 1-2-3 ran three times as fast as VisiCalc and was more 'user friendly.' Despite the fact that 1-2-3 cost $495 compared with $249 for VisiCalc, it massively outsold it.

Lotus 1-2-3's ascendancy over VisiCalc was absolute. By 1985 Visi-Corp (the publisher of VisiCalc, formerly known as Personal Software) was close to bankruptcy, while Bricklin and Frankston's development company Software Arts was taken over by Lotus Development, and they became employees of the firm. For a couple of years Lotus was perhaps America's fastest growing major company, and by 1987 it was the world's largest independent software vendor with over 2000 employees and revenues of $400 million a year, almost entirely derived from sales of 1-2-3.

Lotus 1-2-3 dominated IBM-compatible PC spreadsheet sales until 1990, when the Microsoft Windows operating system achieved market success and was supplied with virtually every new PC. The arrival of Windows was another platform discontinuity. Because Lotus 1-2-3 would not run under Windows, users wanting to take advantage of the new operating system had to buy Microsoft Excel—the only spreadsheet for Windows then

Spreedsheet evolution: Lotus 1-2-3 and Microsoft Excel

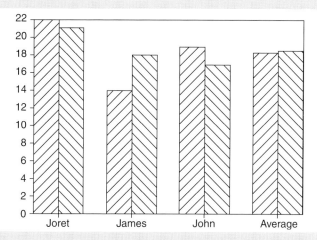

Fig. 12.3 Lotus 1-2-3, introduced in January 1983, integrated three functions—a spreadsheet, graphical data analysis, and a simple database. The figure shows the data from the spreadsheet in the previous text box presented as a bar-chart.

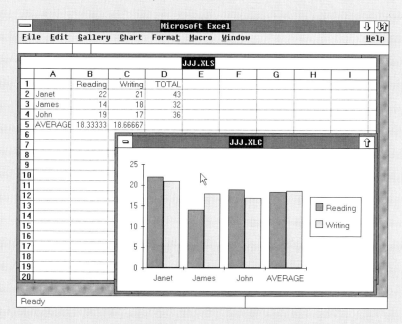

Fig. 12.4 Microsoft introduced Excel 2.0 for the IBM-compatible PC in autumn 1987. The new program made a leap forward in terms of ease-of-use and data analysis. It provided the user with multiple windows, direct manipulation by means of a mouse, and powerful aids for charting. The figure shows the data from the spreadsheet in the previous text box presented as a bar-chart.

available. This change was made as painless as possible by Microsoft offering price discounts for Lotus 1-2-3 users and by the product supporting Lotus-compatible file formats and facilities. It is worth noting that Microsoft's spreadsheet Multiplan had been available since 1980 and in improved form as Excel since 1987. However, it was not until the platform discontinuity of Windows that it was able to capture significant market share.

Product improvement and perfection

As increasing-returns theorists would predict, over the years spreadsheet publishers have made massive investments in product improvements. Compared with today's spreadsheets, early versions of VisiCalc look primitive in the extreme. One index of the advancement is the size of the programs themselves. When it was introduced in 1979, VisiCalc consisted of roughly 10 000 program instructions. By 1990, Microsoft Excel contained about 650 000 lines of code.[13] Although frequently criticised as 'bloatware', the increased code size had been accompanied by vast improvements in data analysis and presentation, and ease of use.

VisiCalc originally manipulated and displayed nothing other than text and numbers on a monochrome screen. However it quickly became apparent that users were keen to do more with their spreadsheets and VisiCorp introduced a number of complementary products, such as VisiPlot and VisiTrend that produced visual representations of spreadsheet data for incorporation in written reports and presentations.

Lotus 1-2-3 eliminated the need for such complementary products by incorporating the same facilities in the basic software package. The tight integration of all these functions within a single program was very attractive to users. Lotus 1-2-3 was systematically enhanced throughout the 1980s. For example, release 2 in 1985 introduced sorting and enabled much larger spreadsheets to be manipulated. Release 3 in 1989 enabled users to manipulate several spreadsheets simultaneously.

Alongside improved data manipulation and presentation, ease of use has been a major selling point of spreadsheets. VisiCalc's user interface now seems archaic. For example, to exit VisiCalc one was required to type the sequence of characters '/SQY'. Users were provided with a 'Pocket Reference' card with many such command strings. Lotus 1-2-3 made an advance in user friendliness by introducing menus in English. Now, to quit a program, one simply had to select a menu and chose the 'Quit' command. With the introduction of Microsoft Windows and Excel it became possible to use a mouse to perform many operations by direct manipulation.

At the time of writing, Lotus 1-2-3 and Excel, and the only other significant spreadsheet Quattro Pro, are all in about their tenth major release. There is a commonly held suspicion that the introduction of new product versions at approximately 18-month intervals is as much about revenue generation as genuine product improvement. Nonetheless there is no question that each product release adds new capabilities to the software, partly to keep up with competitors, and partly to exploit new technological opportunities such as the World Wide Web. There is no subject on which users are so vocal and so at odds with the manufacturers as the phenomenon of 'creeping featurism'. Users find themselves buried beneath an overwhelming weight of wizards, problem solvers, and 'macro' facilities that go far, far beyond the needs of any one individual, and far beyond what any person who is not a full-time developer could master. Indeed, three frustrated authors were even driven to write a book on the topic, *Excel 97 annoyances*.[14] And yet, as Cusumano and Selby have shown in *Microsoft secrets*, their excellent academic study of Microsoft's development processes, enhancements to products such as Excel are largely driven by user research—focus groups, ethnographic field studies, usability laboratories, and feedback from telephone help-desks.[15] It seems that every feature, however apparently obscure, is indispensable to a significant group of users. This is perhaps the inevitable outcome of designing one-size-fits-all software for a community of a hundred million users.

Complementary products

Compared with the population of different mathematical tables, the number of spreadsheets was remarkably small. By the mid-1980s there were less than 20 significant products, and today there are just three. However, in the 1980s there developed a secondary market in complementary programs for the best-selling spreadsheets. These software packages enhanced or adapted the spreadsheet by extending its capability, or provided 'templates' for particular applications. One might say that owning a spreadsheet and some complementary products was rather like owning a book of common logarithm tables together with a few volumes of special purpose tables. While every engineer, scientist, or actuary owned the same standard logarithm tables, each profession had its own set of specialist tables.

Fig. 12.5 One of the most popular complementary products for Lotus 1-2-3, Sideways enabled a spreadsheet to be printed across several pages of tractor-feed stationary, eliminating the need for scissors and glue. (Reproduced with permission of Funk Software.)

The market for complementary products probably began with VisiPlot and VisiTrend, mentioned previously, that enhanced the capability of VisiCalc. Although Lotus 1-2-3 on its first release in 1983 incorporated all that VisiCalc and its complementary products could do within a single package, third party suppliers quickly began to make 'add ons' to enhance 1-2-3 in new and unforeseen directions, a phenomenon that *Forbes* magazine call the 1-2-3 'after market'.[16] For example, a product known as Sideways produced by Funk Software enabled a spreadsheet to be printed along the length of continuous stationery, instead of across, which saved the user from having to use scissors and glue when printing large spreadsheets. Sideways was a real boon for users and it sold hundreds of thousands of copies. In 1988 Funk Software introduced a new product Allways that added desktop publishing capabilities to 1-2-3 to exploit the new generation of laser printers; this was another best seller.

At first Lotus did not actively co-operate with complementary product suppliers, seeing the add-on market as parasitic and something it would like to keep in-house. However, it soon became apparent that the add-on market had two major benefits. First it increased user lock-in, for now a competing spreadsheet would have to supplant not only 1-2-3, but also a user's collection of add-on products. Second, specialized add-ons enabled 1-2-3 to satisfy many different user communities in a way that no mass-appeal product from a large firm could. In 1985 Lotus began to actively co-operate with third-party producers by supplying an 'Add-in Development Kit'. An add-in was different to an add-on. While an add-on program complemented 1-2-3, it was a separate program with all the difficulties that communicating between two programs involved. By contrast an add-in was a program that could be tightly linked with 1-2-3, so that the two behaved as a single integrated program. Indeed it was possible to have several add-ins attached to 1-2-3. The add-in market boomed, and soon there were scores of add-ins from dozens of suppliers. For example, one product SeeMore enabled the spreadsheet to be reduced in size so that more of it could be seen on the screen. Another, Noteworthy, enabled electronic Post-It notes to be attached to the cells of a spreadsheet. D.A.V.E. enabled data to be entered into a spreadsheet by means of a standard form. A package called The Analyst enabled errors in a spreadsheet to be detected. Every one of these products had several competitors. By 1987, Lotus claimed that there were more than a thousand products available from 450 different suppliers.

Now something truly remarkable and unexpected occurred. Almost unwittingly, Lotus 1-2-3 had been transformed from a stand-alone software product into a technological system, by which the market determined the appropriate mix of capabilities through its choice of add-ins. A small number of add-ins (Allways among them) proved extremely popular, and Lotus responded to these strong market signals by acquiring the rights to the best-selling add-ins, or by taking over the producer directly. In the late 1980s Lotus started to sell 1-2-3 bundled with Allways and other popular add-ins, which enabled it to maintain a 70 per cent market share despite intense competition from Excel and Quattro Pro. In the early 1990s, the features of the most popular add-ins were seamlessly integrated into 1-2-3 and the original source of these capabilities was no longer evident. Lotus 1-2-3 now incorporated advanced desktop publishing capabilities, the user could zoom into the spreadsheet with different levels of magnification, one could create dialogs for data entry, one could annotate cells, and there was help to detect errors. All of these features had once been provided by add-ins. And that, rather like a technological *Just so* story, is how the spreadsheet got its modern form. In the 1990s the phenomenon of add-ins vanished as quickly as it had risen.

The most numerous complementary products for spreadsheets, however, were not add-ons or add-ins, but application templates, and these continue to have a market niche. An article in *Software news* in spring 1985 explained:

The software enhancements to Lotus products come in many flavors. We have identified some 200 Lotus-related template, add-on, add-in and standalone software products. By far the largest number of products are the many 'pre-solved problems' or templates as they are called, which are worksheet models developed to solve specific types of problems. These can be loaded into 1-2-3... and then 'filled in' with data.[17]

A typical and very popular type of application template was for tax preparation. While most users could in principle have created a spreadsheet for their tax return, it was time consuming and error prone; $50 for a ready made template was thus money well spent. By mid-1985 there were more than a dozen products in just this category, going under names such as 1-2-3-Tax, TaxAide, TaxCalc, and the wince-inducing TaxPertise. The *Software news* article cited above gave a taxonomy of four template categories, with each sub-divided into different branches:

General Business: financial analysis; sales & marketing; budget; office management; sales tools; asset management; cash management; accounting; project management; production management; business forecasting

Personal Organization and Finance: personnel investment; real estate; tax planning and preparation

Utility Templates: multi-products sets; engineering management

Vertical Markets: television and film; aviation; insurance; real estate; oil production; banking

If one were looking for evidence of the extent to which the spreadsheet has replaced the table, it is in the realm of template publishing. To take just one example, for many years the Craftsman Book Company of America supplied books such as *The practical lumber computer*, *The practical rafter calculator*, and *The national construction estimator*. These books consisted largely of mathematical tables. Today the first two of the these titles have vanished, while the role of the third is filled by *Estimating with Microsoft Excel—unlocking the power for home builders*. The Craftsman Book Company now publishes just one compendium of tables, *Estimating tables for home building*, but several spreadsheet templates.[18]

Intellectual property and plagiarism

The code of a program—whether expressed in binary or symbolic form—has always been protected by ordinary copyright law, but until recent times patent law has explicitly excluded algorithms and mathematical concepts. When Dan Bricklin invented VisiCalc, very few software patents had been granted, and those that were usually relied on hardware-related processes. The publishers of VisiCalc retained a patent attorney in 1979, and the advice given was that applying for a patent would be expensive and unlikely to succeed. VisiCalc was subsequently protected only by copyright and trademark law. Hence Dan Bricklin never became fabulously wealthy, although he has earned a place in history as the inventor of a device that created an industry and changed the working lives of millions of individuals—in his own words, he has 'made a difference'.[19]

It was relatively easy to clone an existing software product by 'reverse engineering'. This could be done at a much lower cost than that of developing the existing program because the development costs of the original consisted not just of code writing, but user research, failed product launches, and the cost of market building. Important protection against such cloning was provided by 'look-and-feel' case law. Look-and-feel protection had been granted in a handful of cases dating from the 1950s for products such as greetings cards and TV quiz-show formats. In 1987, the Lotus Development Corporation pursued a landmark look-and-feel lawsuit against Paperback Software International. This firm had created a spreadsheet similar to Lotus 1-2-3 called VP-Planner, which it sold for $99—a fifth of the price of 1-2-3. In court Paperback Software testified that it had planned to enter the spreadsheet market in 1983, but decided before launching the product in 1984 that it would need to be fully compatible with Lotus 1-2-3 in order to compete with it. After protracted legal argument, the judge ruled in June 1990 that VP-Planner's 'keystroke for keystroke' copying of 1-2-3's interface infringed Lotus's intellectual property rights. Lotus was awarded costs and damages and VP-Planner had to be withdrawn from the market; Paperback Software subsequently ceased trading.[20] Another prominent 1-2-3 clone, The Twin, was quickly pulled from the market by its publisher Mosaic Software.

This decision was of great importance to Lotus (and all other software makers) because for the first time it gave legal protection to an important

Fig. 12.6 Lotus 1-2-3 clones. These spreadsheet programs included menus compatible with Lotus 1-2-3 so that users could immediately make use of the software without learning new commands. The products were withdrawn from the market following legal action by the Lotus Development Corporation. (Copyright Eleanor Robson.)

aspect of user lock-in, familiarity with a product's user interface. A few days after the Lotus–Paperback Software decision, Lotus sued Borland, the publishers of the Quattro spreadsheet. Quattro's case was somewhat different. Quattro had its own distinctive user interface, which was generally considered to be better than that of 1-2-3, but it also allowed the user optionally to select a 1-2-3 style menu and command structure. This was provided so that users of 1-2-3 could painlessly switch to Quattro without learning an entirely new set of commands. Borland initially lost the case, but the decision was reversed on appeal a few months later, the Court of Appeals accepting that the protection of menus would impede user learning and program interoperability. (This public spirited result was analogous to the use of a universal control layout in motor cars, where economic and safety considerations have long transcended any design whims of manufacturers.) Lotus pursued the case to the Supreme Court which confirmed the decision of the Court of Appeals in 1996. The current position appears to be that the look-and-feel of a program is protected only insofar as this does not stifle innovation by impeding a user's reasonable desire to switch to a competing product.

Errors in spreadsheets

When using mathematical tables there were two important sources of error: incorrect values in a table, and mistakes made in hand computation. Incorrect values in tables were generally eliminated by publishing errata sheets and new editions, while errors by users were minimized by disciplined calculating regimens. The situation with spreadsheets is analogous.

There are few if any recorded instances of mathematical errors produced by bugs in a spreadsheet program. One reason for this was that the cost of a product recall was very high so that spreadsheet publishers engaged in extensive internal and external testing before releasing a product. In one well known case, however, Lotus 1-2-3 Release 2, launched in September 1985, had an error in some of its financial functions. This did not produce an incorrect mathematical result (which would have been a major PR disaster) but failed to compute anything at all. Nonetheless:

Since [these] functions were at the heart of many a spreadsheet analysis, Lotus Development hurriedly turned out 1-2-3 Release 2.01 in July 1986. Release 2.01 was offered to Release 2 users as a $15 upgrade (to cover postage and handling).[21]

User spreadsheet error was a much more serious problem. There was at first no easy way to avoid spreadsheet errors, and conscientious users had to evolve a culture of building in check sums to ensure the validity of a spreadsheet. However, then and now, there was no obligation to check for errors, and users stumbled along doing their uninformed best. Most academics, for example, will have encountered spreadsheet errors in the tabulation of student grades. While one presumes that most of the errors eventually come to light and are fixed, one can never be wholly sure that a large-scale spreadsheet is error free, nor does one feel confident in rearranging rows or columns without carefully checking a sample of results afterwards.

The problem of spreadsheet error was evident very early on. For example, the second issue of *Lotus* magazine devoted its cover story to the topic.[22] As the add-in market developed in the mid-1980s, a number of products came onto the market for error analysis, including The Analyst, The Auditor, and Cell-Mate. These products, which all worked in much the same way, enabled users to see behind the spreadsheet, to verify the relationships between cells in a graphical form, to unravel circular definitions, and so on. Another set of add-ins (such as Noteworthy, above) enabled individual cells to be annotated so that the developer could create an audit trail as the worksheet was created.

During the late 1980s a number of empirical investigations of user-created spreadsheets were undertaken. The first major study was conducted by Price Waterhouse in 1987. This revealed that 21 per cent of spreadsheets had serious errors that caused them to produce incorrect results. As one writer noted, by the late 1980s spreadsheets had gone far beyond an informal scratchpad, but had become major business applications in their own right, and that it was really a question of 'not if, but how many' errors they contained.[23] Such reports prompted the manufacturers of spreadsheets to pursue the add-in makers to acquire their expertise. The company that made The Analyst was acquired by Lotus, while The Auditor was bought by the publisher of SuperCalc. The modern spreadsheet now has auditing facilities fully integrated, but they are so buried beneath a morass of features that few users know about them, and even fewer use them.

The literature on spreadsheet errors seems to share the characteristics of the gloomy literature on programming errors. In both cases there is plainly a rhetorical element because if programs and spreadsheets were *really* so bad, it would not be possible to use them. In the case of programs, the problem of reliability has been partly solved by the substitution of reliable software products written under controlled conditions for end-user programming. An analogous role for spreadsheets has been played by application templates.

Users of spreadsheets

As with tables, most historical writing has focused on the production and technology of the spreadsheet. We know comparatively little about their use.

There is considerable anecdotal evidence that VisiCalc was first used by middle-managers for preparing budgets and business plans. According to the journalist Robert Slater:

> Allen Sneider, a partner at the accounting firm of Laventhol and Horwath in Boston, was the first commercial user of VisiCalc. He bought an Apple computer in 1978 and was trying without much success to run financial models on it. A friend showed him a test copy of VisiCalc at the local computer store, and Sneider knew that this was the tool he had been looking for. He wrote a cheque on the spot. The more he used the program, the more impressed he became. He asked senior management to introduce more Apples and VisiCalc, putting the computers and programs to work in different departments. The time it took to run applications on a time-sharing system was thereby cut by eighty per cent. The spreadsheet revolution was on its way.[24]

Dan Bricklin has recalled that he was regularly buttonholed at computer shows by users who claimed that they were now able to do in minutes what had formerly taken hours.

Whether the time spent using spreadsheets was economically useful is moot. For many years, there has been a debate about the so-called 'productivity paradox'—the fact that despite huge investments in information technology there appears to be no concomitant, measurable increase in GNP at the macro level. At the micro level there is a anecdotal evidence of a misapplication of human effort to optimize resources at the margin, on the grounds that two per cent of anything big enough is worth saving. As Thomas Landauer has observed in his polemical book *The trouble with computers*:

> Spreadsheet programs can tempt users into endless puttering—changing this, trying that—of little value. Additional time is stolen by housekeeping: filing results, making disk copies, finding disks and files later, organising and cleaning up overloaded file systems, upgrading programs, learning and teaching how to do new operations, keeping the hardware working. Bookstores stock fat volumes about spreadsheet programs, and community colleges offer courses in their use.[25]

It seems that the spreadsheet initially began as an *ad hoc* tool, rather like the electronic calculator of the early 1970s. However, it quickly became the analytical tool of choice for numerate managers everywhere, and soon many other professions. By the mid-1980s college and university courses had also begun to use the spreadsheet as a classroom tool for teaching numerical methods, the sciences, engineering, economics, and quantitative business school courses of all kinds.

There is a lack of statistical data on the diffusion of the spreadsheet in the various professions and education. Probably the best proxy we have is the textbook literature. Among the 1000-plus spreadsheet texts in the Library of Congress catalogue one can discern three broad categories that had begun to emerge by the mid-1980s. The first, and most populous, genre consisted of general purpose texts for using a proprietary spreadsheet—these typically had titles such as *Lotus 1-2-3 for idiots, Excel in 21 days*, etc. A second genre consisted of books designed for particular professions, especially the management and accounting professions but also in the sciences, engineering, economics, agriculture, geology, and many other specialities. These typically had titles such as *Management accounting in Lotus 1-2-3, Spreadsheet applications for analytical chemists*, etc. Books were sometimes written for a 'generic' spreadsheet, and sometimes tailored to a particular product, with different editions for each of the popular spreadsheets. The third genre was college textbooks that used a spreadsheet as a pedagogical device for learning a particular subject. These typically had titles such as *Numerical analysis with spreadsheets, Financial accounting—a spreadsheet approach*, etc.

Although a deeper analysis of spreadsheet users is beyond the scope of this chapter, there is a huge potential for research, and there are at least two untapped information sources to mine. First, there is the ephemeral periodical literature that surrounded spreadsheets. The best by far was *Lotus* magazine for 1-2-3 users, published monthly from May 1985 for nearly a decade. Before that there was *SpreadSheet* magazine for VisiCalc users, and after there was *Excel user*. The correspondence columns of these magazines tell us much about spreadsheet users and their world. For example the most frequent correspondents were enthusiasts explaining a 'macro' for some obscure process such as turning decimal numbers into Roman numerals. These were not people one would want to spend a long train journey with, but other letters came from actuaries, engineers, doctors, and most other numerate professions one could think of. Second, there were user groups. Lotus 1-2-3 was the best served, with groups in every major US city, and international chapters in Europe and Japan. At user-group meetings a local spreadsheet whiz might explain, for example, how he (usually a he) had analysed sales data for his employer.

There was never a user group for logarithm tables, but that perhaps says more about the times in which we live than about comparative technologies. At all times the writers of letters and attendees of meetings were a minority. The silent majority simply used the tools at hand—whether logarithms or spreadsheets—and got on with the job.

Further reading

An excellent overview of current spreadsheet technology and practice in given by Deane Arganbright's 'Spreadsheet' entry in the *Encyclopedia of computer science* (4th edition, ed. A. Ralston, E. D. Reilly, and D. Hemmendinger, Nature Publishing, London, 2000, pp. 1670–4.) There is vast publishing industry associated with propriety spreadsheets, particularly the market leader Microsoft Excel. Texts are aimed at three levels of user: the beginner, the intermediate or advanced user, and the professional developer. The intended readership is usually explicitly stated on a book's cover, or is implicit in the length of the book and the prose style. Beginners' books usually have a title that hints at the readership level, though not always as annoying as *Excel for dummies*. Books are typically 200–300 pages in length. Texts for seasoned users are much longer—700 or 800 pages being quite usual—in order to cover the myriad features of the software. Finally, there are professional texts for developers of spreadsheet applications, templates, and add-ins. These are usually long and expensive, and a peep between the covers will be enough to convince most readers that this level of expertise is far beyond what they could ever need in everyday spreadsheet use.

The early history of the spreadsheet and VisiCalc is a staple of every popular history of the personal computer. One of the earliest and best accounts is given in Paul Frieberger and Michael Swaine, *Fire in the valley: the making of the personal computer*, McGraw-Hill, New York, 1984 (revised edition 1999). There is no good history of the Lotus 1-2-3 spreadsheet, but the company's magazine *Lotus* gives a superb coverage of the spreadsheet in its heyday. The best account of the history of Microsoft's Excel is given in Daniel Ichbiah and Susan L. Knepper, *The making of Microsoft* (Prima Publishing, Rocklin, Calif., 1991.)

A competitive analysis of the spreadsheet industry appears in Stanley J. Liebowitz and Stephen E. Margolis, *Winners, losers and Microsoft: competition and antitrust in high technology* (The Independent Institute, Oakland, Calif. 1999). A good account of the Lotus Development Corporation's look-and-feel lawsuits is given in Lawrence D. Graham, *Legal battles that shaped the computer industry* (Quorum Books, Westport, Conn., 1999). There is some periodical literature on errors in spreadsheets, but the best starting point is one of the books on writing reliable spreadsheets such as Ronny Richardson, *Professional's guide to robust spreadsheets* (Manning, Greenwich, Conn., 1996). A refreshing and sceptical discussion of spreadsheets and computers generally is given in Thomas K. Landauer, *The trouble with computers* (MIT Press, Cambridge, Mass., 1996).

Notes

1. J. B. Best, *Cognitive psychology*, West Publishing Co., Minneapolis, 1995, p. 95.
2. M. G. Croarken, *Early scientific computing in Britain*, Clarendon Press, Oxford, 1990, p. 103.
3. E. L. Grant and L. F. Bell, *Basic cost accounting*, 2nd edn, McGraw-Hill, New York, 1964, pp. 67–8.
4. Anon., *BBL Basic Business Language: report generation and financial modeling, available on the TYMCOM-X system*, Tymshare Inc., Cupertino, Calif., March 1973.
5. D. A. Krueger and J. M. Kohlmeier, 'Financial Modeling and "What If" Accounting,' *Management accounting*, 1972—reprinted in A. Rappaport (ed), *Information for decision making*, Prentice Hall, Englewood Cliffs, NJ, 1982, pp. 172–81; esp. 172.
6. M. Campbell-Kelly, *From airlines reservations to Sonic the Hedgehog: A history of the software industry*, MIT Press, Cambridge, Mass., 2003.
7. P. Frieberger and M. Swaine, *Fire in the valley: the making of the personal computer*, McGraw-Hill, New York, 1984, pp. 229–30.
8. R. T. Fertig, *The software revolution: trends, players, market dynamics in personal computer software*, North-Holland, New York, 1985, p. 178.
9. S. J. Kaplan, 'Some future trends in spreadsheets', *Lotus*, February 1986, 138.
10. E. Sigel and the staff of Communications Trends, *Business/professional microcomputer software market, 1984–6*, Knowledge Industry Publications, White Plains, NY, 1984, p. 38.
11. *SPA worldwide data program for the year 1996*, Software Publishers Association, Washington, DC.
12. A good discussion of increasing-returns economics is given in B. W. Arthur, 'Positive feedbacks in the economy', *Scientific American,* February 1990, 92–9.

13. M. A. Cusumano and R. W. Selby, *Microsoft secrets*, The Free Press, New York, 1995, p. 246.
14. L. Woody, L. Hudspeth and T. J. Lee, *Excel 97 annoyances*, O'Reilly, Sebastapol, Calif., 1997.
15. Cusumano and Selby, *Microsoft secrets*, passim.
16. D. W. Carroll, 'Lotus enhancements blossom forth', *Software news*, March 1985, 66–71.
17. Ibid.
18. Information taken from the website of the Craftsman Book Company, www.craftsman-book.com—accessed July 2001.
19. Interviews with Dan Bricklin appear in: R. Slater, *Portraits in silicon*, MIT Press, Cambridge, Mass., 1987, pp. 285–94; Susan Lammers, *Programmers at work: interviews with 19 programmers who shaped the computer industry*, Tempus-Microsoft Press, Redmond, Wash., 1986, pp. 130–51; R. Levering, M. Katz, and M. Moskowitz, *The computer entrepreneurs: who's making it big and how in America's upstart industry*, New American Library, New York, 1984, pp. 128–33.
20. L. D. Graham, *Legal battles that shaped the computer industry*, Quorum Books, Westport, Conn., 1999, pp. 56–68.
21. K. R. Conatser, '1-2-3 through the years', *Lotus*, June 1992, 38–45, esp. 40.
22. J. Grushcow, 'Avoid these common spreadsheet errors', *Lotus*, July 1985, 59–62.
23. R. R. Panko, 'What we know about spreadsheet errors', *Journal of end user computing* 10:2 (1988), 15–21.
24. Slater, *Portraits in silicon*, pp. 291–2.
25. T. K. Landauer, *The trouble with computers: usefulness, usability, and productivity*, MIT Press, Cambridge, Mass., 1996, p. 150.

Biographical notes

Martin Campbell-Kelly is reader in computer science at the University of Warwick, where he specializes in the history of computing. His publications include *Computer: a history of the information machine* (Basic Books, 1996), co-authored with William Aspray, and *ICL: a business and technical history* (Oxford University Press, 1989). He is editor of the *Works of Babbage* (Pickering & Chatto, 1989). He has recently completed *From airlines reservations to Sonic the Hedgehog: a history of the software industry*, (MIT Press, 2003).

Mary Croarken gained a Ph.D. in History of Computing from the University of Warwick in 1986. She is currently a Visiting Fellow at Warwick University. Her publications have centred on the history of computing and mathematical tables in the pre-1950 period and include *Early scientific computing in Britain*, (Clarendon Press, 1990). Mary Croarken is on the Editorial Board of the *IEEE Annals of the History of Computing* and is Honorary Secretary of the British Society for the History of Mathematics.

Raymond Flood is University Lecturer in Computing Studies and Mathematics at the Department for Continuing Education, Oxford University, and Fellow of Kellogg College. His main research interests lie in statistics and history of mathematics. He is a co-editor of *The nature of time* (Blackwells, 1986), *Let Newton Be!* (1998), *Mobius and his band* (1993) and *Oxford figures* (2000) published by Oxford University Press.

Ivor Grattan-Guinness is Professor of the History of Mathematics and Logic at Middlesex University. He was formerly editor of the journals *Annals of Science* and *History and Philosophy of Logic*. His books include the two-volume *Companion encyclopedia of the history and philosophy of the mathematical sciences* (Routledge, 1994) and *The search for mathematical roots, 1870–1940: logics, set theories and the foundations of mathematics from Cantor through Russell to Gödel* (Princeton University Press, 2000). He is an Associate Editor (mathematicians and statisticians) for the *New dictionary of national biography*, due in 2004. He is editing for Elsevier a large collection of essays on *Landmark writings in Western mathematics: case studies 1640–1940*, also to appear in 2004.

David Grier is an associate professor of computer science and international affairs at the George Washington University. He is Editor-in-chief elect of the *IEEE Annals of the History of Computing* and has published articles in that journal as well as *Chance*, the *American Mathematics Monthly*, the *Communications of the ACM* and *Inference*. He is currently writing a book on human computers, which will be published by Princeton University Press in 2003.

Edward Higgs is a senior lecturer in history at the University of Essex. He worked for many years as an archivist at the Public Record Office in London. His publications include *Making sense of the census* (HMSO 1989). He was editor of *History and electronic artefacts* (OUP 1998). His latest book, *The information state in England: state gathering of information on citizens since 1500*, is currently in the press.

Graham Jagger is an associate lecturer with the Open University, where he is carrying out research into activities of the seventeenth century British mathematical community. He is the author of 'Joseph Moxon, FRS, and the Royal Society', *Notes and Records of the Royal Society* 51 (1995), 193–208.

Chris Lewin is Head of UK pensions at Unilever plc. An actuary by training, he is an authority on risk management and he received the Finlaison medal of the Institute of Actuaries in 1999. Mr Lewin has contributed several papers on actuarial science to BSHM conferences. His publications include numerous articles on the early history of actuarial science and insurance in *The Actuary* and other actuarial magazines. His book *Pensions and insurance before 1800—a social history* will be published by Tuckwell Press in 2003.

Arthur L. Norberg holds the ERA Land-Grant Chair in History of Technology in the University of Minnesota, where he is also Professor of Computer Science. He is also Director of the Charles Babbage Institute for the History of Information Processing, a research and archival center. He is the author, with Judy E. O'Neill, of *Transforming computing technology: information processing for the Pentagon, 1962–1986* (Johns Hopkins University Press, 1996) and of a number of articles in the history of astronomy and in the history of computing.

BIOGRAPHICAL NOTES

Eleanor Robson is a Fellow of All Souls College, Oxford. She works on the intellectual history of ancient Iraq and teaches ancient Near Eastern archaeology, history, and languages at the Oriental Institute, Oxford. She is the author of *Mesopotamian mathematics, 2100–1600 BC: technical constants in bureaucracy and education* (Clarendon Press, 1999).

Doron Swade is an engineer, an historian of technology, and a leading authority on the life and work of Charles Babbage. Until recently he was Assistant Director & Head of Collections at the Science Museum, London. He has published widely on curatorship and the history of computing. Books he has authored include *The dream machine: exploring the computer age* (with Jon Palfreman, BBC Books, 1991). His most recent book is *The cogwheel brain: Charles Babbage and the quest to build the first computer* (Little, Brown, 2000; Viking Penguin, 2002).

Margaret de Valois gained a BSc in Pure Mathematics at the University of Warwick, where she studied the history of mathematics with David Fowler, specializing in the letters of Sir Isaac Newton. Ms de Valois has written for various industry publications and *The Actuary* magazine, and now works as a professional clarinettist.

George Wilkins joined H M Nautical Almanac Office in 1951 and served as Superintendent from 1970 to 1989. He was Head of the Almanacs and Time Division of the Royal Greenwich Observatory from 1974 to 1989. He was active in the International Astronomical Union and served as president of Commissions 4 (ephemerides) and 5 (documentation and astronomical data); he was also chairman of working groups on numerical data and on the determination of the rotation of the Earth. He is currently an Honorary University Fellow in the School of Mathematics at Exeter and participates in the activities of the Norman Lockyer Observatory at Sidmouth.

Michael R. Williams is professor emeritus of the Department of Computer Science, University of Calgary and Head Curator at the Computer History Museum in Mountain View California. He is an authority on calculating machinery, and his many publications include *A history of computing technology* (1987 and 1995). He is a former editor-in-chief of the *Annals of the History of Computing*.

Index

Abramowitz, Milton 283–284
Académie des Sciences, Paris 107, 111–112, 117, 132, 187
Accounting machines 244, 296
    Burroughs 302
    Ellis, *see* National
    National 137–138, 244, 245, 247, 297, 302–303, 311
Accounting tables 20–27
Actuarial profession 5, 8, 79, 98–99
Actuarial tables 79–99
Adab 29
Adams, John Couch 193
Adelard of Bath 6
Admiralty 193, 300, 302, 305
Admiralty Computing Service 9, 256, 303
Aiken, Howard 138, 245
Airey, John 241–242, 254
Airy, George Biddell 157, 192–193, 197, 198, 202
Akkad 22
Akkadian language 22, 27, 35, 36, 37
al-Battani 6
Alfonsine tables 6
al-Khwarizmi 6
Almanacs
    see ephemerides
    see Connaissance des temps
    see Nautical almanac
American ephemeris
    see Nautical almanac, US
American Mathematical Society 269, 272
Analytical engine 12, 129, 151
Analytical Society, Cambridge University 126
Annuities 87, 89, 90, 94
Anu, god of Uruk 37
Apple II 330
Applied Mathematics Panel 282, 283, 286
Aramaic language 27, 29
Archibald, Raymond Clare 132, 287
Arnott, Neil 221
Ashur 26, 28, 34
Ashurbanipal, king of Assyria 29, 35
Assurance 80, 89, 93–94, 98
Assyria 20, 26, 29, 34–35, 38

Astronomer Royal 8, 10, 156–157, 191, 192, 193, 295
Astronomers, 5–6, 8, 36–41, 42, 52, 60, 177–203, 295–316
Astronomical cuneiform texts 37–41
    diaries 38
    Enuma Anu Ellil 37–38, 41
    ephemerides 39, 41
    lunar tables 38–40
    planetary tables 39–40
    procedure texts (precepts) 39–40
Astronomical observations 178–203
    comparison with tables 181, 183, 184, 189, 181, 193, 198
Astronomical Research Institute, Berlin 135
Astronomical Society, *see* Royal Astronomical Society
Astronomical tables 7, 39–41, 177–203, 295–316
Aubrey, John 60
Auwer, Arthur 202

Babbage, Charles 8–9, 11–12, 15, 69, 106, 110, 123, 126–129, 133, 136, 139, 149–151, 157–163, 222
Babbage, H. P. 139
Babylon 23, 25, 28, 35, 37, 39, 40, 41
Babylonia 13, 20, 27, 28, 34, 37
Bachelder, John 133
Bagay, Valentin 118
Baily, Arthur, 8
Baily, Francis 8
Ballistics Research Laboratory 11
Ballistics tables 283
BASIC 326
Basic Business Language (BBL) 326, 328
Bauschinger, Julius 135–136
Bell Laboratories 124, 278, 282
Bennett, A.A. 270, 287
Berlin Academy 185
Bernoulli, Daniel 96
Bessel function tables 12, 14, 239–240, 241, 243, 244, 245, 258, 286
Bessel, Friedrich W. 191–192, 193, 196, 197, 198, 202
    Tabulae Regiomontanae 192
Bethe, Hans 266, 281

INDEX

Bickley, William 248–249, 254–255
Bills of Mortality (Carlisle) 89
Bills of Mortality (London) 80–86
Binding 145, 147
Binet, Jacques P. M. 188
Biot, J.B. 112
Blagden, Charles 110, 111, 117
Blanche, Gertrude 266–267, 272–289
Blundeville, Thomas 56
Borda, J.C. 111, 117–118
Borland (company) 341
Bothwell, Janet 50
Bourchier, Henry 58
Bouvard, Alexis 189, 198
Bradley, James 189, 191, 202
Brahe, Tycho 7, 52
Bricklin, Dan 327–328, 339, 343
Bridgewater, Earl of: *see* Egerton, John
Briggs, Henry 7, 9, 15, 49–50, 54–62, 64–72
   Arithmetica logarithmica 57, 62, 65–67, 69
   Logarithmorum chilias prima 50, 58–59, 64–65
   Trigonometria Britannica 49–50, 65–66, 68, 69
Briggs, Lyman 270–271, 278, 284
British Association for the Advancement of Science 8, 196, 197, 235–249
British Association Mathematical Tables Committee 8, 10, 12, 235–261, 279
   Bessel Function Sub-Committee 245
   Chairmen 241
   Members 246
   Secretaries 241
Brookhaven National Laboratory 286–287
Burckhardt, Johann K. 188, 192, 197
Bureau de Cadastre 10, 69, 107, 108
Bureau des Longitudes 107–109, 111–112, 189, 197
Burg, Johann T. 188
Bush, Vannevar 269

Calculating machines 11, 12, 98, 133, 145–170, 225, 240, 244, 247, 259, 270, 279, 280, 283, 296, 297, 301, 308
   Arithmometer 146–147, 224–225
   Brunsviga 136–137, 146, 300, 302
   Brunsviga-Dupla 137, 244
   Burroughs Adding Machine 300–302
   Comptometer 131
   Edmondson 240
   Leyton Arithmometer 300
   Mercedes Euklid 135
   Monroe 280, 281
   Nova Brunsviga 244
   Odhner 137

   Sunstrand 281
   see also accounting machines
   see also punched card machines
Calculation *see* Computation
Calculators, electronic 50, 98, 315, 323, 344
Calculators, human
   see computers, human
Cambridge University 52, 126, 137, 193, 247, 250, 257
   King's College 71
   Mathematical Laboratory 248, 250
   Observatory 193
   St John's College 54
   Trinity College 66
Carlini, Francesco 197
Carlisle table 93
Carnot, Lazure 108
Cayley, Arthur 236–238, 239, 253
Census, England and Wales 4, 5, 11, 87, 209–214, 225, 229
   of 1861 209
   of 1901 213
   of 1911 12, 219, 225
Census, US 12
Centennial International Exhibition, Philadelphia, (1876) 132–134
Centesimal system 111, 117
Chadwick, Edwin 221
Charles II, king of England 49, 70
Checking tables 109, 151–153, 296, 299, 302, 304
Church, William 147
Cleaver, Frank 247
Clement, Joseph 127–8, 139
Coast and Geodetic Survey 268
CODATA 314
Colebrooke, Henry Thomas 8
Columbia University 271, 279
Compound interest 79–80
Compound interest tables 80–83
Computation 3, 20, 24, 25–26, 31, 32, 33, 39, 50–52, 54, 56–58, 66, 70, 96–97, 105, 118, 145–147, 150–159, 179, 180, 188, 189, 190, 194, 195, 201, 202, 203, 213, 215, 221–222, 224–225, 227, 326, 328–330, 341
   see also computers, electronic
   see also computers, human
Computers, electronic 5, 11, 12, 94, 145, 203, 225, 247, 252, 257–260, 284, 296, 297, 302, 305–307, 310, 315, 316, 323–345
Computers, human 1, 10, 11, 108–110, 150–152, 179, 180, 190, 192, 195, 201,

INDEX 355

224–225, 238, 240, 247, 248, 265–266, 273–289, 296
see also hairdressers
Comrie, Leslie John 12, 137–138, 243–247, 251, 255, 279, 284, 297, 300–303, 304, 307–310, 325
Conant, James 286
Condon, Edward 284
Connaissance des temps 7, 108, 196–197, 203
see also ephemerides
Continuous Mortality Investigation Bureau 94
Cowell, P. H. 300
Craftsman Book Company 339
Cugerus 52
Cuneiform script 22, 27, 29, 41
Cuneiform tablets 14, 18–42
    A 681 29
    AUAM 73.0400 23–24
    AUAM 73.0639 21, 23–24
    BM 34083 40
    CBS 2124 32
    CBS 3323 18–19
    Plimpton 322 33, 34
    VAT 12593 27, 30–31
Cunningham, Allan 240–242, 244–245, 250, 254
Curtiss, John 277, 285, 287

Dale, William 90
Dartmouth College 132–133
Darwin, Charles 249, 252
Davis, Harold 269
Davy, Humphrey 127
De Colmar, Thomas 146, 224
De Moivre, Abraham 87
De Morgan, Augustus 99, 148–149, 153–156
De Prony, Gaspard Riche 4, 10, 69, 105–118, 127, 150–152, 265–266
Deacon, Alfred 136
Death certificate 227
Decimalization 3
Delambre, Jean Baptiste Joseph 111, 117–118, 189, 190, 192
Delaunay, Charles 198
Deparcieux, Antoine 88
Department of Oriental Pyrenees 107
Dépôt Générale de la Guerre 111, 117
Design of tables 307–310
Difference engines 11–12, 15, 110, 123–140, 159, 160, 163–170, 222–224, 297
    Babbage's 1833 demonstration piece 128
    Babbage's Difference Engine No. 1 159, 163
    Babbage's Difference Engine No. 2 129, 139, 160, 163–170

    Grant's 132–135
    Hamann's 135–136
    Minor, 136–137
    Scheutz' 129–131, 139, 159, 222–224
    Wiberg 131–132, 159
Differences 11, 53, 59, 66, 67, 68, 72, 108–110, 113–116, 124–125, 150, 163, 277, 296, 302, 304, 308–310
Digital Equipment Corp. (DEC) 327
Disease classification systems 216–219
Distribution media for tables 314–315
Distribution of tables 147
Dodson, James 89
Doodson, Arthur 241
Downing, A. W. M. 300
Dudley Observatory, Albany, New York 130–131
Dunkin, Edwin 297, 299, 300
Dunkin, William 300

Early Dynastic period (*c.* 3000–2400 BCE) 22, 27, 29–30
East India Company 52
Eckert, Wallace 279
École de Géographes 107, 108
École des Ponts et Chaussées 106, 107, 112
École Polytechnique 106, 112, 186, 194
Edmonds, Thomas Rowe 91–92
Egerton, John, first Earl of Bridgewater 60–61, 62
Elderton, W. Palin 93
Elliptic Functions 238, 240, 244
Emden Functions 243
Encke, Johann F. 197
English life tables 131, 222–224
ENIAC 288
Enlil, god of Nippur 19, 23, 37
Ephemerides 5, 7, 181, 183, 190, 194, 195–198, 203, 295–316
    comparison of British, French, and US 196–197
    see also Connaissance des temps
    see also Nautical almanac
    see also astronomical cuneiform texts
Equitable Life Assurance Society 89, 93
Errors in tables 14, 72, 87, 90–92, 97, 111, 114, 117, 118, 138, 146, 149, 151, 153–157, 159–160, 184, 192, 222, 224, 228, 238, 274–277, 307, 310–312, 315, 325, 341–343, 346
Esarhaddon, king of Assyria 29
Euler, Leonhard 115–116, 179–180, 184, 185, 188, 194
Excel user magazine 344
Exponential tables 238

Factor tables 239, 240
Fair, William 139, 212, 220, 224, 229
Felt, Dorr. E. 131
Filon, Louis 241
Financial analysis software 326–328
Finkelstein, Samuel 287
Finlaison, John 89–90, 91, 95
Fippard, Richard C. 93
Firmin Didot 111–112
First World War 269
Fisher, Ronald 241, 244, 245
Forbes magazine 337
Frankston, Bob 327
French Revolution 185, 186
Friendly Societies 94–95
Funk Software (company) 336–337

Galton Eugenics Laboratory, UCL 228, 245, 247
Galton, Francis 227
Garnier J.G. 108
Gauss, C.F. 156
Gellibrand, Henry 55–56, 66, 68
General Board of Health 222
General Registrar Office 12, 13, 208–230
    Annual report 216, 227
    Census report 210
    Dictionary of occupational terms 208–209
    Search Room 211
    Statistical review 216, 228
    and Treasury 213–215, 224
    and Stationery Office 224
Gibbs, Wolcott 133
Gifford, Emma 10
Glaisher, James 238–239, 253
Glaisher, James Whitbread Lee 155, 156, 236–240, 244, 253
Gompertz, Benjamin 8, 91–92
Goodwin, Charles "E.T." 252, 255–258
Goulburn, Henry 157
Graham, George, 224
Grant, George Bernard 12, 132–135
Graunt, John 83–85
Great Exhibition, London (1851) 224
Greeks 29, 38
Greenhill, Alfred 239
Gresham College 54–56, 60, 61–62, 66, 68, 72
Gresham, Sir Thomas 54–55
Groombridge, Stephen 8
Guard digits 151
Gunter, Edmund 7, 50, 55–56, 59, 60–65, 66
    Canon triangulorum 50, 60–65
    De sectore et radio 60, 62
    Gunter's Sector 60, 61

Hahn, Philipp Matthäus 124
Hairdressers 109, 114–115, 118
Halley, Edmund 8, 86–87, 90, 96
Halley's Comet 39
Hamann, Christel 12, 135–136
Hammurabi, king of Babylon 23
Hansen, Peter Andreas 193, 194–195, 197, 198, 199–200
Hart, Andrew 49
Harvard Mark I 12, 138, 245
Harvard University 12, 124, 133, 138, 327
Heliocentric position co-ordinates 179, 180, 184
Henderson, John 243
Herschel, John 8, 126–127, 157–158
Herstmonceux Castle, *see* Royal Greenwich Observatory
Heysham, John 89
Hicks, William 239
Highland Society of Scotland 91, 95
Hill, George William 201
Hill, Micaiah 241
Hind, John 300
Hipparchus 13
Hitchins, Malachy 299, 300
Hollerith
    Census Machine 12, 225–226
    see also punched card machines
Hudson, T. C. 300–301
Hydrographic Office, US 282

IBM Corp. 281, 282, 303–306, 312–313
IBM-compatible PC 331, 333
Imperial College, London 305
Imperial Observatory, Strasbourg 135
Inana, goddess of Uruk 22
Increasing returns theory 331–332, 334
Indiana University 269, 281
Initial values 274
Institut de France 107
Institute for Numerical Analysis 287
Institute of Actuaries 98–99
Institute of Advanced Study, Princeton 271
International Astronomical Union 243, 314
Interpolation 113, 118, 150, 308, 310
Iraq
    see Mesopotamia
Isin 23, 26, 33

James VI, king of Scotland (James I, king of England) 50
Jet Propulsion Laboratory 306, 314
Jones, C.W. 258

Jones, Jenkin 93
Jones, William 62, 67, 83

Kalhu (Nimrud) 26, 29, 37
Kanesh 34
Kaplan, Jerrold 328
Kassite period (c. 1600–1100 BCE) 18–19, 26, 28–29, 34
Kelly, Patrick 8
Kepler, Johannes 7, 52, 179
Kersseboom, William 88
Kish 35
Klingenberg, Poul 80
Klipstein, Philipp Engel 124, 126
Königsberg 202

Laderman, Jack 284
Lagrange, Joseph Louis 106, 107, 182, 184–189, 195
   Method of variation of arbitrary constants 184, 194, 199, 200
   Three-point interpolation formula 72
Lalande, Jerome 7
Language of tables
   English 52, 69–72
   Latin 60, 69, 72
Laplace, Pierre Simon 182, 185–189, 199
   Laplacian programme (*see* also precepts: search for unified set) 188, 189, 191, 198, 201, 202, 203
   Mécanique céleste 188, 189, 199
Lardner, Dionysius 129–130, 136, 152, 157, 158
Larsa 23, 25, 33, 34
Law of mortality 91–93, 98
Lee, Wyman 97
Lefort, Pierre Alexandre Francisque, 112–113, 117
Legendre, A.M. 108, 111, 116
Legendrian Functions 238
Leibniz, Gottfried Wilhelm von 10, 124, 146, 150
Leverrier, Urbain J. J. 193, 194–195, 198, 199–200, 203
Leybourn, R. and W. 70
Library of Alexandria 6
Life assurance tables 5, 80, 83–91, 93–94, 95, 98 131
Linear programming 288
Linear zigzag function 38, 41
Lister, Thomas 214
Liverpool University 247, 250
Lodge, Alfred 239–240, 241–242, 253
Lodge, Eleanor 240

Logarithms 5, 7, 10, 11, 14, 48–72, 90, 98, 105–118, 127, 135–136, 151, 157, 236, 296, 297, 299, 300, 301, 323, 336, 345
   artificial numbers 52, 61, 63
   differences 53, 59, 66, 67, 68, 72
   examples of use 52, 63, 65, 69, 71–72
   of trigonometric functions 49, 52–53, 57–58, 66, 70, 71
   of trigonometric functions to base 10 50, 59, 62–64
   of trigonometric functions with decimal division of the degree 59, 66, 68, 70
   of trigonometric functions with decimal division of the quadrant 66
   of natural numbers to base ten 50, 57–59, 65, 67, 69–70
   popularization of 56, 65, 69–72
   problems with Napier's definition 54, 56–57
   rules for use 63, 71–72
London Mathematical Society 244
Longstreth, Miles 197
Look-and-feel 340–341, 346
LORAN, *see* navigation tables
Los Alamos 283, 286
Lotus 1-2-3 331–335, 337–338, 340–341, 344–345
Lotus Development Corp. 328, 332, 334–335, 340–341, 346
Lotus magazine 342, 344, 345
Lowen, Arnold 8, 266–267, 271–289
Lunar motion, tables of 188, 189, 193, 196, 197, 198
   see also astronomical cuneiform texts
   see also astronomical tables
   see also ephemerides

Makeham, William 91
Manchester Royal Infirmary 228
Manchester Unity Friendly Society 95
Manchester University 247, 250
Manly, Henry W. 97
Marduk, god of Babylon 23, 35, 37
Mari 33
Marshak, Robert 281
Maskelyne, Nevil 7, 10, 156, 189, 191, 202, 295–296, 299, 315
Massachusetts Institute of Technology (MIT) 258, 269, 280, 282
Mathematical cuneiform texts
   coefficient lists 33, 34
   metrological lists and tables 33–35, 39–40, 42
   multiplication lists and tables 27, 29–34, 39–40, 42

Mathematical cuneiform texts (*contd.*)
  reciprocal lists and tables 31–35, 40
  tables of inverse cubes 33
  tables of squares and square roots 27, 29–30, 33, 35, 40
  tabular calculations 33
Mathematical symbolism 70
Mathematical tables and other aids to computation 138, 247, 252, 256
Mathematical Tables and Other Aids to Computation, Committee 270, 279, 287
Mathematical Tables Committee
  see British Association Mathematical Tables Committee
  see Royal Society Mathematical Tables Committee
Mathematical Tables Project, New York 8, 11, 258, 265–289
Mayer, Tobias 7, 188
Mechanization 123–140, 145, 158–170, 224–225, 301–313
Menabrea, Luigi 151
Merchiston, Lord: *see* Napier, John
Mesopotamia (ancient Iraq) 10, 13–14, 18–42
Microsoft Corp. 334, 335
Microsoft Excel 9, 330–335, 338, 344, 345
Microsoft Windows 332, 333, 335
Milan 197
Miller, Jeffrey 242–244, 247–249, 252, 255, 257–258
Milne, Joshua 89
Monetary functions 96–98
Moore School of Electrical Engineering 11
Morrow, Malcolm 267–270
Morse, Phillip 280, 282, 286
Mortality tables 92–94
Mosaic Software (company) 340
Mouton, G. 115–116
Moxon, Joseph 70–72
Müller, Johann Helfrich 124, 126, 146
Multiple decrement tables 95–97
Multiplication tables 236, 300

Napier, Archibald, seventh Lord Merchiston 50
Napier, John, eighth Lord Merchiston 7, 9, 48–54, 56–58, 59, 61–63, 65, 72
  Constructio (Mirifici logarithmorum canonis constructio) 52
  Descriptio (Mirifici logarithmorum canonis descriptio) 48–49, 52, 56–58, 61–62, 65, 69
  Napier's bones 50
  Promptuary 51
  Rabdologiae 50–51

Napier, Robert 52
National Academy of Sciences 267–270, 274, 287
National Accounting Machine, *see* accounting machines
National Bureau of Standards 257, 266, 270, 278, 281, 284–285, 288
National Cash Register (company) 137
National Mercantile Life Assurance Society 93
National Physical Laboratory 248, 249, 250, 256, 257, 303, 305, 313
Nature 252
Nautical Almanac Office, British 9, 11, 137–138, 225, 243–245, 279, 295–316
  Superintendents 301
Nautical Almanac Office, US 176–177, 195, 197, 198, 201, 203, 310, 315
Nautical almanac, British 7, 10, 12, 152, 156–157, 189, 192, 196–198, 200, 203, 295–316
  see also ephemerides
Nautical almanac, US 176–177, 189, 196–198, 200, 201, 203
  see also ephemerides
Navigation and navigation tables 5, 11, 52, 56, 60, 177, 188, 193, 196, 203
  LORAN 266, 282–283
Nebuchadnezzar II, king of Babylonia 29
Neo-Assyrian period (*c.* 900–610 BCE) 27, 29, 34–35, 37, 38
Neo-Babylonian and Persian periods (*c.* 540–320 BCE) 27, 29, 37
Neumann, Caspar 86
New York University 267
Newcomb, Simon 180, 193, 195, 198–203
Newton, Isaac 7, 178–179
  Newtonian gravitation 187
  Newtonian mechanics 188
Newton, John 69–70
Nimrud: *see* Kalhu
Nineveh 27, 29, 35, 37
Ninurta, god of Nippur 25
Nippur 18–19, 23, 24, 25–26, 32–3, 35, 41
Non-polynomial functions 150
Northampton life table 88–89, 93
Nosology 216–219
Number theory tables 239, 240–241, 244
Number-divisor tables 244

Occupational classification systems 216, 219–221
Occupational dictionary 208–209
Office for Scientific Research and Development 282–284, 286
Ogle, William 227

INDEX

Olbers, Wilhelm 191
Old Babylonian period (*c.* 2000-600 BCE)
    25–26, 31–34, 36, 41
Orbital elements 180–181, 182–183, 184, 185,
    192, 193, 203
Oughtred, William, 60–61, 70–72
    Trigonometria (Trigonometry) 70–71

Paper 148, 163
    Flong 148
Paperback Software (company) 340–341
Paris Exposition (1855) 130, 222
Paris Observatoire 112–113, 194, 195, 198, 202
Parseval, A.M. 108
Parthians 29, 37
Pascal, B. 146
Pattern wheels 161
Pearson, Karl 9, 227–228
Pell, John 66, 72
Pennsylvania University 11, 134
Pension funds 95–97
Personal Software (company) 327, 332
Perturbations of the planets, perturbation theory
    178, 179, 181, 183, 185–188, 190, 192,
    193, 194–195, 198, 199–200, 201, 202
Peters, Johann Theodor 135–136
Philip II, king of Spain 50
Phillips, Arthur 313
Piazzi, Giuseppe 191, 202
Pierce, Benjamin 133, 197, 198, 199
Pitiscus 65
Pivotal values 150, 152
Plana, Giovanni A. A. 197
Planetary motion, tables of 180, 189, 196, 198,
    200, 201
    corrections to 184, 190, 192
    equations of 178–181, 186, 191, 200, 202
    short-term usefulness of 181, 184, 191, 193, 198
    see also astronomical cuneiform texts
Poisson, Simeon 188
Polynomial functions 150, 163
Pond, John 156, 202, 299
Poulkova Observatory 198
Powers, tables of 244
Precepts 179, 180, 181, 188, 189, 195
    search for unified set 178, 179, 185, 188, 191,
        195, 198, 202
    see also Laplace: Laplacian programme
Precession, theory of 191
Present value 81–82
Price Waterhouse (company) 342
Price, Bartholomew 239
Price, Richard 88–89, 95

Prieur de la Côte d'Or, C.A. 108
Printing of tables 124, 126, 129–130, 132–134,
    136–139, 145, 147, 161, 165–170, 224,
    297, 311–313
Probabilities of survival 80
Probability function tables 274
Proof reading 145, 147–149, 154–158, 310–311
Protector Assurance Society 9
Ptolemy, Claudius 6, 13, 39
    Almagest 6, 39
    Handy tables 6
    Syntaxis 6
Public Health Act (1848) 222
Punched card machines 12, 146, 247, 279,
    281, 282, 296, 297, 302–304, 311,
    313, 314
Puzrish-Dagan 21, 23–24
Pythagoras' theorem 33, 34

Radio Shack TRS-80 330
Rayleigh, Lord 239
Ready reckoners 1, 3–4
Recorde, Robert 97
Regiomontanus 64
Registrar General 131
Registration and Marriage Acts (1836) 211
Rheticus 10, 53, 64, 66
    Canon doctrinae triangulorum 53
    Opus Palatinum 10, 64, 66
Rhodes, Ida 284
Richards, John 97
Riche de Prony, Claude 106
Roosevelt, Franklin 268, 279, 282, 289
Rothamsted Agricultural Experimental Station
    244, 247
Royal Astronomical Calculating Institute, Berlin
    135
Royal Astronomical Society 8, 158, 300
Royal College of Physicians, London 214
Royal Exchange 54, 55
Royal Greenwich Observatory 189, 191, 192,
    193, 196, 197, 198, 202, 238, 296, 297,
    300, 303, 306
Royal patronage 35, 37
Royal Society 55, 86, 127, 222, 240, 249–260
Royal Society Mathematical Tables Committee
    249–260
    Bessel Function Panel 258
    Depositary of Unpublished Tables 252
    General Sub-Committee 251
    Mathematical Tables Series 252
Royal Swedish Academy of Sciences 130
Royal Technological Institute, Stockholm 130

360 INDEX

Sadler, Donald 243–244, 249, 252, 255, 297, 303, 310
Salzer, Herbert 265–266, 284
Samsu-iluna, king of Babylon 23
Sang, Edward 69, 113–114, 117
Sargon II, king of Assyria 29, 37
Sargon, king of Akkad 22
Savile, Sir Henry 55, 60
Scheutz, Pehr Georg and Edvard 12, 129–131, 132, 133, 136, 222, 224
Schickard, W. 146
Science Museum, London, 128, 139
Scientific Computing Service Ltd., London 138, 245
Scribal schooling 20, 23, 27, 29–35, 36, 42
Scribes 14, 24, 36
  Bêl-ban-apli 35
  Nammah 31
Second World War 247, 279, 282–283, 303, 314
Secular variations, investigation of 185–187, 195, 197
  see also astronomical tables
Seleucid period (*c.* 320–120 BCE) 39–42
Sennacherib, king of Assyria 29
Service Géographique de l'Armée 117
Sexagesimal place value system 31, 41, 111
Shaduppum 33
Sheppard, W. F. 245
Shulgi, king of Ur 22
Shuruppag 27, 29–30
Sickness tables 94–95
Simpson, Thomas 87
Sippar 26, 33, 35
Slide rule 225
Smart, John 87
Smith, Adam 10, 108
Smith, David Eugene 9
Smith, Henry 236
Smith, Thomas Southwood 221
Smithsonian Institution, Washington D.C. 131
Snyder, Virgil 272
Society of Apothecaries, London 214
Software Arts (company) 327, 332
Software news 338–339
Solar motion, tables of 189, 190–191, 192, 197, 203
  see also ephemerides
Speidell, John 69
Spreadsheet 2, 5, 14, 15, 322–345
    anatomy of 329–330
    clones 339–341
    errors 325, 341–343, 346
    etymology 325–326
    macros 344
    user groups 344–345
Spreadsheet (complementary products) 334, 335–339
  Add-ons and add-ins 337
  Allways 337–338
  application template 338
  Cell-Mate 342
  D.A.V.E. 337
  Noteworthy 337, 342
  SeeMore 337
  Sideways 336–337
  The Analyst 337, 342
  The Auditor 342
  VisiPlot 334, 337
  VisiTrend 334, 337
Spreadsheet (products); *see* also VisiCalc, Lotus 1-2-3, Microsoft Excel
  Microsoft Multiplan 330, 334
  Quattro Pro 335, 338, 341
  SuperCalc 342
  The Twin 340
  VP-Planner 340
SpreadSheet magazine 344
Standard Tables Program 94
Stegun, Irene 284
Stereotyping 148–149, 154, 160–161, 163, 165–170
Stevenson, T. H. C. 219
Stevin, Simon 81
Stocks, Percy 228
Stokes, George 236
Stokes, Richard 71–72
Stratford, William 300
Struyck, Nicolaas 88
Subtabulation 150, 152, 275–277, 296, 300, 302, 304, 308–310
Sumer 20–22
Sumerian language 21–22, 30, 35, 36
Susa 33
Syllabaries and lexical lists 35–36

T. J. Watson Computing Laboratory 279
Tabling sheets 213
Taxonomy of tables 2–5
Taylor's theorem 116
Temples 18–20, 21, 22, 25–29, 37
Tetens, Johannes Nikolaus 91
Thomas, Ernest C. 97
Thompson, Alexander 137, 252
Thomson, William (Lord Kelvin) 236, 239
Three-body problem 179
Ticking method 213, 225

Todd, John 257
Toledan tables 6
Tonti, Lorenzo 80
Tontine 80, 89
Transcription 145–146, 149–151
Trenchant, Jean 80–82
Trigonometric functions and tables 10, 53,
    63–64, 66, 68, 70, 105–118, 151, 179,
    236, 299, 300
  definition of 53
  logarithms of: *see* logarithms
  spherical trigonometry 61–64, 70
Typesetting 145, 147–149, 153–154, 160, 163, 166
Typography of tables 2, 14, 15, 49, 52–54, 59,
    61, 62–64, 65, 66, 67, 71–72, 147, 244,
    307–308, 311
  in cuneiform 18–19, 20, 21, 23–24, 26, 27,
    34, 36–37, 40, 41
  tabular lists (cuneiform) 25, 32–33, 36, 37,
    38–39

University College, London 137
  see also Galton Eugenics Laboratory
Ur 21, 22, 23–24, 31, 33, 35
Uruk 20–21, 22, 33, 35, 37, 39
US Army 266, 281, 282
US Naval Observatory 298, 305, 306, 307, 312,
    314
US Navy 266, 282–283
Ussher, Archbishop James of Armargh 56, 58

Van Orstrand, C.E. 285
Veblen, Oswald 9, 269–270, 284
VisiCalc 322–323, 325, 327–328, 330–332, 334,
    337, 339
VisiCorp (company) 332

Vlacq, Adrian 65–66
von Lindenau, Bernhard A. 189, 198
von Neumann, John 280, 285, 286, 288
von Zach, Franz X. 189

Walker, Sears 198
Ward, John 64
Wargentin, Pehr 88
Watson, Alfred 95
Weaver, Warren 286
What-if analysis 326
Wiberg, Martin 12, 131–132
Wilkes, Maurice 250, 252, 256, 258–260
Wilkins, George 310
William Thomas 60
Wingate, Edmund 64
Winlock, Joseph 197
Witt, Richard 79, 82–83, 97
Wittstein, Theodor 96
Woolhouse, W.S.B. 91
Works Project Administration (WPA) 8, 11,
    267–289
  see also Mathematical Tables Project, New York
Worksheet (alternative term for spreadsheet)
    325–326
Worksheets 275–277, 278
World Health Organisation 228
World Wide Web 335
WPA *see* Works Project Administration
Wright, Edward 52, 56, 61, 66, 69
Writing boards 29, 38, 41

Yale University Observatory 305
Yeshiva University 271, 285
Young, Thomas 91–92, 299–300
Yule, George Udny 227–228

DISCARD

DATE DUE

DEMCO, INC. 38-2931